Advances in Geographical and Environmental Sciences

Series Editor

R. B. Singh, University of Delhi, Delhi, India

Advances in Geographical and Environmental Sciences synthesizes series diagnostigation and prognostication of earth environment, incorporating challenging interactive areas within ecological envelope of geosphere, biosphere, hydrosphere, atmosphere and cryosphere. It deals with land use land cover change (LUCC), urbanization, energy flux, land-ocean fluxes, climate, food security, ecohydrology, biodiversity, natural hazards and disasters, human health and their mutual interaction and feedback mechanism in order to contribute towards sustainable future. The geosciences methods range from traditional field techniques and conventional data collection, use of remote sensing and geographical information system, computer aided technique to advance geostatistical and dynamic modeling.

The series integrate past, present and future of geospheric attributes incorporating biophysical and human dimensions in spatio-temporal perspectives. The geosciences, encompassing land-ocean-atmosphere interaction is considered as a vital component in the context of environmental issues, especially in observation and prediction of air and water pollution, global warming and urban heat islands. It is important to communicate the advances in geosciences to increase resilience of society through capacity building for mitigating the impact of natural hazards and disasters. Sustainability of human society depends strongly on the earth environment, and thus the development of geosciences is critical for a better understanding of our living environment, and its sustainable development.

Geoscience also has the responsibility to not confine itself to addressing current problems but it is also developing a framework to address future issues. In order to build a 'Future Earth Model' for understanding and predicting the functioning of the whole climatic system, collaboration of experts in the traditional earth disciplines as well as in ecology, information technology, instrumentation and complex system is essential, through initiatives from human geoscientists. Thus human geosceince is emerging as key policy science for contributing towards sustainability/survivality science together with future earth initiative.

Advances in Geographical and Environmental Sciences series publishes books that contain novel approaches in tackling issues of human geoscience in its broadest sense—books in the series should focus on true progress in a particular area or region. The series includes monographs and edited volumes without any limitations in the page numbers.

More information about this series at https://link.springer.com/bookseries/13113

Mukunda Mishra · R. B. Singh ·
Andrews José de Lucena · Soumendu Chatterjee
Editors

Regional Development Planning and Practice

Contemporary Issues in South Asia

Springer

Editors
Mukunda Mishra
Department of Geography
Dr. Meghnad Saha College
Uttar Dinajpur, West Bengal, India

R. B. Singh
Department of Geography
University of Delhi
New Delhi, Delhi, India

Andrews José de Lucena
Federal Rural University of Rio de Janeiro
Rio de Janeiro, Brazil

Soumendu Chatterjee
Department of Geography
Presidency University
Kolkata, West Bengal, India

ISSN 2198-3542 ISSN 2198-3550 (electronic)
Advances in Geographical and Environmental Sciences
ISBN 978-981-16-5680-4 ISBN 978-981-16-5681-1 (eBook)
https://doi.org/10.1007/978-981-16-5681-1

This Springer imprint is published by the registered company Springer Nature Singapore Pte Ltd.
The registered company address is: 152 Beach Road, #21-01/04 Gateway East, Singapore 189721, Singapore

Foreword

I am delighted that Springer Nature Singapore Pte Ltd. has published this book, *Regional Development Planning and Practice—Contemporary Issues in South Asia,* edited by Mukunda Mishra, R. B. Singh, Andrews José de Lucena, and Soumendu Chatterjee. The editors are to be congratulated for their work to prepare a volume that is so timely and appropriate.

Geographers have long stressed the need to regard "development" as a "regional process," Development planning is invariably more effective when responsive to, and customized to, a particular region's needs. Of course, this demands attentiveness to locale. Successful regional development planning also requires the right tools, methods, conceptual frameworks, and data. Paying particular attention to regional development in South Asia this book sheds light on these key elements in chapters by academics and planning practitioners from Bangladesh, Brazil, India, Mexico, Nepal, Poland, and Sri Lanka.

Collectively, the chapters of this book aspire to contribute to better understandings of, and practices for, sustainable regional development. They do so through coverage of a wide range of contemporary issues including climatic and other environmental changes, human health, gender inequality, accessibility, challenges associated with displaced populations, and tribal livelihood strategies in addition to more conventional themes such as regional inequalities and regional urban problems.

Notably, the editors have emphasized the importance of experiences in South Asia, a diverse region which confronts critical economic, demographic, environmental, and societal challenges. There is much to be learned both within the region and further afield from the experiences of the contributing scholars.

I thank the editors and each of the contributors for the thought and care they have put into this work and I acknowledge everyone associated with the publication process for the attention they have given to the delicate work of bringing such a book to print. I do hope these efforts will be rewarded by worldwide reading of this volume by audiences who seek to support regional development that is conducted in ways that are locally nuanced, effective, and sustainable.

March 2021

Iain Hay, Ph.D., LittD
First Vice-President of the International Geographical Union
Matthew Flinders Distinguished Emeritus Professor of Geography
College of Humanities
Arts and Social Sciences
Flinders University
Adelaide, Australia

Preface

Conceptualizing "development" as a "regional process" is geographers' brainchild. The "sense" of "region" has long been rooted deeply in the fundamental research practices to which the geographers are accustomed. The geographical perspective of regions entails the conceptualization of the formal region to be nested horizontally, whereas the functional region is by hierarchical relationships on space with spatial flows or interactions. In geographical researches, the region works as a tool by serving as a statistical unit of analysis. However, what is more important that regions serve as the fundamental spatial units of management and planning by specifying a territory or a part of it for which a certain spatial, development, or regulatory plan is sought.

The first decade of the new millennium was eventful. The 2008s Global Economic Crisis was a severe alert to the policy researchers and policymakers. Shreds of evidence in Economic Sciences, Trade Theories, and Economic Geography speculate that the regional crises are becoming more important than the national crisis for many parts globally, especially in developing areas. The scholarly world witnesses that some regions are becoming poorer than their national economies. Lagging regions need to see their economic base renewed if they are to stay competitive, and this requires all actors—natural, economic, social, political, and education—to work together proactively. The development and management practices to be focused on regions which is the need of the time.

Climate change issues have stormed the laboratories of almost all fields of knowledge—may it be natural science, biological science, mathematical science, or social science. The complex interaction between climate change and human development can hardly be explored without considering how the regional version of "deflected" climatic parameters excerpt impacts regional society or economy to shape human development of that region. Besides, there will be no practical global policy framework that will be equally effective for every region to mitigate climate change and its impact on local society or economy.

The last year of the second decade of the millennium has been witnessing the pandemic outbreak of the Novel COVID 19, spreading violently at almost all parts of the globe. Loss of human lives is the primary threat of a pandemic. Beyond a severe health hazard, it has long-term social, economic, and behavioral implications that mankind will be facing during the decade to come. Like the global phenomena of the climate change, the COVID 19 pandemic also carry its impact differently on the different region depending upon its economic strength, health, and medical systems' tenability, government policies and probably the most powerful will be the health behavior of the residing population of the region. All these things have been emerging as the newer research dimensions under the purview of Regional Development Planning.

There is a sea change in the whole of South Asia. The socio-economic–political ups and downs in this vast region are almost coterminous with the Indian subcontinent. It continues to have mass poverty with its attendant evils of ill-governance, impoverished public health, technological backwardness, and territorial disputes in India, Bangladesh, Pakistan, Myanmar, and Afghanistan. The South-Asian states suffer from internal divisions, deep mistrust, and internal incoherence. It has prevented the region from realizing its potential of economic growth, political power, and the cultural excellence of its millions of population. Within this geopolitical setup, there are critical regional issues at different corners of South Asia, which seek investigation and regional policy prescriptions.

In India, the population grows from 1,000 million to over 1,500 million; ecosystem and land-cover changes take an ever-larger share of Net Primary Production; the slow transition from rural agrarian to secondary and tertiary economic sectors; growth of non-farm rural livelihoods, informal urban services; "joblessness"; formalization in the industrial sector; slow decline in poverty, with rapid increases in differential wealth creation by new entrepreneurial classes, urban elites, entrenched dominant caste, and trading groups; mobilization of lower caste (Dalit) groups and women in society, with significant consequences for the social and political dynamics—all these contemporary issues in India need to be studied, analyzed, and managed regionally.

This book, within its folds, accommodates systematic and analytical scientific commentary on the emerging methods of regional development, measurement of the development, regional development models, and policy measures that are found instrumental for South Asia. The broad spectrum of discussion on the planning and practice of regional development issues in the South-Asian countries is discussed in 13 chapters from the scholars from different parts of the South Asia, where the first three chapters in Part I speculate the sustainable regional development process from three different angles.

Part I, entitled "Sustainable Regional Development," is self-explanatory regarding its content. The first chapter (Chap. 1) on "Sustainable Regional Development with Environmental Practices" is written by scholars from the renowned Polish Academy of Sciences who speculates the importance of environmental goals in defining the development direction through the multidimensional concept of sustainable development. They have also explained the role of natural capital as a factor of human development and well-being; traces the integrated approaches and regional instruments for achieving environmental goals. The following chapter (Chap. 2) visualizes the twentieth and twenty-first centuries to bring to the fore some interesting discussion on the geopolitics of sustainable development. The scholars from the Federal Rural University of Rio de Janeiro present the challenges of using and appropriating nature from a prism of Western modernity, which influences the dualistic conception between preservation policies and the dynamics of a capitalist development agenda. The authors exemplify some important Brazilian regions that are part of border zones with other countries and are part of the expansion of the agribusiness economic frontier. They speculate how Brazil's insertion in the world economic system and the adoption of a policy of re-primarization of its economy has placed the geopolitics of sustainability in this Latin Amerian country. The third chapter (Chap. 3) is a collaborative venture of the authors from Brazil and Mexico. The authors have presented the scientific and political cooperation between these two countries regarding the territory region relationship, food sovereignty, agroecology, and they explained how these cooperations result in sustainable development.

Part II is comprised of four chapters that envisage the regional development practices in four South-Asian countries—Sri Lanka, Nepal, Bangladesh, and India. Under this Part, Chap. 4 speculates the regional development from the standpoint of the sustainable responses to climate change phenomena and other environmental issues in the South-Asian country of Sri Lanka. The chapter, authored by a group of practitioners in Geography in Sri Laka, has made meaningful discussion on how the regional approach in development explores promising avenues for the vast rural population in this country through the critical review of the Accelerated Mahaweli Development Programme, IRDP, Southern Development Plan and other premier government-sponsored schemes. The following chapter (Chap. 5) is from the landlocked nation Nepal, which has ensured a balanced development sustainably between three regions—plains, hills, and mountains, through the regional development planning policy implemented in the country in the early 1970s. The renowned Trbhuvan University's practicing geographers have presented a broad spectrum of the country's regional planning policies with detailed discussions and some magnificent illustrations. Chapter 6 brings to fore a burning regional issue in South Asia, i.e., the dislocation and involuntary migration, and the author has made a detailed study on the communities displaced by the rapid river bank erosion by Teesta in Bangladesh. The study assesses the sociocultural and economic change in the displaced communities, and the author has made detailed observations on the psychological consequences of dislocation and migration. The chapter mentioned some recommendations and sought a national guideline for the rehabilitation of the displaced. The following chapter (Chap. 7) deals with how the physical and relative accessibilities are becoming essential for rural people, ensuring a balanced development by enhancing the basic social and administrative services at the country's furthest rural corners. Two practicing spatial scientists bring forth their unique connotation of development and extend their efforts to measure the level of accessibility which are correlated to the development parameters. The result of the study is instrumental to framing micro-spatial regional planning strategies in terms of connectivity.

Part III of the book deals with the very pertinent and long-debated issue of regional imbalance in India. Chapter 8, under this Part, vividly discusses the "Transformation of Aspirational District Program (TADP)", launched by the Government of India in 2018. The authors have critically reviewed the working principles of the TADP; its efforts in linking the government's development scheme with the Sustainable Development Goals (SDGs); its prospect to accelerate the development process of the backward districts through utilizing local resources for development; time-bound workflow; real-time data capturing and disseminating efficiencies. The following chapter (Chap. 9) is people-centric. The author speculates that the burden of disability in India falls disproportionately across geographic regions and socio-economic groups, which should be carefully considered in the public health policies. The chapter vividly presents the distribution of the disabled population and discusses how providing access to basic capabilities may be strategically important to reduce the challenges for the disabled in India. The following chapter (Chap. 10) deals with another critical issue: how far gender inequality suppresses India's human development. Authors from the Population Research Centre of the Institute for Social and Economic Change has revealed considerable disparities in the level of human resource development among the different states of India, and interestingly, the higher-income states, being constrained by the unequal gender development, fail to hold themselves at the top of the HDI table.

Part IV consists of three chapters that deal with the urban dimensions of Regional Development. Chapter 11, the first chapter in this Part, examines whether the prevalence of child morbidity in urban India has any association with housing, water, and sanitation condition based on NHFS-4 datasets. The study succeeds to find out that diarrhoea and Acute Respiratory Infection (ARI) are significantly diseased among urban India; children below 1 year of age are more affected than the reference category; male children are more susceptible than females in both diseases; ST and SC children are more vulnerable to ARI and diarrhoea. The study has recommended community engagement in the planning and implementation of water sanitation programs in urban India. The following chapter (Chap. 12) examines the association between land-use and land-cover (LULC) pattern and the urban heat islands (UHI) in the urban-industrial landscape. The study confirms that the rise in the LST pattern is spatially associated with land-use change. Chapter 13 deals with another urban issue—the housing condition concerning seepage. The authors have done intensive field study in the dilapidated buildings in different pockets of chawls, slums, and slum rehabilitated houses in Mumbai to extract the actual scenario. The author sought urgent engineering practices of building construction and renovation work to ensure the safety of life and wealth at vulnerable condition if the meteorological aspects of the region are concerned.

Part V is interesting that comprises three chapters on the challenges and options of sustainable livelihood. Chapter 14, under this Part, deals with the livelihood of the population displaced from Bangladesh and set up their habitat in India, which has been a critical discourse since the Independence in India geared up by the 1971's Bangladesh liberation war. By capturing field data, the authors try to cross-validate whether cross-border migration complies with the classical migration theories. The authors also attempt to evaluate how the immigrant families have been living in an unfamiliar social and cultural setup at the present place. The study is concerned with these displaced populations' critical social dynamics, which often reinforce political debates in the country, while the root causes of their plight remain unattended. The following chapter (Chap. 15) envisages the socio-economic dimension of the climate change phenomena by investigating how climate change increases the vulnerability at the household level to the tribal communities. The authors have conducted an extensive survey on 650 tribal households in the Dooars region to examine how the Oraon, Munda, and Santal families are affected by climate change using the IPCC prescribed dimension of adaptability, sensitivity, and exposure. In the last chapter of this volume (Chap. 16), the authors have studied the residents of the unique ecosystem of Sunderban regarding their degree of dependency, awareness about sustainability of renewable resources, and the knowledge-base for a sustainable livelihood through the protection of the ecosystem. All these have been examined through a valuation exercise to conserve the overall ecosystem of this area using the Contingent Valuation Method (CVM). The study concludes with entrusting the "local model" of development as a result of the CVM reflects that the people of the Sunderbans are aware of the conservation of the ecosystem for their long-run benefit.

We will feel our efforts effective if the book works for the practitioners of geographical sciences, socio-economic planners, public policymakers, and policy researchers who desire to understand and address the complex processes of a region for the purpose of sustainable regional development through rational planning.

Uttar Dinajpur, India — Mukunda Mishra
New Delhi, India — R. B. Singh
Rio de Janeiro, Brazil — Andrews José de Lucena
Kolkata, India — Soumendu Chatterjee

Acknowledgments

Sincere collective efforts have made this book possible. We thank all of our friends, colleagues, students, and all others concerned. Without their active supports, this book would not have seen the light of day. They all have executed their respective roles perfectly.

We express our most profound sense of indebtedness to Professor Iain Hay, Matthew Flinders Distinguished Emeritus Professor of Geography in the College of Humanities, Arts and Social Sciences at Flinders University, and the First Vice-President, International Geographical Union (IGU). We are thankful to him for his kind advice and valuable time writing the "Foreword" for this volume.

Authors are the key resources in an edited volume. We convey our sincere thanks to all contributors for offering us the opportunity to include their valuable works in this volume. Their prompt response and active cooperation have made this volume successful on time. It incorporates a good representation from different countries in South Asia.

We are grateful to the administrative authorities of our respective institutions for their administrative supports.

We also thank all the reviewers for their constructive comments and advice that substantially improved this book's contents and quality.

Constructive editorial advice and constant support from Yosuke Nishida, *the Editor in Earth Sciences, Geography, and Environment of Springer Japan*, remains unparalleled. We acknowledge the support of the entire team of the Springer Nature associated with the publishing process, disseminating their respective roles as perfectly as ever.

Mukunda Mishra
R. B. Singh
Andrews José de Lucena
Soumendu Chatterjee

Contents

Editors and Contributors

About the Editors

Dr. Mukunda Mishra is an Assistant Professor in the Department of Geography and designated Vice Principal of Dr. Meghnad Saha College, affiliated with the University of Gour Banga in West Bengal, India. He completed his postgraduate studies in geography and environmental management at Vidyasagar University and holds a Ph.D. in geography from the same university. He was selected for the National Merit Scholarship by the Ministry of Human Resource Development, Government of India. His research chiefly focuses on analyzing unequal human development and creating multi-criteria predictive models. He has, so far, authored one monograph and edited 3 contributory volumes from the house of Springer Nature and published more than 25 research articles and book chapters. He has more than 10 years of hands-on experience in dealing with development issues at the ground level in various districts of eastern India.

Dr. R. B. Singh is a Professor of geography at the Delhi School of Economics, University of Delhi; the Secretary General and treasurer of the International Geographical Union (IGU); and the Chair of the Council for Scientific and Industrial Research (CSIR)–Central Food Technological Research Institute of the Government of India. He is also a member of the International Council for Science (ICSU) and the Scientific Committee for Urban Health and Well-Being. He was awarded the Japan Society for the Promotion of Scientific Research Fellowship and has presented papers and chaired sessions in more than 45 countries. He has published 14 books, 36 edited research volumes, and more than 240 research papers. He has supervised 36 Ph.D. and 81 M.Phil. students. In 1988, the UNESCO/International Social Science Council awarded him research and study grants in social and human sciences.

Dr. Andrews José de Lucena is an Associate Professor, the Federal Rural University of Rio de Janeiro, Brazil. He has completed his masters in geography from the State University of Rio de Janeiro and holds a Ph.D. in atmospheric sciences from the

Alberto Luiz Coimbra Institute for Graduate Studies and Research in Engineering (COPPE/UFRJ). His research interest is chiefly concerned with the urban climate and its research methods, climate change, and the natural environment. He participated in and integrated research projects with several professionals and institutions in the areas of climatology and urban meteorology, atmospheric hydrological modeling, remote sensing in urban areas, and urban development in the metropolitan region of Rio de Janeiro. He coordinates the Integrated Laboratory of Applied Physical Geography (LIGA/UFRRJ) and integrates the Environmental Satellite Applications Laboratory of the Meteorology Department (LASA/UFRJ). He manages the website *Climatologia* with information about land surface temperature (LST) of the metropolitan area of Rio de Janeiro from 1984 to the present.

Dr. Soumendu Chatterjee is a Professor and the Head of the Department of Geography at the Presidency University in Kolkata, India. He has been teaching geographical science at the undergraduate and graduate levels for more than 20 years. His primary research interest is creating scientific models to predict complex physical and human processes on the Earth's surface. He has more than 100 publications in national and international journals and has headed several research projects funded by the University Grants Commission (of India), the Department of Science & Technology (GOI), the Indian Council of Social Science Research (ICSSR), and other agencies in India and abroad.

Contributors

Ram Babu Bhagat Department of Migration and Urban Studies, International Institute for Population Sciences (IIPS), Mumbai, India

Nilendu Chatterjee Department of Economics, Bankim Sardar College, South 24 Parganas, West Bengal, India

Soumen Chatterjee Department of Geography, Visva Bharati University, Santiniketan, India

Mamta Chaudhary School of Health Systems Studies, Tata Institute of Social Sciences, Mumbai, India

André Santos da Rocha Laboratory of Economic Geography, Policy and Planning(LAGEP), Federal Rural University of Rio de Janeiro (UFRRJ), Seropédica, Brazil

Leandro Dias de Oliveira Laboratory of Economic Geography, Policy and Planning(LAGEP), Federal Rural University of Rio de Janeiro (UFRRJ), Seropédica, Brazil

Pamela Deb Department of Geography, Visva-Bharati, Santiniketan, West Bengal, India

Bożena Degórska Institute Geography and Spatial Organization, Polish Academy of Sciences, Warsaw, Poland

Marek Degórski Institute Geography and Spatial Organization, Polish Academy of Sciences, Warsaw, Poland

M. D. K. L. Gunathilaka Department of Geography, University of Colombo, Colombo, Sri Lanka

Krishnendu Gupta Department of Geography, Visva Bharati University, Santiniketan, India

Pablo Ibanez Laboratory of Economic Geography, Policy and Planning(LAGEP), Federal Rural University of Rio de Janeiro (UFRRJ), Seropédica, Brazil

N. C. Jana Department of Geography, The University of Burdwan, Burdwan, West Bengal, India

Sk. Karim Department of Migration & Urban Studies, International Institute for Population Sciences, Mumbai, India

Bappaditya Koley Department of Geography, Bankim Sardar College, South 24 Parganas, West Bengal, India

C. M. Lakshmana Population Research Centre, Institute for Social and Economic Change, Bangalore, India

L. Manawadu Department of Geography, University of Colombo, Colombo, Sri Lanka

Rahul Mishra International Institute for Population Sciences (IIPS), Mumbai, India

Mukunda Mishra Dr. Meghnad Saha College, Itahar, West Bengal, India

Rameswar Mukherjee Department of Geography, Samsi College, Malda, West Bengal, India

Somnath Mukherjee Bankura Christian College, Bankura, West Bengal, India

Anindita Nath Department of Geography, Bankim Sardar College, South 24 Parganas, West Bengal, India

Kalosona Paul Department of Geography, Sidho-Kanho-Birsha University, Purulia, West Bengal, India

Pushkar K. Pradhan Central Department of Geography, Tribhuvan University, Kathmandu, Kirtipur, Nepal

Subham Pramanick Department of Geography, University of Calcutta, Kolkata, India

César Adrián Ramirez-Miranda Laboratory of Rural Regional Development, Chapingo Autonomous University (UaCh), Texcoco, Mexico

Guilherme Ribeiro Laboratory of Economic Geography, Policy and Planning(LAGEP), Federal Rural University of Rio de Janeiro (UFRRJ), Seropédica, Brazil

Baikunth Roy PG Department of Economics, College of Commerce, Arts and Science, Patliputra University, Patna, India

Utpal Roy Department of Geography, University of Calcutta, Kolkata, India

Marcos Aurelio Saquet Integrated Laboratory of Territorial Studies (GETERR), State University of Western Paraná (UNIOESTE), Cascavel, Brazil

Raju Sarkar Population Research Centre, Institute for Social and Economic Change, Bangalore, India

Tanmoy Sarkar Department of Geography, Malda, West Bengal, India

Puspa Sharma Central Department of Geography, Tribhuvan University, Kathmandu, Kirtipur, Nepal

R. B. Singh Department of Geography, Delhi School of Economics, University of Delhi, Delhi, India

Mst. Rebeka Sultana Department of Geography and Environmental Studies, University of Rajshahi, Rajshahi, Bangladesh

Syfujjaman Tarafder Gour Mahavidyalaya, Malda, West Bengal, India

V. P. I. S. Wijeratne Department of Geography, University of Colombo, Colombo, Sri Lanka

Arun Kumar Yadav Evidence Action's Technical partner in India, Patna, Bihar, India

Vidya Yadav P.G. Department of Geography, College of Commerce Arts & Science, Patna, Bihar, India

Abbreviations

ABAD	Apni Beti Apna Dhan (India, Development Scheme)
ADB	Asian Development Bank
APP	Agricultural Perspective Plan (Nepal, Development Scheme)
APY	Atal Pension Yojana (India, Development Scheme)
ARI	Acute Respiratory Infection
BBBP	Beti Bachao Beti Padhao (India, Development Scheme)
BLHI	Boundary Layer Heat Island
BSF	Border Security Force (India)
BSY	Balika Samriddhi Yojana (India, Development Scheme)
CAGR	Compound Annual Growth Rate
CBS	Central Bureau of Statistics (Nepal)
CDC	(U.S.) Centers for Disease Control and Prevention
CDG	Central Department of Geography (of Tribhuwan University)
CENVyTT	Centro de Validación y Transferencia de Tecnología (Center for Validation and Technology Transfer, Brazil)
CHC	Community Health Center (India)
CLHI	Canopy Layer Heat Island
COMPASS	Comparative Analysis of Territorial Governance and Spatial Planning Systems (Project)
CONACyT	Consejo Nacional de Ciencia y Tecnología (Council for Science and Technology, Brazil)
CPR	Common Property Resources
CSR	Child Sex Ratio
CVM	Contingent Valuation Method
DALY	Disability-Adjusted Life Years
DFID	Department for International Development
DOLIDAR	Department of Local Infrastructure Development and Agricultural Roads (Nepal)
DOR	Department of Roads (Nepal)
DPIIT	Department for Promotion of Industry and Internal Trade (India)
EAG	Empowered Action Group

EAPS	Administrative (Public) Service Accessibility
EASS	Social Service Accessibility
EEA	European Environment Agency
ESPON	European Spatial Planning Observation Network
EU	European Union
FAO	Food and Agriculture Organization
FDI	Foreign Direct Investment
FGD	Focus Group Discussion
FY	Financial Year
GDP	Gross Domestic Product
GFP	Garment Factory Program (Sri Lanka, Development Scheme)
GIAP	Grupos de Investigación Acción Par-ticipativa (Participatory Action Research Groups)
GIS	Geographical Information System
GOI	Government of India
GP	Gram Panchayat (India)
GPI	Gender Disparity Index
GPI	Gender Parity Index
GUS	Growing Up in Scotland (data product)
GVA	Gross Value Added
GWR	Geographically Weighted Regression
HDI	Human Development Index
HDR	Human Development Report
HRBH	High-rise building households
HVI	Household Vulnerability Index
IBAMA	Instituto Brasileiro do Meio Ambiente e dos Recursos Naturais Renováveis (Brazilian Institute of the Environment and Renewable Natural Resources)
ICDS	Integrated Child Development Services (India, Development Scheme)
ICIMOD	International Centre for Integrated Mountain Development (Nepal)
ICMBIO	Instituto Chico Mendes de Conservação da Biodiversidade (Chico Mendes Institute for Biodiversity Conservation)
IDI	In-Depth interview
IICA	Instituto Interameri-cano de Cooperación para la Agricultura (Inter-American Institute for Cooperation on Agriculture)
IPCC	Intergovernmental Panel on Climate Change
IRDP	Integrated Rural Development Program (Sri Lanka)
LEADER	Liaison Entre Actions de Développement de l'Économie Rurale (Link between Actions of Development of Rural Economy, Brazil)
LMIC	Low and Middle Income Country
LPG	Liquid Petroleum Gas
LST	Land Surface Temperature
LULC	Land-Use and Land Cover (LULC)
LWE	Left Wing Extremism (India)
MDG	Millennium Development Goal

MDP	Mahaweli Development Project (Sri Lanka)
MESMIS	Marco para la Evaluación de Sistemas de Manejo incorporando Indicadores de Sustentabilidad (Framework for the Evaluation of Management Systems incorporating Sustainability Indicators, Brazil)
MHADA	Maharashtra Housing and Development Authority (India)
MHUPA	Ministry of Housing and Urban Poverty Alleviation (India, Development Scheme)
NAIi	Network Accessibility Index
NDBI	Normalized Difference Built Index
NDVI	Normalized Difference Vegetation Index
NGO	Non-Governmental Organization
NHFS	National Family Health Survey
NORAD	Norwegian Agency for Development Cooperation
NPP	National Population Policy (India)
NPP	Nature Protection Potential
NRC	Nepal Resettlement Company
NTFP	Non-Timber Forest Produce
OAI	Overall Accessibility Index
OBC	Other Backward Classes (India)
OECD	Organization for Economic Co-operation and Development
OLS	Ordinary Least Square
ORS	Oral Rehydration Salt
PCA	Principal Component Analysis
PHC	Primary Health Center (India)
PMJJBY	Pradhan Mantri Jeevan Jyoti Bima Yojana (India, Development Scheme)
PMSBY	Pradhan Mantri Suraksha Bima Yojana (India, Development Scheme)
PROSA	Regional Prototypes for Food Security and Sovereignty and the Fight against Poverty (Brazil)
PWD	Person with Disability
RCC	Reinforced Cement Concrete
RCM	Replacement Cost Method
Red GTD	Red de Gestión Territorial para el Desarrollo Rural Sustentable (Territorial Management Network for Sustainable Rural Development, Brazil)
RGSEAG	Rajiv Gandhi Scheme for Empowerment of Adolescent Girls-SABLA (India, Development Scheme)
RMK	Rashtriya Mahila Kosh (India, Development Scheme)
SC	Scheduled Castes (India)
SCP	Semi-automatic Classification (Plugin in QGIS)
SDA	Southern Development Authority (Sri Lanka)
SDG	Sustainable Development Goal
SGRY	Sampoorna Gramin Rozgar Yojana (India, Development Scheme)
SIDA	Swedish International Development Cooperation Agency

SIRGI	Sistema Regional para la Gestión de la Innovación (Regional System for Innovation Management, Brazil)
SREC	Students Research Ethics Committee
SRH	Slum Rehabilitated Houses
SRTM	Shuttle Radar Topography Mission
ST	Scheduled Tribes (India)
SUHI	Surface Urban Heat Island
TADP	Transformation of Aspirational Districts Programme (India)
ToA	Top of the Atmosphere
TSS	Transport Service Score
UHI	Urban Heat Island
ULB	Urban Local Bodies
UMC	Urban Municipal Corporation
UN	United Nations
UNDP	United Nations Development Program
UNEP	United Nations Environment Programme
UNESCO	United Nations Educational, Scientific and Cultural Organization
UNHCR	United Nations High Commissioner of Refugee
UNICEF	United Nations International Children's Emergency Fund
VIF	Variance Inflationary Factor
VPI	Village Priority Index
WASH	Water Supply, Sanitation, and Hygiene (conditions)
WFI	Weighted Facility Index
WHO	World Health Organization
WPA	World Programme of Action
WTP	Willingness to Pay

List of Figures

List of Tables

Part I
Sustainable Regional Development

Chapter 1
Sustainable Regional Development: Environmental Practices

Marek Degórski and Bożena Degórska

Abstract In Poland, as in the entire European Union (EU) and in many countries of the world, sustainable development, as an evolving and multidimensional concept, is widely used in defining development directions, in which environmental goals are considered increasingly important. The aim of this study is to indicate integrated approaches and instruments in the environmental dimension responding to challenges of great importance for the sustainable development of Polish regions and to determine the interdependence between natural capital and the achievement of sustainable development by regions. On the basis of the ranking of Polish regions in terms of the level of sustainable development achieved in 2005–2010 and 2011–2015 and changes in the place of regions in the analyzed periods, regions of success, and weaker regions were identified. The chapter also deals with natural capital as a factor of human development and well-being, integrated approaches, and regional instruments of high importance for achieving environmental goals.

Keywords Sustainable development · Regional development · Natural capital · Ecosystem services · Circular economy · Poland

1.1 Introduction

In view of the limited natural resources at the Earth's disposal, climate change, decline in biodiversity, environmental pollution, and rapid demographic growth in connection with the processes of urbanization and growing demand for space and consumer goods, it becomes crucial to use solutions that ensure the maintenance of the ability of ecosystems to provide services without risk for the preservation of the stability of the natural system with development corresponding to the present challenges and without detriment to future generations. Therefore, sustainable development (SD) as an evolving and multidimensional concept, based on a set of values and positive principles in achieving social, environmental, and economic goals, taking into account

M. Degórski (✉) · B. Degórska
Institute Geography and Spatial Organization, Polish Academy of Sciences, Warsaw, Poland
e-mail: m.degor@twarda.pan.pl

M. Mishra et al. (eds.), *Regional Development Planning and Practice*,
Advances in Geographical and Environmental Sciences,
https://doi.org/10.1007/978-981-16-5681-1_1

the integration of multi-faceted activities, from the end of the twentieth century has become a leader in creating policies and solutions adopted in Poland, in the EU and in the world. More and more attention is paid to taking into account the mutual relations that give synergistic effects of SD (Kroll et al. 2019). Thus, the synergy effect adds value to such a development. It seems very important to distinguish between the terms "sustainability" and "sustainable development." Shaker (2015) suggests that "the term 'sustainability' should be viewed as humanity's target goal of human—ecosystem equilibrium (homeostasis), while 'sustainable development' refers to the holistic approach and temporal processes that lead us to the endpoint of sustainability."

Despite its ambiguity and an evolving approach, SD is a positive vision of development at present and in the future, but often due to conflicts of interest, it requires the reconciliation of different, sometimes contradictory values and goals toward a new synthesis (Kates et al. 2005). The most common conflicts occur between economic development and environmental protection.

Therefore, in the face of unfavorable phenomena, the environmental goals of SD play an increasingly important role, although the perception of their importance over the last quarter century has fluctuated and differed in approach, even within the EU. According to experts from EU countries, SEA and the Habitats and Birds Directives are indicated as the most influential on spatial planning and territorial management in the context of sustainable development, as the basis for the creation of the Natura 2000 network, as well as the Water Directive (Komornicki et al. 2018b).

Due to the implementation of the law and the implementation of the directions and objectives of the EU policy to national systems, more and more emphasis is placed on integration between social, environmental, economic, and institutional orders, and broadly understood environmental objectives accompany currently the majority of policies, programs, and plans of general and sectoral character, at all levels of development planning and territorial management. It should be emphasized that SD was adopted as the basis for the development of Poland, which is clearly stated in Article 5 of the Constitution of the Republic of Poland of 2nd April 1997. "The Republic of Poland shall safeguard the independence and integrity of its territory and ensure the freedoms and rights of persons and citizens, the security of the citizens, safeguard the national heritage and shall ensure the protection of the natural environment pursuant to the principles of sustainable development."

Strategic documents (policies, strategies, programs) prepared at the national level have adopted various approaches toward taking into account the SD environmental objectives. Currently, the basis of the state policy is the Strategy for Responsible Development until 2020 (with a perspective until 2030) adopted in 2017 (MIiR 2017). Its major aim includes creating conditions for the growth of income of Polish residents with a simultaneous increase in social, economic, environmental, and territorial cohesion, although environmental objectives are less emphasized. Already in the period before Poland's accession to the EU structures, there were first documents drawn up that included environmental issues and activities of integrated nature in the achievement of goals of sustainable development (Poland's Sustainable Development Strategy until 2025, National Environmental Policy and the Concept of the

National Spatial Development Policy). A similar approach characterizes the National Spatial Development Policy, adopted in 2019.

A binding document of the regional policy conducted by the state, setting out the priorities of regional development, is the National Strategy of Regional Development 2030, adopted in 2019, although its priority, i.e., territorially balanced development, is perceived primarily as reducing disproportions in the level of socio-economic development of various territories. On the other hand, strictly with regard to ecological issues, it is the political framework that determines the successive editions of the Environmental Policy. The last one (State Environmental Policy 2030—development strategy in the area of environment and water management), adopted in 2019, sets the main goal as: "Development of the potential of the environment for the benefit of citizens and entrepreneurs," and the objectives were defined in response to the most important trends in the area of environment, adopting the integration of environmental considerations with economic and social needs.

The development strategies of voivodeships are the basic and most important strategic documents of regional self-governments that define the areas, goals, and directions of development policy interventions carried out in the regional space. SD is usually the overriding goal in most regional strategies, which also define numerous goals relating to the broadly understood environment. A reference to other regional policy instruments of high importance for the environment is provided later in this chapter.

The aim of the study is to indicate integrated approaches and instruments in the environmental dimension, responding to challenges of great importance for the sustainable development of Polish regions, and to determine the interdependence between natural capital and the achievement of SD by regions, as well as to present the ranking of regions in terms of SD.

The research referred to the division of Poland into administrative units in the rank of voivodeships (NUTS 2), in which the Warsaw Capital Region and the Mazowiecki Regional Region were treated jointly (Mazowieckie Voivodeship). In this analysis, voivodeships were treated as regions. The results of the Kłóska ranking analysis (2017) for the years 2005–2015 were used to evaluate the regions in terms of achieving SD. The index of the averaged ranking place was calculated for the periods 2005–2010 and 2011–2015 together with the place of the region in these periods. In order to indicate positive or negative directions of changes in the implementation of sustainable development, changes in the place of regions between these periods were determined. High-ranking places and promotion by at least two positions were treated as the region's success in achieving SD. With regard to natural capital, a ranking analysis was performed for the following features: percentage of: (i) forests, (ii) meadows and pastures, (iii) surface waters, (iv) national parks and reserves, and also for (v) state of the population of important protected animals without the beaver (Castor fiber), which is common in Poland, as well as (vi) soil quality for agriculture. They served as the basis to define the ranking of regions with respect to each variable, to calculate the Averaged Ranking Index, and to determine the final ranking places of the regions. The ranking of regions in terms of natural capital was made on the basis of GUS data from 2018 (GUS 2019). The strength of the relationship between SD and natural capital was analyzed by use of Spearman's rank correlation.

1.2 Natural Capital as a Factor of Human Development and Well-Being

In the opinion of some authors, in economic theories of the end of the twentieth century, the problem of the role of the environment in economic growth formed the basis of the concept of sustainable development (Harding 2006). This means that economic and social development in many countries of the world should take place with parallel respect to the stability of the natural environment, in such a way that human interference does not violate the limits of nature tolerance, the principles of social justice, and human well-being. Taking into account the above conditions, the natural potential of the environment, constituting the basis for ecosystem services, can be treated as an equal factor of sustainable development, especially of regional development, which together with economic and social factors significantly affects the directions of activities and methods of their implementation (Degórski 2007, 2009).

One of the recently developed concepts providing the methodological basis for a multi-directional assessment of environmental benefits is the so-called ecosystem services (Dale and Beyeler 2001; Barbier 2007; Boyd and Banzhaf 2007; Scholte et al. 2015). In research conducted in the field of human-environment interdependence, the concept of ecosystem services describes the relationship of the natural system with the socio-economic system, focusing at the same time on the benefits that humans derive from the environment (Fig. 1.1). It also takes into account an attempt to evaluate these benefits economically (Balmford et al. 2008; de Groot et al. 2010, 2012; Burkhard et al. 2012, 2018; Alam 2018), as well as activities supporting the implementation of SD assumptions (Vihervaara et al. 2010).

Pioneering research on the valuation of ecosystem services was carried out at the end of the twentieth century by the team of Costanza et al. (1997), which raised interest in the topic and sparked a lively discussion. On their basis, many critical comments were formulated, which became the driving force for further work, both methodical and organizing (MEA 2005; TEEB 2010; Haines-Young and Potschin 2018), as well as verifying of applicable nature (Agbenyega et al. 2009; Boerema et al. 2016). Much attention is devoted to the indicative role of services at various levels of spatial organization, or in various functional areas (Bolund and Hunhammar 1999; De Groot et al. 2012), mainly in the context of supply services (satisfying basic human needs) and regulatory services (making use of biogeophysical and biogeochemical processes in improving the state of the environment and the quality of human life). Also, the cultural ecosystem services play an important role in the development of regions (Sagie et al. 2013; Scholte et al. 2015). Additionally, they can play an

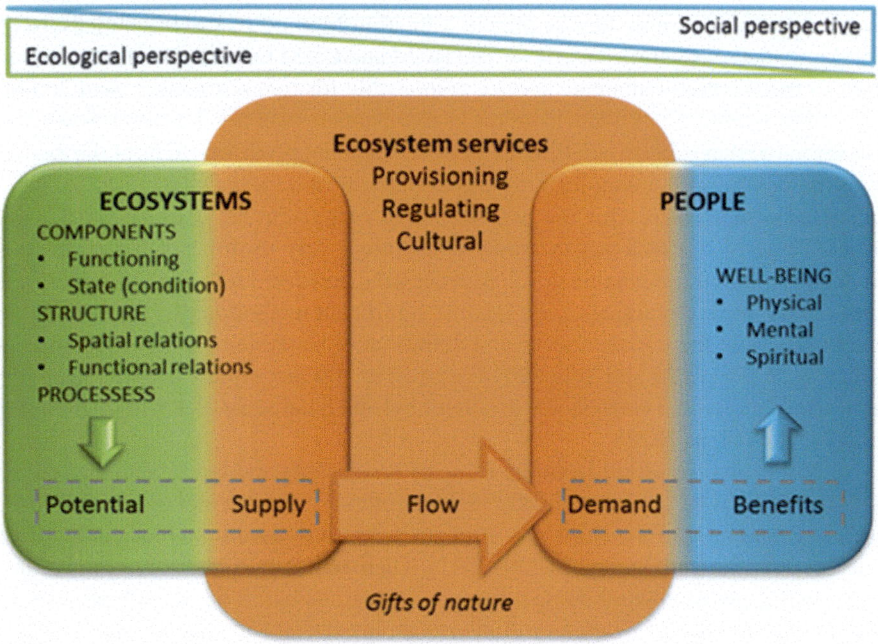

Fig. 1.1 The relationships between nature and society within the ecosystem services concept (Prepared by the authors after Affek et al. 2020)

important role in integrating science with sector policies and regional management (Degórski 2012b).

It was attempted in Poland to use ecosystem services to shape sustainable development at the regional level (Mizgajski and Stępniewska 2009a), as well as to quantify and evaluate individual services, their practical use in spatial planning (Raszka and Hełdak 2013), and in the optimization of functioning of some activities of the water-sewage system (Mizgajski and Stępniewska 2009b). These works may also be of great importance for creating directions of development and spatial management using the potential of the environment and landscape, especially in zones of strong spatial urbanization (Degórski 2013, 2017), as well as agricultural, tourist, and recreational regions (Affek et al. 2020). The services of water ecosystems are also very important for sustainable development, particularly in regions with a high share of inland waters or with access to the sea, especially that as a result of increasing human pressure, their proper functioning is severely limited. According to the Helsinki Commission, among 24 identified ecosystem services provided by the Baltic Sea, only ten functions well, while seven are threatened, including the ability of the ecosystem to regenerate and buffer anthropogenic waste input (HELCOM 2010). For example, in order to assess the possibility of using ecosystem services to optimize protective activities related to the waters of the Gulf of Gdańsk and the development of coastal regions, the Replacement Cost Method (RCM) was used as a measure of revealed preferences and the Contingent Valuation Method (CVM) as a measure of replacement costs (Kamińska 2019). The RCM and CVM have shown that the value of the eutrophication balancing service responsible for removing more than 20,000 tons of nitrogen through denitrification in sediments is over $ 0.5 trillion, while the average value that society is willing to bear in financing programs to fight the causes of eutrophication is a few dollars per capita in the form of a tax (Kamińska 2019).

The sets of measures for the assessment of ecosystem services already at the present stage of research advancement constitute a very useful tool for evaluating the achievement of sustainable development goals; however, in Poland, they have not been widely used in practice so far. Despite significant methodological progress, the lack of clarity in the understanding and definition of ecosystem services constitutes still an issue (Degórski and Solon 2014) as well as still incomplete knowledge about the value of individual ecosystems on a regional and local scale, determining natural capital (Affek et al. 2020).

Natural capital resulting from the value of ecosystems and landscape, the size of natural resources, i.e., the potential of the environment, is more and more often indicated as a generator of development, strengthening, or determining the development trajectory of regions (Degórski 2010, 2011). There are studies undertaken in Poland to assess the environmental competitiveness of regions based on the potential value of the environment and its quality. Analyzes are performed at various spatial scales, from local to national (Bródka 2010; Kasztelan 2011; Bielska and Krupińska 2016). In deliberations on the analysis of the geographic diversity of potentials, spatial econometric models (Ratajczak 2008; Domański 2008; Degórski 2010) and factor analysis (Borys 2005; Rozmus and Trzęsiok 2018) are used with great success. The results of these and similar analyzes, including the SD index assessment, ecosystem services, analysis of the potential of municipalities in the area of nature protection (Degórski 2012a), and delimitation of areas constituting strategic intervention space for the state (Śleszyński et al. 2017), were used in formulating conclusions and expert recommendations for national strategic documents and above all, for the National Spatial Development Concept 2030 (Degórski 2008; Korcelli et al. 2010). One of such key analyses in the area of natural heritage protection was the determination of the nature protection potential (NPP) for all communes in Poland (Fig. 1.2).

The specific spatial differentiation of the value of the indicator of nature protection potential of communes demonstrated that mountain and lake regions have the highest protection potential. In both cases, these are areas with relatively high forest cover and a large share of permanent meadows and pastures. The most valuable areas include: the Carpathian and foothills mountain ranges extending in the south of Poland (1), areas located in the latitudinal lake district—South Baltic Lake District (2), Eastern Baltic Lake District (3), Podlasie Region, (4) with extensive wetlands of the Biebrza Basin, the large forest complex of the Białowieża Primeval Forest, constituting area inscribed on the UNESCO heritage list, the Augustów and Knyszyńska Primeval Forests. There are also Roztocze (5) extensive ranges of hills of attractive landscape, (6) Barycz Valley with the largest complex of ponds in Europe (Stawy Milickie), and (7) Bory Dolnośląskie—the largest compact forest complex in Poland.

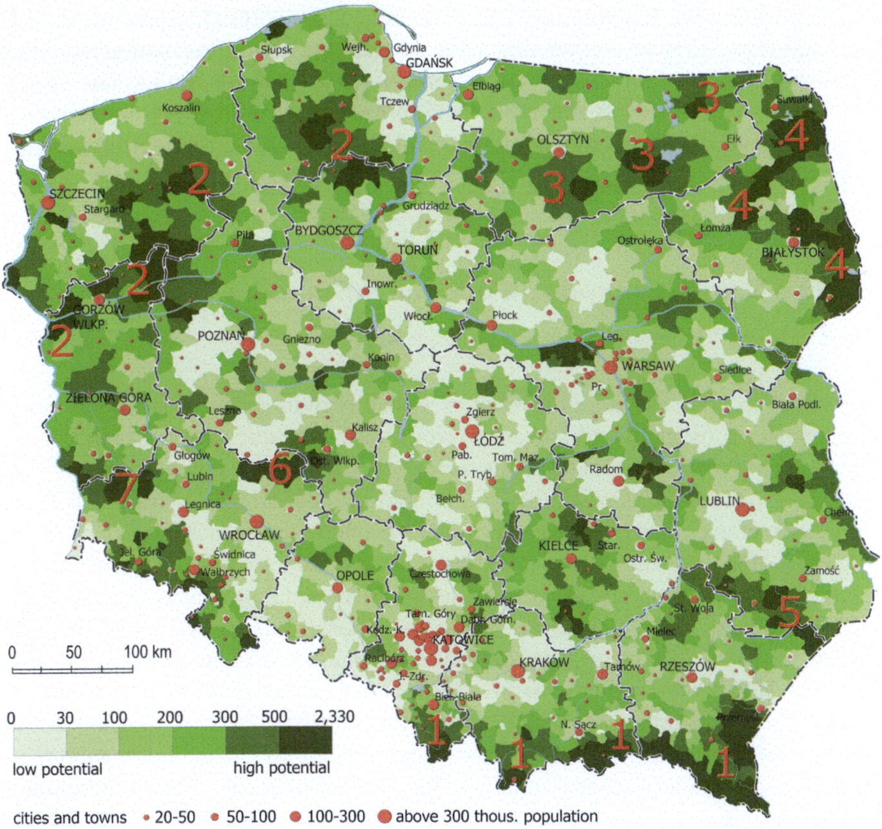

Fig. 1.2 Indicator of the nature protection potential of communes (prepared by the authors after Degórski 2012a, 2020) with the most valuable areas from protection potential marked with the numbers 1 to 7

The optimal use of the resources and potential of the natural environment together with the protection of its most valuable fragments, apart from human resources and economic potential, constitute currently one of the pillars of the regional development strategy in the European Union countries. Also, in Poland, it gains more practical application on a broader scale. The emphasis on environmental issues varies at the regional level between general documents adopted by local governments, such as Voivodeship Development Strategies and some sectoral documents. It is usually the strongest in areas focused on developing tourism based on natural heritage. The research results so far indicate that resources and endogenous potential of the environment of a given region with similar economic and social conditions may determine the strength and directions of its development and the quality of human life to various degrees (Morris and Therivel 2000; Degórska and Degórski 2003). However, it seems that in regional development policy, natural capital is not always perceived to an adequate degree as an essential factor of human well-being and as an important factor influencing the functioning of the socio-economic system. However, the EU policy and legislation in conjunction with the support within the cohesion policy increasingly impose the inclusion of SD objectives in socio-economic and spatial development, which triggers that more and more attention is focused on the mutual relations between the environment and sectoral policies, such as agricultural, urban, energy, and transport policies.

1.3 The Nature Component and the Sustainable Regional Development

Taking into account the large diversity of Polish regions in terms of biodiversity, and especially ecosystem and landscape biodiversity, the question arises as to how the natural component affects the achievement of SD objectives by the regions. Two things were important to answer this question. First concerning determination of the ranking of regions in terms of natural capital, second in terms of SD.

In terms of the natural environment, Poland is a highly diversified country due to the large landscape diversity. In the system similar to the latitudinal, there are mountain, piedmont, upland, lowland, lake-side, and coastal landscapes occurring from the south to the north. The high natural value of Poland is evidenced by the fact that as much as 30% of the country is covered by areas subject to legal protection, approx. 20% of the country's area are habitat areas of the Natura 2000 network, and about 30% constitute forests. Obtained results of the ranking regarding the natural components indicate that the leaders are mountain regions of the Polish part of the Carpathians, i.e., Podkarpackie (1st place), Małopolskie (3rd place) as well as the coastal and lake region—Pomorskie (2nd place). Agricultural regions are placed further in the ranking, with the last place taken by the Łódzkie Voivodeship constituting one of the most deforested in the country (Fig. 1.3).

Most regions with high natural value in Poland are characterized by peripheral location, depopulation processes, low level of urbanization, and economic weakness. In these regions, the potential of endogenous development resulting from large natural capital may therefore stimulate development and facilitate the achievement of SD, e.g., by focusing the development on tourism, based on the regional natural and cultural heritage, clean environment, and attractive natural landscape. Regions of high-ranking places concentrate on numerous forms of national and international nature protection, mainly national parks, which constitute key Natura 2000 sites and landscape parks. This result corresponds to the location of areas with a high potential for nature protection (Fig. 1.2). Additionally, such regions are characterized by qualified human resources in the area of environmental protection, planning, and management, by the local community with higher ecological awareness, than in other regions and have a priority in obtaining EU financial support in the implementation of SD objectives.

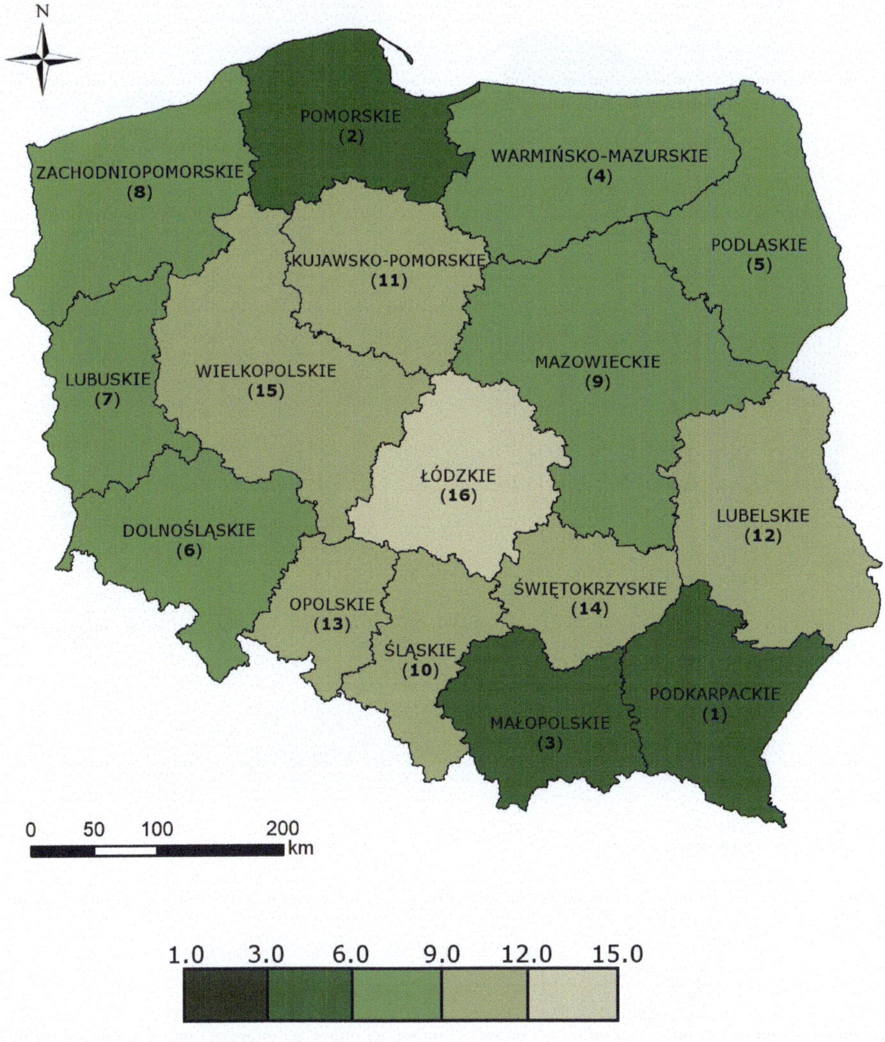

Fig. 1.3 Natural capital ranking of Polish regions (*Source* Authors' own study based on GUS 2019)

Referring to the area of sustainable development, the region that achieved the highest place in the ranking in 2005–2015 was Mazowieckie, as indicated by the total sustainable development indicator determined for individual regions of Poland (Klóska 2017). While performing a comparison of the ranking places determined on the basis of the value of the average ranking position index for the periods 2005–2010 and 2010–2015, initially Mazowieckie was the vice-leader after Pomorskie voivodeship (Fig. 1.4). However, after 2010 for Mazowieckie, it decreased slightly, which means an increase in the place and moving to the leader position in the SD ranking (Fig. 1.5). Thus, the first place in the ranking was taken by the region with the highest

Fig. 1.4 Ranking of regions (numbers in parentheses) and the average place of region in years 2005–2010 (*Source* Authors' own study based on Klóska 2017)

indicators of socio-economic development, concentrating the capital (Warsaw) and metropolitan functions (Warsaw Metropolitan Area), with the highest GDP in Poland (160% of the national average). However, the region occupies a much weaker position in terms of nature conservation (Fig. 1.2) and natural capital (Fig. 1.3).

The next regions in the ranking in terms of SD are Pomorskie and Małopolskie, i.e., regions with high economic and social potential, but unlike Mazowieckie, also with a very high natural potential (Fig. 1.2). It should be emphasized that in some regions, the basic determinants of SD are socio-economic conditions. Such regions

Fig. 1.5 Ranking of regions (numbers in parentheses) and the average place of region in years 2011–2015 (*Source* Authors' own study based on Kłóska 2017)

are often affected by social and economic problems, with examples of Warmińsko-Mazurskie and Podlaskie, which despite the great natural potential, are placed further in terms of the overall SD level. However, in case of a large number of regions with a relatively low level of socio-economic development, they are also characterized by a low place both in the overall assessment of SD (Fig. 1.4) and in the ranking of natural capital (Łódzkie, Świętokrzyskie, Lubelskie—Fig. 1.3).

Similar conclusions can be found in the literature on the subject. It is emphasized that taking into account the division of SD indicators into the social, economic, and environmental order, and there is a differentiation of leaders between individual dimensions. According to E. Roszkowska and R. Karwowska (2014), the best parameters of sustainable development in economic terms are characteristic of Dolnośląskie, Mazowieckie, and Pomorskie voivodeships, i.e., the richest regions with the best-developed infrastructure and service sector. In terms of social dimension, the leaders are Pomorskie, Wielkopolskie, and Mazowieckie. In the environmental dimension, the leading places in the ranking are occupied by Podlaskie, Podkarpackie, Małopolskie, Warmińsko-Mazurskie, and Pomorskie, which corresponds to the ranking of regional leaders in terms of natural capital (Fig. 1.2). According to K. Drabarczyk (2017), in terms of implementing the pro-ecological solutions from the functioning of the social perspective, the highest standards are characteristics of the Mazowieckie, Łódzkie, Opolskie, and Wielkopolskie voivodeships. Summing up, in all dimensions of SD, the Mazowieckie Voivodeship is in the forefront, as already mentioned, the leader in terms of SD. Performed statistical analysis proves that in practice, the level of regions' SD is determined by two fundamental issues to a similar extent, i.e., the natural capital of the region and economic factors. This is confirmed by the results of two statistical analyses (Spearman's rank correlation). Both indicate a statistically significant relationship. First one, between the region's place in the ranking in terms of level of sustainable development and the ranking according to the average position of the studied characteristics of the environment ($r = 0.63$). Second, between the place of the region in the ranking according to the level of sustainable development and the place defined on the basis of the level of gross domestic product per capita of a given region ($r = 0.61$).

Referring to individual characteristics of the environment (percentage share of forests, meadows and pastures, surface waters, national parks and reserves, as well as state of the population of important protected animals without common in Poland beaver (Castor fiber) and the quality of soils for agriculture) the strongest relationship of the places of regions in the ranking in terms of the level of sustainable development was determined toward the state of the population of important protected animals ($r = 0.90$), while the weakest toward the share of national parks ($r = 0.49$) and forests ($r = 0.47$). No statistically significant correlation was found against other environmental characteristics.

The success or failure of regions in pursuing sustainable development policy is in practice evidenced not only by leading places in the ranking but also by changes in places of regions. The assessment of these changes between 2005–2010 and 2011–2015 (Fig. 1.6), based on changes in the value of the index of an average place of the region, showed that the increase in the place concerned seven regions, and therefore also seven regions were affected by a decrease in the ranking. Two regions (Świętokrzyskie and Lubelskie) did not change their place. It should also be emphasized that in 9 out of 16 regions, only slight changes were recorded, namely, increase or decrease by one place.

Changes by two or more places were treated as an indicator of the success or failure of the region. Similarly, high places in the ranking of regions in terms of SD are treated as a success; on the contrary, the last as a weakness. Comparing

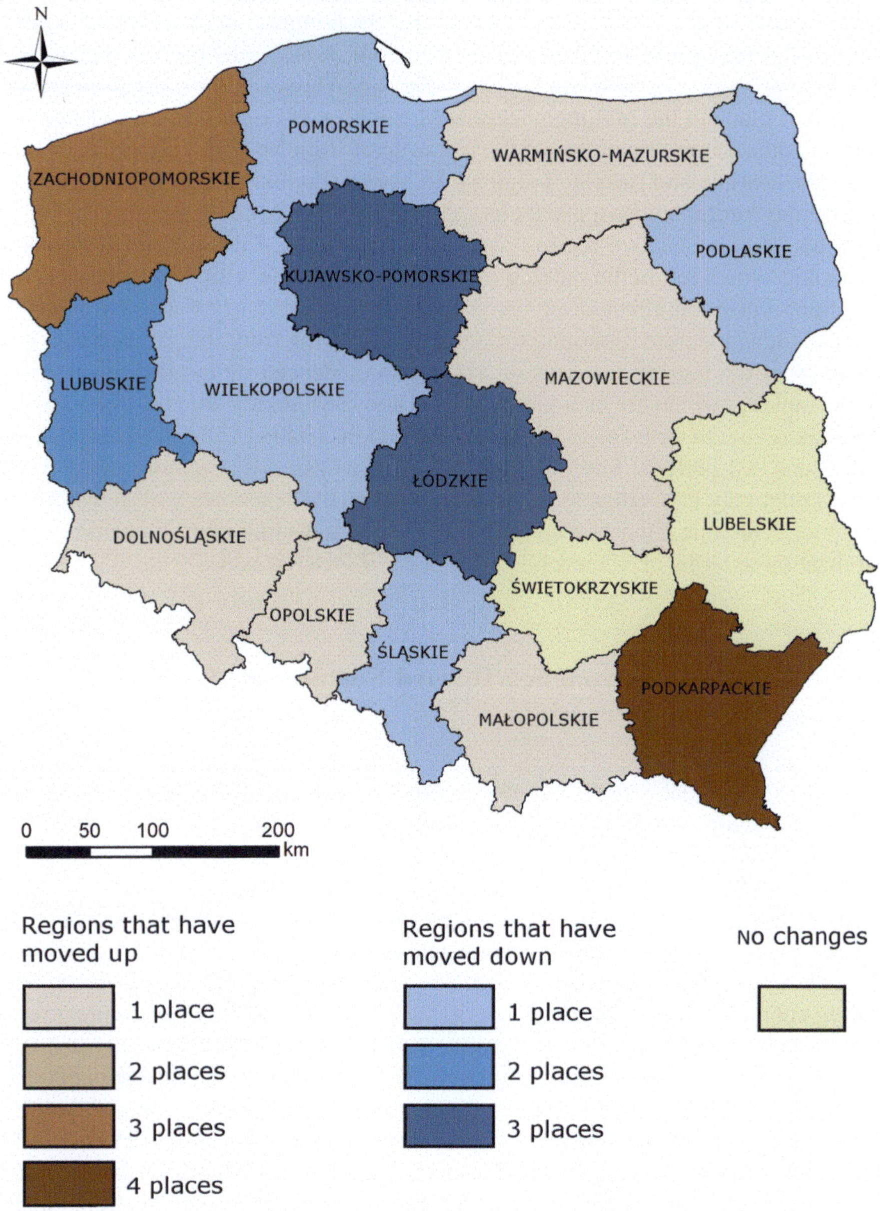

Fig. 1.6 The change of the place in the ranking determined for the years 2005–2010 and 2011–2015 (*Source* Authors' own analysis)

the two analyzed periods, we can conclude that the regions, which achieved significant success are regions such as Podkarpackie, which advanced by four places, and Zachodniopomorskie—up by three places (Fig. 1.6). However, the success measured by the first place in the ranking concerns Mazovia as the richest region of Poland. In conclusion, the improvement of the place, apart from Mazovia, concerns mainly mountain, seaside, and Lakeland regions, i.e., generally tourist regions, attractive in terms of the natural landscape and characterized by high natural capital.

In this assessment, two regions scored the worst, i.e., Łódzkie and Kujawsko-Pomorskie, which fell in the ranking by three positions (Fig. 1.6). Thus, the greatest reduction in position affected regions with the lowest and fairly low environmental potential, of a moderate economic situation, but suffering from social problems. The weakness of the regions in terms of SD is also evidenced by the last positions in the SD ranking, which are occupied by Lubelskie, Świętokrzyskie, i.e., the weakest and weak region in terms of economy. These regions had not changed their position in the analyzed periods. Summing up, it can be assessed that mainly agricultural regions relatively less valuable in terms of biodiversity resources with a weaker or very poor economic condition are characterized by a decline in their position or its stabilization in terms of SD level.

1.4 Integrated Approaches Toward Sustainable Growth and Regional Practices

In achieving SD goals, more and more attention is paid to the coherence between the environmental dimension and the social, economic, political, and institutional ones. Integrated measures are implemented, which are particularly important in the process of implementing multi-faceted and innovative projects with complex interdependencies between the environment and development. These include mitigation and adaptation to climate change and green economy, as well as the related low-carbon economy, resource-efficient economy, including circular economy and sharing economy. These issues operate in the direct feedback loop or indirect relationships and therefore require a systemic decision-making approach (e.g., income and employment growth should be driven by public and private investments that reduce carbon emissions and pollution, increase the energy and resource efficiency of the economy, and prevent loss of biodiversity and ecosystem services) (UNEP 2011).

The Europe 2020 Strategy provides the basis for such solutions. The three main priorities are: sustainable growth (this issue is discussed in more detail later in the chapter), smart growth, as well as inclusive growth. Sustainable growth means building a more resource-efficient, greener, and more competitive economy. Such an approach requires a transformation focused on adapting to functioning in a low-carbon economy and with limited resources while preventing environmental degradation, biodiversity loss, and unsustainable use of resources. Currently, it is also important to combat climate change, increase resilience to climate risks, and improve disaster prevention and response. (EUROPE 2020). The Voivodeship Development Strategies play a key and integrating role in determining the directions of sustainable development of regions, while the Voivodeship Spatial Development Plans and the Spatial Development Plans of Functional Urban Areas (Metropolitan Areas) in the coordination of spatial development. Their inclusion of environmental issues corresponds to the potential of endogenous development of the region related to the significant potential of the environment for tourism, resources of raw materials, or environmental pollution. The strategies are periodically assessed, taking into account the achievement of SD objectives with the use of SD indicators.

In order to achieve the ecological goals adopted in the national strategic documents, mainly in the National Environmental Policy, and to integrate activities aimed at improving the condition of the environment, the voivodeship self-government prepares an Environmental Protection Program for the voivodeship. The special role of the Program is to refer to the specific ecological problems of the region and the scale in which they occur and to indicate the most important tasks that should be performed in the region in a specific time perspective. Apart from improving the condition of the environment in the region, other important issues include reducing pollution, protecting biodiversity and developing environmental values, rational resource management, energy security, and adapting to climate change. The Environmental Protection Program is also used to support the decision-making process and harmonize the activities undertaken to introduce an ecological order in the voivodeship and to ensure that the inhabitants continuously improve their environmental conditions of living.

Environmental objectives also include numerous sectoral policies, strategies, plans, and programs, which are referred to later in this chapter. Therefore, it should be emphasized that the documents, apart from the integration of activities, may additionally contribute to achieving the synergy effect between ensuring socio-economic development with parallel creation of conditions for human well-being, and broadly understood protection and stability of the environment. Established priorities and objectives usually include new approaches focused on integrating socio-economic development with environmental improvement and adaptation to climate change.

The enormous role of the green economy is increasingly emphasized in achieving sustainable development. UNEP has defined the green economy as "one that results in improved human well-being and social equity, while significantly reducing environmental risks and ecological scarcities, which is a low carbon, resource-efficient, and socially inclusive" (UNEP 2011). It is similarly described by Green Economy Coalition (2011) as "a resilient economy that provides a better quality of life for all within the ecological limits of the planet." In general, this economy can be defined as a set of pro-ecological activities in the development processes of regions and countries (EEA 2016). It is indicated that in the green economy, growth in income and employment should be driven by public and private investments that reduce carbon emissions and pollution, enhance energy and resource efficiency, and prevent the loss of biodiversity and ecosystem services (UNEP 2011). The basis is the integration of sectoral economic policies and environmental policy, and the goal is to create green growth. At the same time, such a transformation in the perception of regional development has the potential to strengthen social capital and the already mentioned more equitable access to environmental services. Additionally, it is in line with the European Green Deal adopted by the European Commission as an action plan for a sustainable EU economy, the aim of which is to achieve by 2050 an improvement in the quality of the environment and the neutrality of the economy's impact on the climate on a continental scale. On the one hand, green economy and green growth can be perceived as one of the main pillars of sustainable development (Degórski 2013, 2017, 2018), but on the other, the question arises whether, due to the high convergence with the concept of SD, it will not replace the concept of SD commonly accepted by international bodies and all EU countries (Szarmej-Baran and Baran 2014). Regardless of the definition of green economy and green growth, the most important thing is that the ultimate goal is sustainable development.

In Poland, the key challenge for the implementation of the green economy is related to the achievement of the low-carbon economy goals, as the energy economy is still based on non-renewable solid fuels (hard coal 47.1%; lignite 29.1%) (GUS, 2019). Reasons for this unsatisfactory energy balance of Poland are related to both the historical conditions resulting from the availability of domestic hard coal and lignite resources and to the currently unstable energy policy of Poland, concerning mainly the unspecified future of coal, wind, and nuclear energy. Additional problems are created by a very low share of renewable energy sources (only about 11%), the high energy consumption of production, growing demand for energy resulting from technological progress, where the use of devices powered by electricity is becoming common, despite introducing solutions reducing their energy consumption. For example, as a result of civilization development, electricity consumption per capita has more than quadrupled in the last 50 years from around 900 KWh in 1960 to over 4000 KWh in 2019 (GUS 2020).

Nationwide, one of the significant goals of improving economic and ecological balance is the transition to a low-carbon economy, in which framework is set out in the National Energy and Climate Plan for 2021–2030 and the National Low-Emission Strategy. These strategic documents set out the visions and directions for the transformation of the Polish economy to a low-carbon economy and the achievement of climate neutrality by 2050. One of the significant goals of improving economic and ecological balance is the development of renewable energy sources. In this regard, the regional government is developing a Program for the Use of Renewable Energy Sources. Silesia, as a heavily polluted region, has prepared a draft of regional energy policy until 2030 (Low-Emission Economy Policy). It aims, inter alia, at defining the framework for the conduct of the region's development policy, which includes improvement of the quality of the environment, the increase in the competitiveness of the voivodeship, and the improvement of the quality of life of its inhabitants in the area of the energy economy, especially low-emission economy. For zones where air pollution levels are exceeded, regional governments develop Air Protection Programs, most frequently due to exceeded permissible levels of particulate matter—PM10, PM1.5, benzo (a) pyrene, and ozone.

Such activities are particularly important for the quality of the environment and the quality of life in industrialized regions and large urban agglomerations. According to the EEA Report (2017), air pollution contributes to about 48,000 premature deaths in Poland every year (in Warsaw, about 3,000). Forty-six thousand premature deaths are attributed to PM2.5 exposure; 1,700 is NO2; and 970 to O3 exposure. PM2.5 is considered the most harmful to human health among suspended components in the atmosphere. According to WHO report (2016), in 2012–2015, out of 50 UE cities and towns with the highest annual concentration of PM2.5, 33 are located in Poland, and 16 are in the top 20. This affects mostly the Silesia Region, which encompasses the most polluted cities in Poland and the EU.

The poor condition of air, which deteriorates significantly during the occurrence of smog episodes, is mainly caused by road transport and by burning low-quality solid fuels. Additionally, some heating stoves or small boilers of single houses and estates are outdated and poorly operated heating devices. Another factor refers to the common phenomenon of congestion, i.e., a greater volume of car traffic than the capacity of the road infrastructure, triggering road paralysis and generating large amounts of pollution. For example, currently, additional 0.5 million motor vehicles are poured into Warsaw every day, which in relation to the number of vehicles registered in the capital, nearly 1.3 million (GUS, 2016), significantly increases the traffic load and the scale of pollution. Dense urban development with the lack or inefficient ventilation system of many cities, especially with light winds, facilitates the concentration of pollutants. Moreover, current urban ventilation reveals the adverse phenomenon of building over the air exchange and regeneration systems. An example of losing the ventilation function by such systems is Warsaw, which had a model system of wedges providing ventilation designed in the 1980s, which to a large extent have already been built-up. Although spatial planning tries to integrate the planning of green and blue infrastructure with the city's ventilation system, the current legal system makes it difficult to act in this direction.

In order to improve air quality, and above all, to reduce the hazards associated with smog, in addition to the already mentioned Air Protection Programs, regional self-governments undertake the so-called Anti-smog Resolutions. They introduce restrictions and bans in the regions, mainly regarding heating installations and used fuels. Regions establish also support programs for clean air matters. For example, the Management Board of the Mazowieckie Voivodeship adopted for implementation in 2020 the "Mazowieckie Instrument for Air Protection Support MAZOWSZE 2020." Reduction of the negative impact of transport on the environment is achieved by the Sustainable Development Plans of Public Collective Transport prepared for regions, agglomerations, and cities. They tend to include on an ever wider scale the introduction of zero-emission vehicles.

An important pillar of the green economy is the resource-efficient economy. The stocks of non-renewable resources are depleted and can become unavailable within a measurable period of time or maybe drastically limited. An example is copper, in which resources with the current rate of consumption will end in about 40 years (Piłat 2017). In Poland, this raw material has been the basis for the functioning of the south-western macroregion, which encompasses major areas of copper ore mining and processing. In the longer term, a similar situation will also apply to other raw materials, especially coal, which is currently the basis of the Polish energy sector and the development of south-western regions (Śląskie, Dolnośląskie).

The circular economy concept gives some hope for a more resource-efficient economy and reduction of environmental burdens. Its assumptions result from new solutions in the economics of production and ecological economy. A circular economy minimizes the use of non-renewable resources and reduces the generation of waste and pollution, which also protects the environment. It is also important to reduce the environmental impact of the overproduction of consumer goods. According to Piłat (2017), we are currently irretrievably throwing away about 92% of all raw materials obtained for the production of products that we have used only once. It is assumed that materials used to produce goods can be recycled into production after their end of life (a production method called from cradle to cradle). From the economic point of view, this concept is a contradiction to the linear economy, which has currently served as a common production model, constituting the following pattern of conduct: to obtain raw materials, process them, and then freely dispose of the product (take, make, dispose). A circular economy supports a resource-efficient, energy-saving, and low-carbon economy and enables the safe introduction of waste into the biosphere by implementing innovative technologies and production solutions. According to the forecasts of Accenture Strategy, the circular economy supported by new business models and technologies may be worth up to USD 4500 billion by 2030 (Lacy and Rutqvist 2015).

In Poland's regional policy, more and more attention is paid to rational waste management, which refers to preventing and reducing the amount of industrial and municipal waste deposited in landfills, increasing recycling, increasing the safety of waste disposal, and the sustainable use of biomass, with particular stress put on minimizing biodegradable municipal waste sent to landfills. Regional perspective encompasses both weaknesses and slightly better sides of waste management. Although the majority of industrial waste is generated by regions with the mining industry located mainly in Dolnośląskie and Śląskie (54% in total), they have also seen the largest quantitative decline. However, large landfill areas constitute still an issue. In recent years with constant GDP growth, the amount of industrial waste generated per year has remained at a similar level (115.3 million tons, 2018, Central Statistical Office). However, the significant percentage increases in the amount of industrial waste in some regions (in Łódź by 42%) may be alarming, in Pomorskie by 25% compared to 2017. Concurrently, significant declines in the regions included in the Green Lungs of Poland (Warmińsko-Mazurskie by 28%, Podlaskie by 18%) can be indicated as a good side (GUS, 2019).

With regard to municipal waste, which accounts for almost 10% of all waste, there is a dichotomous division of Poland into the eastern and western parts. In general, eastern regions generate small amounts of municipal waste, including mixed waste going to landfills (the least in Lubelskie region is 107.8 kg per capita), which significantly distinguishes it from large amounts in the western regions (Dolnośląskie

247.3 kg per capita, Lubuskie 226.1 kg per capita) and seaside ones (Zachodniopomorskie 206.7 kg per capita, Pomorskie 190 kg per capita), (Environmental Protection, GUS, 2019). There is a visible link between a significant amount of generated municipal waste and GDP and the location of tourist regions. However, limitations in their production may be associated with a higher environmental awareness of the inhabitants of developing ecotourism regions with a large share of legally protected areas (eastern Poland). The amount of mixed municipal waste from households collected during the year per capita is also growing (from 158.4 kg in 2012 to 186 kg in 2019). The positive side constitutes a significant increase in the share of municipal waste collected separately during the year in relation to the total municipal waste (from 10.5% w 2012 to 31.2% in 2019).

Regions and municipalities are also struggling with an alarmingly high increase in waste imports and a particular threat of an increase in the number of fires of waste gathering sites, the largest number of which was recorded in the Śląskie (28) and Mazowieckie (28) voivodeships in 2018, while at the beginning of the twenty-first century they were sporadic situations (Environment, 2018). Unfortunately, in the area of municipal waste management, a further increase in the generated waste can be expected, because with a result of 325 kg per person, Poland is in the penultimate position in the EU, which produces an average of 486 kg per person, and neighboring Germany produces 633 kg.

Polish regions have a tool for organizing a waste management system—the Regional Waste Management Plan. According to EU requirements, it is a mandatory document in the planning, coordination, management, and monitoring processes. Thanks to multidimensional activities and support for investments with EU funds with a large participation of local governments, which have been responsible for municipal waste since 2012, including the introduction of selective collection and fees, regions manage to reduce the negative impact of waste on the environment, and therefore better implement the principles of sustainable development.

Sharing economy, also known as collaborative economy, as the same phenomenon was named so by the European Commission in 2016, or the "peer-to-peer economy," is becoming more and more popular due to new environmental-friendly social trends. Sharing economy equals sharing of assets, resources, time, skills, or capital. Unlimited consumption or property accumulation partially loses their importance to the exchange and lending of goods. The forms of collective consumption can take the form of one for profit and one non-for-profit and different business models. These models include collaborative economy, access-based consumption, collaborative consumption, peer-to-peer economy, on-demand services-applications (platforms) connecting consumers and suppliers of goods/services (Koźlak 2017) directly. Forms of shared consumption include cloth swap, toy swap, cohousing, room-sharing, couchsurfing, crowdfunding, coworking, and carpooling (Wardak and Zalega 2013; Ziobrowska 2017). According to the TNS Polska (2015) report, the sharing economy in Poland is gaining momentum, as 20% of residents made use of such services in 2015, i.e., only 8 percentage points less than the EU average, but together with the interested people, it was already the highest percentage in the EU. Due to high air pollution in most large urban agglomerations, car-sharing becomes one of particular importance. Shared use of passenger cars (car-sharing) reduces the number of vehicles and their kilometer mileage and thus reduces CO_2 emissions (Clewlow 2016). Moreover, access to shared cars results in their more optimal use (Martin 2016). In many large Polish cities, vehicles, cars, scooters, bicycles, and scooters are shared. Long-distance mobility is also developing. It includes traveling together in one car (ride-sharing), which in turn supports the development of tourism while reducing the traffic load (Degórski and Degórska 2019). Although car-sharing constitutes a partial alternative to car ownership and supplement to public transport and can result in environmental benefits, including reduction of congestion and smog problems, its prerequisite is a relatively high replacement rate for private cars (Tchorek et al. 2020).

The sharing economy certainly supports SD. However, this is a major challenge that requires a revolutionary change in business models and the functioning of the market due to constituting a threat to many traditional models, operating in opposition to the emerging, new, more "eco-friendly" consumer model. Online services contribute to a partial reduction of real mobility (Degórski and Komornicki 2017), and the sharing or circulation of things supports a resource-efficient economy, concurrently reducing the burden on the environment.

In the context of reducing the negative effects of climate change, mitigation and adaptation measures are very important (Degórska 2012a, b, 2014, 2015). The intensification of unfavorable climatic phenomena and extreme weather phenomena results in losses related to the economy of society and biodiversity. Increasingly higher air temperatures and the increasing number and length of periods with high temperatures are one of the dangerous manifestations of climate change. The threat is additionally reinforced by the urban heat island effect. Such situations are especially dangerous for young children and the elderly, who often suffer from coronary heart diseases or respiratory failure (Degórska 2013; Błażejczyk et al. 2014; Kuchcik 2017). The elderly population is therefore particularly sensitive to thermal stress, and its number is currently increasing (Degórska and Degórski 2018). Among Polish cities, the most difficult demographic situation in this respect is in Warsaw, which has the largest number of inhabitants over 60 years old, amounting to 405,000 (23.6%), which is also the largest percentage among cities over 50,000 (Degórska 2014; Degórski 2020). Additionally, in conditions of high air temperatures and high pollution, the concentration of atmospheric ozone increases, posing a great threat to health. The number of days with exceeded ozone levels in cities with over 100,000 inhabitants is significant and amounts to several days a year, for example, in the Silesia Region, it is 18, and in Warsaw—16. This means that for over half a month during 1 year, the inhabitants of these cities are exposed to a strong impact of toxic ground-level ozone (Degórska and Degórski 2018). This state forced the authorities of the regions to take mitigation measures. For example, the local government of the Mazovia Region implemented in 2016 the Short-Term Action Plan for the Warsaw Agglomeration Zone, where there is a risk of exceeding the alarm level of ozone in the air.

Another problem related to climate change is the increasing extreme hydrological phenomena: droughts and floods (Degórska 2012a). Due to the increasing water deficit and droughts, the regions develop Small Retention Programs and are also obliged to take into account Drought Effects Counteracting Plans in the River Basin Districts. These issues are particularly important due to the critically low water resources in Poland and the high risk of drought. Regarding flood risk, regions have to take into account Flood Risk Management Plans. These instruments support the process of management and spatial planning in regions in terms of mitigation and adaptation to climate change.

According to the conceptual assumptions of regional development, the maintenance of the hinterland of the open areas for the Warsaw agglomeration, as for other agglomerations, was to be ensured by green rings, introduced into the spatial development plans of most voivodeships. However, in practice, it is not possible to protect such areas from urban sprawl, i.e., because they are attractive landscape areas. In the vicinity of Warsaw, a lot of development pressure concerns the planned ring, and in it, among others, the surroundings of the national park, landscape parks, forests, and water reservoirs (Degórska 2012a; Degórska and Degórski 2018). In the Warsaw Agglomeration, the pressure covers the Warsaw Area of Protected Landscape.

A very important measure for the implementation of the SD concept and improvement of the quality of the environment is increasing the country's forest cover. This is achieved by the program of increasing forest cover of regions, the aim of which is not only to increase their share in the total area of a given region but also to improve the efficiency of forest ecosystems. Equally important is the protection of biodiversity, which includes all activities, especially legal measures aimed at preserving the biological potential of regions.

1.5 Summary

Adoption of the concept and goals of sustainable development is now considered a very important premise for Poland's development. It has contributed to a significant strengthening of environmental issues in regional development. Significant EU support played an important role for SD, mainly within the cohesion policy, which contributed to strengthening the inclusion of environmental aspects in economic, spatial, and social development at all levels of territorial management and spatial planning, as well as broadly understood environmental protection. It should be noted that the multi-faceted nature of the sustainable development policy, due to the need to take into account both mutual relations and synergy effects, requires a partial change. The change should be in the approach in the management and development planning process, especially at the regional level, as the main level of coordination of policy implementation and SD objectives achievement, as an approach focused on new integrated development planning. SD, and especially the green economy, creates opportunities for synergy effects from integrated activities for sustainable development.

Based on the analysis of activities for the implementation of the principles of sustainable development in the regions of Poland, it can be concluded that they include new tools and solutions that aim to optimize society's life and environmental protection. In recent years, such activities have included the increasing use of resources and the potential of the environment to generate regional development trajectories, including the use of ecosystem services or the implementation of circular economy and sharing economy mechanisms. The effect of implementing these solutions is an ever-stronger connection of regional development with the principles of sustainable development. Demonstrated differences in the potential and resources of the environment of the regions on the national scale, together with the socio-economic conditions, have a large impact on the level of implementation of the concept of sustainable development and the quality of society's life. It should be emphasized that despite the differences between regions, Poland is characterized by a very low Gini index compared to other countries, and GDP growth in the regions does not increase the value of the index, which, in the opinion of many economists, suggests that the inhabitants live better and better (Smith 2020).

The regions suffering from social and economic problems demonstrate the worst results, despite the fact that they have a large natural and landscape potential used in accordance with the assumptions of the sustainable development concept, such as Warmińsko-Mazurskie or Świętokrzyskie voivodeships, which represent a high level of environmental development, a medium-low level of economic development, and a low level of social development.

In Poland, it is also particularly important in the implementation of the concept of sustainable development to develop a green economy, including the low-carbon economy and the circular economy, especially in regions characterized by very high urbanization and industrialization rates, which are at the same time very sensitive to changes in the quality of the environment and social problems (Gill et al. 2007; Degórski 2017). Achieving stability between the socio-economic system and the natural system in these regions is a challenge of the twenty-first century, considering that since the end of the twentieth century, there has been rapid urban growth, mainly in some large cities, in which negative side is the strong urban sprawl process and the exposure of residents to numerous threats (Degórski 2017; Degórska and Degórski 2018).

Referring to the key issues of implementing SD in Polish regions, especially in urbanized areas, the attention is also drawn to the negative impact on the environment and human well-being of three factors: energy-based mainly on coal, the threat resulting from poor air condition in urban agglomerations, and poor preparation of the economy and cities to climate change. This indicates that there are still areas and sectors of the economy in Poland that require "deep" transformation, such as energy, water management, and urban policy. Unfortunately, research within the ESPON/COMPASS project showed that implementing the principles of sustainable development worked better in strategic documents than in practice in Poland (Komornicki et al. 2018a).

References

Affek A, Degórski M, Wolski J et al (2020) Ecosystem Service potentials and their indicators in postglacial landscapes: Assessment and Mapping, Elsevier, Amsterdam (The Netherlands); Oxford (United Kingdom); Cambridge (United States)

Agbenyega O, Burgess P, Cook M et al (2009) Application of an ecosystem function framework to perceptions of community woodlands. Land Use Policy 26(3):551–557

Agenda 2030(2015) Transforming our world: the 2030 Agenda for Sustainable Development, United Nations. A/RES/70/1. https://www.un.org/ga/search/view_doc.asp?symbol=A/RES/70/1&Lang=E. Accessed 12 Nov 2020

Alam M (2018) Ecological and economic indicators for measuring erosion control services provided by ecosystems. Ecol Ind 95(1):695–701

Balmford A, Rodrigues A, Walpole M et al (2008) The economics of biodiversity and ecosystems: scoping the science. European Commission, Cambridge, UK

Barbier E (2007) Valuing ecosystem services as productive inputs. Econ Policy 22(49):177–229

Bielska A, Krupińska I (2016) Ocena potencjału środowiska przyrodniczego i kierunków rozwoju jako element planowania przestrzennego gminy wiejskiej (Assessment of the Natural Environment Potential and Directions of Development as an Element of Spatial Planning of a Rural Commune). Infrastruktura i Ekologia Terenów Wiejskich 2(1):473–484 (in Polish)

Błażejczyk K, Kuchcik M, Milewski P et al. (2014) Miejska wyspa ciepła w Warszawie. Uwarunkowania klimatyczne i urbanistyczne (Urban heat island in Warsaw. Climatic and urban conditions). Instytut Geografii i Przestrzennego Zagospodarowania PAN, Wydawnictwo Akademickie Sedno, Warszawa (in Polish)

Boerema A, Rebelo A, Bodi M et al (2016) Are ecosystem services adequately quantified? J Appl Ecol 54(2):358–370

Bolund P, Hunhammar S (1999) Ecosystem services in urban areas. Ecol Econ 29:293–301

Borys T (2005) Indicators for sustainable development—polish experiences. Wydawnictwo Ekonomia i Środowisko, Warszawa-Białystok

Boyd J, Banzhaf S (2007) What are ecosystem services? The need for standardized environmental accounting units. Ecol Econ 63(2–3):616–626

Bródka S (2010) Ekofizjograficzne aspekty oceny potencjału rekreacyjnego (Eco-physiographic aspects of evaluating recreational potential). Problemy Ekologii Krajobrazu 27:59–64 (in Polish)

Burkhard B, Kroll F, Nedkov S et al (2012) Mapping ecosystem service supply, demand and budgets. Ecol Ind 21:17–29

Burkhard B, Santos-Martin F (2018) An operational framework for integrated mapping and assessment of ecosystems and their services (MAES)

Clewlow R (2016) Carsharing and sustainable travel behavior: Results from the San Francisco Bay Area. Transp Policy 51:158–164

Costanza R, D'Arge R, de Groot R et al (1997) The value of the world's ecosystem services and natural capita. Nature 387:253–260

Dale V, Beyeler (2001) Challenges in the development and use of ecological indicators. Ecol Ind 1(1):3–10

de Groot R, Brander L, van der Ploetheroleg S et al (2012) Global estimates of the value of ecosystems and their services in monetary units. Ecosyst Serv 1:50–61

Degórska B (2012a) Problemy planowania struktur przyrodniczych Obszaru Metropolitalnego Warszawy związane z żywiołową urbanizacją przestrzeni (Planning problems of the natural structures of the Warsaw Metropolitan Area related to the spontaneous urbanization of space). Mazowsze Studia Regionalne 10:89–106 (in Polish)

Degórska B (2012b) Spatial growth of urbanised land within the Warsaw Metropolitan Area in the first decade of the 21st century. Geogr Pol 85(3):77–96

Degórska B (2014) Wrażliwość i adaptacja dużych miast do zmian klimatu w kontekście wzrostu temperatury powietrza (Sensitivity and adaptation of large cities to climate change in the context of rising air temperaturę). Biuletyn Komitetu Przestrzennego Zagospodarowania Kraju PAN 254:27–46 (in Polish)

Degórska B, Degórski M (2003) The environmental dimension of European space according to the concept of trajectory. Europa XXI 8:37–44

Degórska B, Degórski M (2014) Ocena wpływu zmian klimatu na gospodarkę przestrzenną (Assessment of the impact of climate change on spatial planning). Przestrzeń i Regiony 3:99–126 (in Polish)

Degórska B, Degórski M (2018) Influence of climate change on environmental hazards and human well-being in the urban areas–Warsaw case study versus general problems. In: Mal S, Singh RB, Huggel Ch (Eds) Climate change, extreme events and disaster risk reduction. Springer International Publishing AG, Heidelberg, New York, London, pp 43–57

Degórska B (2013) Adaptacja miast do zmian klimatu—nowe aspekty wykorzystania geografii w zarządzaniu środowiskiem i planowaniu przestrzennym (Adaptation of cities to climate change—new aspects of using geography in environmental management and spatial planning). In: Janicki G (ed) Wykorzystanie badań geograficznych w zarządzaniu środowiskiem i planowaniu przestrzennym. UMCS Lublin:24–37 (in Polish)

Degórska B (2015) Koncepcja kształtowania struktury przestrzennej sieci ekologicznej Krakowa (The concept of shaping the spatial structure of the ecological network of Kraków). In Baścik M, Degórska B (ed) Środowisko przyrodnicze Krakowa. Zasoby, Ochrona, Kształtowanie. IGiGP UJ, Kraków:363–376 (in Polish)

Degórska B (2017) Urbanizacja przestrzenna terenów wiejskich na obszarze metropolitarnym Warszawy. Konspekt ekologiczno-krajobrazowy (Spatial urbanization of rural areas in the metropolitan area of Warsaw. Ecological and landscape outline. IGiPZ PAN, Prace Geograficzne, Warszawa, p 262 (in Polish)

Degórski M (2003) Environmental matters in the legal aspect to the EU's spatial planning and its eastern European dimension, Geopolitical Studies 11:167–178

Degórski M (2012a) Integrowanie systemu przyrodniczego z systemem społeczno-gospodarczym podstawą nowoczesnego zarządzania regionem (Integrating the natural system with the socio-economic system is the basis for modern management of the region). Przestrzeń i Regiony 1:7–32 (in Polish)

Degórski M (2012b) Quality of life and ecosystem services in rural-urban regions. Europa XXI 22:137–148

Degórski M, Degórska B (2019) Potencjał środowiska przyrodniczego i świadczenia kulturowe ekosystemów jako czynniki kreujące markę turystyczną (The potential of the natural environment and cultural benefits of ecosystems as factors creating a tourist brand). Biuletyn KPZK PAN 275:121–124 (in Polish)

Degórski M, Solon J (2014) Ecosystem services as a factor strengthening regional development trajectory. Ekonomia i Środowisko 51(4):48–57

Degórski M, Komornicki T (2017) Circular economy—social capital and sharing economy, Report for ESPON Project, Territorial Futures, Warszawa

Degórski M (2007) Environmental conditions as a driving force of regional development in Poland. In: Kovacs A (ed) Regionality and/or Locality, Discussion Papers, Special Issue, Center for Regional Studies of Hungarian Academy of Sciences, Pecs:67–80

Degórski M (2008) Przyrodnicze aspekty zagospodarowania przestrzennego kraju—przesłanki i rekomendacje dla KPZK (Natural aspects of the country's spatial development—premises and recommendations for KPZK). In: Saganowski K, Zagrzejewska-Fiedorowicz M, Zuber P (ed) Ekspertyzy do Koncepcji Przestrzennego Zagospodarowania Kraju 2008–2033, Tom IV, Ministerstwo Rozwoju Regionalnego, Warszawa, pp 39–63 (in Polish)

Degórski M (2009) Are environment conditions among factors behind new spatial pattern. In Kovacs A (ed) Old and new borderlines frontiers margins. Discussion Papers, Special Issue, Hungarian Academy of Sciences, Centre for Regional Studies, Pecs, pp 29–39

Degórski M (2010) Wielofunkcyjność przestrzeni przyrodniczej szansą zwiększenia potencjału rozwoju regionów poprzez grawitację atraktorów i wzrost lepkości (The multifunctionality of natural space as an opportunity to increase the development potential of regions through gravity of attractors and increase in viscosity). In: Strzelecki Z, Legutko-Kobus P (ed) Oblicza współczesnego kryzysu a polskie regiony. Ministerstwo Rozwoju Regionalnego, Warszawa, pp 280–287 (in Polish)

Degórski M (2011) Środowisko przyrodnicze jako kreator atraktorów oraz czynnik rozwoju i wzrostu lepkości regionu (Natural environment as a creator of attractors and a factor of development and increase of the region's viscosity). In: Maik W, Rembowska K, Suliborski A, (ed) Geografia regionalna—scalanie i synteza wiedzy geograficznej. Podstawowe Idee i Koncepcje w Geografii 6:83–92 (in Polish)

Degórski M (2013) Zagrożenia zrównoważonego rozwoju obszarów zurbanizowanych w Polsce z punktu widzenia czynników środowiskowych (Threats to the sustainable development of urbanized areas in Poland from the environmental factors point of view). In: Gaczek W (ed) Dynamika, cele i polityka zintegrowanego rozwoju regionów. Aspekty teoretyczne i zarządzanie w przestrzeni. Bogucki Wydawnictwo Naukowe Poznań:163–174 (in Polish)

Degórski M (2017) Jakość życia a potencjał środowiska i świadczenia ekosystemów (Quality of life and the potential of the environment and ecosystem services). In: Okrasa W (ed) Jakość Życia i Spójność Przestrzenna—interakcje rozwoju i dobrostanu w kontekście lokalnym. Wydawnictwo UKSW Warszawa:225–240 (in Polish)

Degórski M (2018) Gospodarka o obiegu zamkniętym circular economy—nowe podejście w rozumieniu relacji człowiek-środowisko (Circular economy—a new approach in terms of the human-environment relationship). In: Churski P (ed) Teoretyczne i aplikacyjne wyzwania współczesnej geografii społeczno-ekonomicznej. Komitet Przestrzennego Zagospodarowania Kraju, Polska Akademia Nauk, Studia 183:27–35 (in Polish)

Degórski M (2020) Środowisko, jego potencjał i zdolności adaptacji do zmian globalnych w kontekście rozwoju kraju (The environment, its potential and ability to adapt to global changes in the context of the country's development). In Śleszyński P (ed), Badania geograficzne dla rozwoju niepodległej Polski, Wydawnictwo Sejmowe, Warszawa:73–86 (in Polish)

Domański R (2008) Przyczynek do modelowania rozwoju zrównoważonego w długim okresie, (Contribution to the modeling of sustainable development in long term). In: Stryjakiewicz T, Czyż T (ed) O nowy kształt badań regionalnych w geografii i gospodarce przestrzennej, Biuletyn KPZK, 237, Polska Akademia Nauk, 237:203–224 (in Polish)

Drabarczyk K (2017) Zrównoważony rozwój województw—analiza porównawcza (Sustainable development of voivodeships—comparative analysis). Zeszyty Naukowe Politechniki Częstochowskiej, Zarządzanie 25(2):23–34

EEA (2016) Green economy. Europe's environment—An Assessment of Assessments, Copenhagen

EEA Report (2017) Air quality in Europe – 2017 report. EEA Report 13/2017. European Environment Agency

EUROPE (2020) A strategy for smart, sustainable and inclusive growth, European Commission https://ec.europa.eu/eu2020/pdf/COMPLET%20EN%20BARROSO%20%20%20007%20-%20Europe%202020%20-%20EN%20version.pdf. Accessed 12 Sept 2020

European Union (2019) Sustainable development in the European Union. Monitoring report on progress towards the SDGs in an EU context. Imprimerie Bietlot in Belgium

Eurostat (2020) Your key to European Statistics. European Commission https://ec.europa.eu/eurostat/web/main/home. Accessed 11 August 2020

Gill S, Handley J, Ennos Pauleit A (2007) Adapting cities for climate change: the role of the green infrastructure. Built Environ 33(1):115–133

Green Economy Coalition (2011) Nine Principles of a Green Economy. Working towards a Balanced and Inclusive Green Economy. International Institute for Environment and Development

de Groot R, Fisher B, Christie M et al (2010) Integratingthe ecological and economic dimensions in biodiversity and ecosystem servicevaluation. In: Kumar P (ed), The Economics of Ecosystems and Biodiversity: Ecological and Economic Foundations. Earthscan, London and Washington, pp 1–40

GUS (2011) Wskaźniki zrównoważonego rozwoju Polski (Sustainable development indicators for Poland). USK Katowice (in Polish)

GUS (2016, 2019, 2020) Rocznik Statystyczny (Statistical Yearbook). Główny Urząd Statystyczny, Warszawa (in Polish)

Haines-Young R, Potschin M, (2018) Common International Classification of Ecosystem Services (CICES) vol 5, issue 1, Guidance on the Application of the Revised Structure, Barton in Fabis, Nottingham, UK

Harding R (2006) Ecologically sustainable development: origins, implementation and challenges. Desalination 187(229)

HELCOM (2010) Ecosystem health of the Baltic Sea 2003–2007. In: HELCOM Initial Holistic Assessment, Baltic Sea Environment Proceedings 122

Kamińska I (2019) Oszacowanie wartości ekonomicznej procesów równoważenia skutków eutrofizacji—regulacyjnej usługi ekosystemowej Zatoki Gdańskiej (Economic valuation of balancing the effects of eutrophication processws—regulating ecosystem services of the gulf of Gdansk). PHD Thesis, Uniwersytet Gdański, Gdynia (in Polish)

Kasztelan A (2011) Ocena środowiskowej konkurencyjności regionów w Polsce (The assessment of environmental competitiveness of regions in Poland). Prace Naukowe Uniwersytetu Ekonomicznego We Wrocławiu 166:258–268 (in Polish)

Kates R, Parris T, Leiserowitz A (2005), What is Sustainable Development? Goals, Indicators, Values, and Practice. Environ Sci Policy Sustain Develop 47(3):8–21, DOI: https://doi.org/10.1080/00139157.2005.10524444

Klóska R (2017) Rozwój zrównoważony regionów w Polsce w ujęciu statystycznym (Sustainable development of individual regions in Poland in terms of statistics). Progr Econ Sci 4:159–175 (in Polish)

Komornicki T, Bednarek-Szczepańska M, Degórska B et al (2018a) COMPASS–Comparative analysis of territorial governance and spatial planning systems in Europe applied research 2016–2018. Final Report: Additional Volume 6 Case Studies Report. ESPON Luxembourg

Komornicki T, Siłka P, Bednarek-Szczepańska M et al (2018b) COMPASS—Comparative Analysis of Territorial Governance and Spatial Planning Systems in Europe Applied Research 2016–2018. Final Report: Additional Volume 3 Phase 1 Questionnaires. ESPON Luxembourg

Korcelli P, Degórski M, Drzazga D et al (2010) Ekspercki projekt Koncepcji Przestrzennego Zagospodarowania Kraju do roku 2030 (Expert project of the National Spatial Development Concept until 2030). Studia 78, Polaka Akademia Nauk, Komitet Przestrzennego Zagospodarowania Kraju, Warszawa (in Polish)

Koźlak A (2017) Sharing economy—as a new socio-economic trend. Res Pap Wrocław Univ Econ 489:171–182. https://doi.org/10.15611/pn.2017.489.15

Kroll C, Warchold A, Pradhan P (2019) Sustainable Development Goals (SDGs): Are we successful in turning trade-offs into synergies? Palgrave Commun 5(140) https://doi.org/10.1057/s41599-019-0335-5

Kuchcik M (2017) Warunki termiczne w Polsce na przełomie XX i XXI wieku i ich wpływ na umieralność (Thermal conditions in Poland at the turn of the 20th and 21st centuries and their impact on mortality), Prace Geograficzne, 263, Instytut Geografii i Przestrzennego Zagospodarowania PAN, Warszawa (in Polish)

Lacy P, Rutqvist J (2015) Waste to wealth: The circular economy advantage. Springer

Lewandowska A (2014) Wdrażanie zrównoważonego rozwoju w Polsce i w krajach Unii europejskiej, (Implementation of sustainable development in Poland and in countries of European Union). Acta Elbingensia 23(4):9–22 (in Polish)

Martin C (2016) The sharing economy: A pathway to sustainability or a nightmarish form of neoliberal capitalism? Ecol Econ 121:149–159

MEA (2005) Ecosystems and human well-being: current state and trends, vol. 1. In: Findings of the Condition and Trends. Working Group of the Millennium Ecosystem Assessment.Island Press, Washington, Covelo, London

MIiR (2017) Strategia na rzecz Odpowiedzialnego Rozwoju (Strategy for Responsible Development) Warszawa, https://www.gov.pl/web/archiwum-inwestycje-rozwoj/strategia-na-rzecz-odpowiedzialnego-rozwoju. Accessed 15 Sept 2020 (in Polish)

Mizgajski A, Stępniewska M (2009a) Koncepcja świadczeń ekosystemów a wdrażanie zrównoważonego rozwoju (The concept of ecosystem services and the implementation of sustainable development) In: Kiełczewski D, Dobrzańska B (ed) Ekologiczne problemy zrównoważonego rozwoju. Wydawnictwo Wyższej Szkoły Ekonomicznej w Białymstoku: 12–23 (in Polish)

Mizgajski A, Stępniewska M (2009b) Zastosowanie koncepcji świadczeń ekosystemów i koncepcji świadczeń infrastrukturalnych w programowaniu gospodarki ściekowej (Application of the ecosystem services concept and the concept of infrastructure services in sewage management programming) Ekonomia i Środowisko 35(1):125–139 (in Polish)

Morris P, Therivel R (2000) Methods of environmental impact assessment. Spon Press, London, New York

Piłat B (2017) Gospodarka okrężna albo powtarzanie starych błędów (The circular economy or repeating old mistakes). Kultura Liberalna: 422(6) https://kulturaliberalna.pl/2017/02/13/gospodarka-okrezna-pilat-ekologia-wegiel/ Accessed 4 Aug 2020 (in Polish)

TNS Polska (2015) Polska jest mobi (Poland is mobi) https://www.benchmark.pl/aktualnosci/sharing-economy-w-polsce-nabiera-rozpedu.html. Accessed 22 Aug 2020 (in Polish)

Raszka B, Hełdak M (2013) Świadczenia ekosystemów w polityce przestrzennej gmin powiatu wrocławskiego (Ecosystem services in the spatial policy of the communes of the Wrocław poviat). Wydawnictwo Uniwersytetu Przyrodniczego, Wrocław (in Polish)

Ratajczak W (2008) Modele ekonometrii przestrzennej w analizie regionalnej (Models of spatial econometrics in regional analysis). In: Stryjakiewicz T, Czyż T (ed) O nowy kształt badańregionalnych w geografii i gospodarce przestrzennej, Biuletyn KPZK, Polska Akademia Nauk 237:86–202 (in Polish)

Roszkowska E, Karwowska R (2014) Wielowymiarowa analiza poziomu zrównoważonego rozwoju województw Polski w 2010 roku (Multidimensional analysis of the level of sustainable development of the Polish voivodships in 2010). Econ Manag 1:9–27 (in Polish)

Sagie H, Morris A, Rofè Y et al (2013) Cross-cultural perceptions of ecosystem services: a social inquiry on both sides of the Israeli-Jordanian border of the Southern Arava Valley Desert. J Arid Environ 97:1–11

Scholte S, van Teeffelen A, Verburg P (2015) Integrating socio-cultural perspectives into ecosystem service valuation: a review of concepts and methods. Ecol Econ 114:67–78

Shaker R (2015) The spatial distribution of development in Europe and its underlying sustainability correlations. Appl Geogr 63:304–314. https://doi.org/10.1016/j.apgeog.2015.07.009. Accessed 22 Aug 2020

Śleszyński P, Bański J, Degórski M et al (2017), Delimitacja obszarów strategicznej interwencji państwa: obszarów wzrostu i obszarów problemowych (Delimitation of strategic state intervention areas: areas of growth and problem areas). Prace Geograficzne 260, Instytut Geografii i Przestrzennego Zagospodarowania PAN, Warszawa (in Polish)

Smith L (2020) The gini index: measuring income distribution, investopedia. http://www.investopedia.com/articles/economics/08/gini-index.asp. Accessed 4 Sept 2020

Szamrej-Baran I, Baran P (2014) Wdrażanie zasad zrównoważonego rozwoju. Polska na tle Unii Europejskiej (Implementation of the sustainable development principles. Poland on the background of the European Union) Ekonomia i Środowisko 50(3):176–190 (in Polish)

Tchorek G. Brzozowski M, Dziewanowska K (2020) Social Capital and value co-creation: the case of a polish car sharing company. Sustainability 12(4713) DOI: https://doi.org/10.3390/su12114713

TEEB (2010) The Economics of Ecosystems and Biodiversity. Ecological and Economic Foundation. Earthscan, London and Washington

UNEP (2011) Towards a green economy: pathways to sustainable development and poverty eradication. A synthesis for policy makers, United Nations, New York, http://www.unep.org/greeneco. Accessed 12 Sept 2020

Vihervaara P, Kumpula T, Tanskanen A et al (2010) Ecosystem services—a tool for sustainable management of human e environment systems: case study finnish forest lapland. Ecol Complex 7(3):410–420

Wardak P, Zalega T (2013) Kolaboratywna konsumpcja jako nowy trend konsumencki. Studia i Materiały Wydział Zarządzania UW 16:7–32. https://doi.org/10.7172/1733-9758.2013.16.1

WHO (2016) Database. http://www.who.int/phe/health_topics/outdoorair/databases/cities/en/

Ziobrowska J (2017) Sharing economy jako nowy trend konsumencki (Sharing economy as a new consumer trend) In: Kalina-Prasznic U (ed) Własność w Prawie i Gospodarce, W-Wydawnictwo Prawnicza i Ekonomiczna Biblioteka. Wydział Prawa Administracji i Ekonomii Uniwersytetu Wrocławskiego, 281–289 (in Polish)

Marek Degórski is the Full Professor, Director at the Institute of Geography and Spatial Organisation, Polish Academy of Sciences (2010–2017), Head of the Department of Geoecology and Climatology (since 2004) as well as Chairman of the Scientific Council (since 2019). Marek Degórski is a member of the Polish National Committee for cooperation with International Council of Sciences and Committee for ICSU-SCOPE and UNESCO-MAB, Committee for International Year of Planet Earth and Committee for Cooperation with International Council of Sciences, Chairman for the Committee of Geographical Sciences Polish Academy of Sciences, and Chief of the National Committee for the cooperation with IGU. Marek Degórski is a Member of Executive Committee of IYPE in IGU (2007-2009), performer and member of Advisory Board or Task Forces for some European Projects and Programs as Landscape Tomorrow, Integrated Environmental Monitoring of Europe, and Vistula ECONET Development and Implementation. Working on some projects concerning six and seven Framework Program of European Commission: (PLUREL e-LUP), ESPON, INTERREG, and Curie-Skłodowska Funds. Head of some projects supported by Polish Ministry of Sciences, National Science Centre. Marek Degórski is Executive Secretary of four projects supported by USDA Forest Service and US Environment Protection Agency. Prof. Degórski is interested in comprehensive geography, landscape analysis as well as spatial planning and management with long experience in the study of the assessment of environmental and human relationship, hazards, and risk, recently in circular economy and smart cities. Marek Degórski is a Fulbright Scholar at the State University of Washington, Seattle and State University of Connecticut, Storrs in 1991, as well as Visiting Professor at the Harvard University, Northern Arizona University, Hawaii State University, and Michigan Technical University (USA), and also Ghent University (Belgium), Oulu University (Finland), Roma University (Italy), and others. Marek Degórski is an expert in the Ministry of Regional Development on environmental and landscape issues, as well as a member of the State Council for Spatial Development, designed by Prime Minister of Poland (2010–2012) and State Council for Nature Protection (2014–2016) and is author of over 400 scientific publications.

Bożena Degórska is a comprehensive geographer, specialist in the protection and development of the environment and landscape, as well as the environmental basis for spatial planning and sustainable development. Bożena Degórska worked for the Institute of Geography and Spatial Organization PAS since 1987, and is currently working as a professor of IGSO PAS. Bożena Degórska is the author or co-author of over 150 scientific publications (including 10 books) and numerous studies of expertise. Scientific research mainly concerns the environmental base of spatial planning and sustainable development, environmental and landscape transformations, identification of mechanisms and correlations between changes of land use and the process of suburbanization (mainly for the Warsaw Metropolitan Area, but also in other regions and at a national scale). Bożena Degórska is participant for domestic and international research projects: KBN, NCM, PHARE/FIESTA, PHARE/OMEGA, POLONIUM, ESPONs. Bożena Degórska is performer of some key projects for practice, among others: Scientific and Research Bases for the Concept of Spatial Development Policy for the Country (1995-1997), Adaptation of Spatial Management of Poland to Climate Change (KLIMADA project), Adaptation of Large Cities to Climate Changes (KLIMADA project), Initial Concept for the Development of the Mazowieckie Voivodeship in variant forms (1999), Development Strategy for the Mazowieckie Voivodeship (2000-2001), Warsaw Development Strategy to 2020 Perspective (2004), Mazovia Development Trends (2012), and Development Strategy of the Gdańsk Metropolitan Area until 2030: Sectorial diagnosis-Energy security and the natural environment (2015). Bożena Degórska is member of designers of conditioning studies and directions for spatial development (among others for the city of Kraków). Bożena Degórska is Lecturer at the Warsaw University of Technology (2008–2015) and at the University of Economics and Management in Warsaw (1997–2008). Bożena Degórska is member of the Urban and Architectural Commission of the Mazowieckie Voivodeship (2004–2008) and also member for the Committee of National Spatial Development at the Presidium of Polish Academy of Sciences, since 2011.

Chapter 2
Geopolitics of Sustainable Development During the Twentieth and Twenty-First Centuries

André Santos da Rocha, Guilherme Ribeiro, Leandro Dias de Oliveira, and Pablo Ibanez

Abstract The contemporary debate on sustainable development is permeating issues that cross economic and geopolitical interests at different scales, mainly because their bases reverberate in possession and appropriation of territory and development policies. Since the second half of the twentieth century, environmental issues have emerged at the center of the international geopolitical debate, materialized in numerous conferences on the environment and the proposed agendas derived from these meetings. The challenges of using and appropriating nature from a prism of Western modernity influence the dualistic conception between policies of preservation and the dynamics of a capitalist development agenda. This conception is still present in the organization of development policies in many countries, such as Brazil. The Brazilian Geopolitics has always been directed to think about the security of its borders and the guarantee of its sovereignty in the territory. This territory is marked by important ecosystems (Amazon, Cerrado, and Pantanal), located in the North and Center-West Regions of the country. These regions are part of border zones with other countries and are part of the expansion of the economic frontier of agribusiness. With the process of Brazil's insertion in the world economic system and the adoption of a policy of re-primarization of its economy, it has placed the geopolitics of sustainability in Brazil at a crossroads.Thus, in this text, we present a theoretical and conceptual review on geopolitics and sustainable development and present current data on the areas of fire occurrence in Brazil. We use data from INPE (Space Research Institute) that indicate the North and Center-West Regions of the country as the largest in outbreaks of fire. These, in turn, are directly and indirectly associated with the expansion of agricultural activities. These impacts on ecosystems affect not only biomes but traditional populations and peoples living in an organized relationship with nature. Thus, these data reinforce the need to rethink the development matrices so that the rights to a genuinely just environment is realized.

Keywords Geopolitical · Sustainable development · Political economy · Brazil

---

A. S. da Rocha (✉) · G. Ribeiro · L. D. de Oliveira · P. Ibanez
Laboratory of Economic Geography, Policy and Planning (LAGEP), Federal Rural University of Rio de Janeiro (UFRRJ), Seropédica 23890-000, Brazil

M. Mishra et al. (eds.), *Regional Development Planning and Practice*,
Advances in Geographical and Environmental Sciences,
https://doi.org/10.1007/978-981-16-5681-1_2

2.1 Introduction

Brazil is one of the nations with the most incredible biodiversity in the world. For this reason, the country is part of a small group of nations called Megadiverse Countries by COP-10,[1] a United Nations conference in which the problems of caring for nature are dealt with. The Amazon forest alone occupies more than four million square kilometers in our territory, more extensive than many countries. Other biomes, such as the *Cerrado,* that occupy a large area in the Center-West Region or the Atlantic Forest,[2] occupying vast areas, have been more altered by anthropic action throughout history. Today, only 12.4% of the total Atlantic Forest remains in Brazil. One of the leading causes of its deforestation in this biome was the occupation by crops, especially coffee.

The devastation by exploratory activities of natural resources was the tone of our development until the middle of the twentieth century, which depended largely on activities that were not intensive in technology. Even with the industrialization process that occurred with greater force after the 1950s, the use of biomes continued to be a relevant part of the economy, which started to occupy previously uninhabited areas.

At the national level, the idea of the occupation and use of these areas have gained significant notoriety with geopolitical studies since at least the 1930s (VLACH 2003). Travassos (1938), for example, already pointed to the need to integrate the Amazon more actively into the Brazilian and South American economies. Later, with the creation of the Escola Superior de Guerra, in 1949, Brazilian geopolitical thought left the paper and became central to national political decisions. To mention the name of Golbery do Couto e Silva, one of the leading scholars of this institution, who, in addition to writing popular books on geopolitics, had a substantial career as policymakers during the Brazilian Military dictatorship (1964–1985). His planning was "to flood Amazon (…) with civilization" (Couto e Silva 2003) through demographic displacement from Center-West Region to Amazon River Channel. In the 1960s and 1970s, the occupation of the Amazon Region was based chiefly on infrastructure projects, mining, and industrialization. Becker (2015) argues that the main intentions of these ideas and policies for that region were to avoid the economic concentration in the southeast region, national integration, and national security.

The Center-West also received attention from the federal government, but the main project focused on agriculture issues. There was a deliberate policy for its occupation based on financial and land incentives. The key objectives were to supply the domestic market and guarantee foreign exchange from exports, especially grains (Frederico 2015). This process has ascribed the region as the largest grain producer in the country today and also as, unfortunately, one of the largest areas deforested.

In these years, both regions have been receiving so much attention because the fires took place in many locals, causing enormous environmental and economic damages.

---

[1] https://www.mma.gov.br/informma/item/6143-megadiverse-countries-discuss-benefit-sharing.

[2] https://agenciabrasil.ebc.com.br/geral/noticia/2020-05/desmatamento-na-mata-atlantica-cresce-272-diz-relatorio.

It is important to pinpoint that in the international scenario, at least since the 1960s, environmental issues are increasingly highlighted, especially the topic of sustainable development. The Amazon Region, since then, has been having a central place in this debate. That is why the questions about the "internationalization" of Amazon have been gaining such an interest in different forums. According to Medeiros (2015), several Brazilian studies related to this topic point out that environmental preservation objectives, continued demarcation of indigenous lands, and reserve units in border areas are at the service of foreign national states' interests. They believe that there is a deliberate articulation of intelligence services, interest groups, lobbies, and national states to halt Brazilian development in favor of a weakening of the Brazilian capacity to realize its potentials.

In other words, there is a huge interrelation between the geopolitical debate and the sustainable development in the curse in Brazil, and this is the central objective of this article. To achieve this goal, in the first part of this chapter, we focus on presenting the geopolitical ideas that oriented most of the occidental intellectuals and, in particular, the Brazilian scholars. This was crucial to delimit the geopolitical thoughts that the Brazilian government served to follow in its policies and discourses. The next section discusses the historical perspective of the sustainable development debate and what we called as environmental geopolitics. It is critical with regard to the environmental issues when Brazilian topics are on the table. The final section will show the recent events in Brazil that call international attention for its impact, dimension, and governmental response.

2.2 The Political Nature of Geography: History and Concepts

According to Michel Foucault, the emergence of a field of knowledge is more than an intellectual issue, and also a political one (Foucault 1971[1970]). This is evident in the case of geopolitics, in which the word itself reveals its aims. However, suppose the Sweden Rudolf Kjéllen coins the rising of this concept at the end of the nineteenth century in Europe in order to understand territorial conflicts and agreements (Kjéllen 1899 and Holdar 1992). In that case, it is not difficult to realize a sort of "epistemological delay" between those rising and ancient practices for controlling and mastering space since the Roman Empire.

In the same way, and although the name still did not exist, how many words are as appropriate as geopolitics to explain the Atlantic Revolution, the conquest of America, and the birth of modernity? Perhaps the absence of the expression itself is one of the reasons by which Carl Schmitt has used the Greek term *nomos* in his masterpiece *The Nomos of the Earth in the International Law of Jus Publicum Europaeum* to show how a certain definition of "space" is part of the Occident *modus operandi* as a whole. Therefore, borders, limits, and frontiers not just composed of a geopolitical discourse in a strict sense but are inseparable from Western instances such as property, law, and sovereignty (Schmitt 2003 [1950]; see also Mezzadra and Neilson 2013).

At service of the nation-state building, the imperialist race, and the capitalist concurrence, geopolitics is a crucial factor in understanding the spatial crisis provoked by the European Empires since at least the Franco-Prussian War and whose aftermath was the First World War. It represents the different facets of the same coin; at this point, national and international issues merged with each other. Hence, an "external danger" could be recognized as such only if a previous sense of territorial pertaining has been forged, as well as an "overseas colony" situated thousands of kilometers from Europe will be accepted as "ours" just if soldiers, missionaries, teachers, and public opinion were convinced that they belong to an "imagined community" (Anderson 2006). Simultaneously, an outer threat ends up masking inner social contradictions (Cowen and Smith 2009).

So, we should reconsider the rising of geopolitics from a mere starting point for political-geographic approaches to the outcome of a long-run process where *nature* and *territory* have become vital for the development of modernity and capitalism (Ribeiro 2015). That is why modern Europe would not exist without environmental resources from the New World and the African slave labor (Quijano 2000). From religious to scientific camouflages, geopolitics is a multi-faced intervention whose *leitmotiv* is creating an unequal zone of "negotiation." It is an addicting game where the will of the few becomes the rule for many, but nobody takes its "necropolitics" consequences (Mbembe 2015).

Much more than a subfield of geography or a German menace such as hegemonic narratives wanted to persuade us (Febvre 1922; Demangeon 1932, 1939; Bowman 1942), geopolitics is a critical window (Ó Tuathail 1996, 1998) through the "darker side of modernity" (Mignolo 1995) might be visualized in a double meaning: on the one hand, it discloses the inhumanity of all those which do not fit into the Enlightenment. As non-humans, non-Europeans could be exploited and massacred until the civilization flourishes from them. On the other hand, it unveils the need to conceive a plethora of discourses legitimating the *status quo*. In other words, every geopolitics (in an empirical sense) is also "geopolitics of knowledge" (Mignolo 2005).

Put differently, geopolitics and nature are unseparable. Elaborated by the Scientific Revolution, the modern version of nature has become one of the pillars of the abyssal thinking (Santos 2007), namely, the human–environment dichotomy. Albeit geographers like Ratzel, Vidal de la Blache, and Mackinder had studied how the history of societies cannot be grasped without the benefits and constraints from the milieu. They had also contributed to fixing nature as an economic resource, unevenly distributed over the globe, which consequent in a matter of clash among Empires (Ratzel 1987 [1897]; Mackinder 1904; Vidal de la Blache 1994 [1917]).

This mindset provoked grave aftermaths. From a Brazilian point of view, four centuries of Portuguese dominion (1500–1889) have shown how politics and economy work together. From mercantilism to the Industrial Revolution, as far as capitalist development requires increasing doses of nature, Brazil was becoming a subordinated piece within the international division of labor. Draining the main gains to metropolitan markets, the agrarian-export model brought substantial ecological impacts not only on the Atlantic Forest [*Mata Atlântica*] (Dean 1995), but also on the whole territory. This structure was also disapproved by national abolitionists like José do Patrocínio and Joaquim Nabuco (Pádua 2002).

However, after the Second World War, the Brazilian geographer Josué de Castro (whose intellectual and political engagement would leave him to work at international organisms such as FAO and UN in the 1960s [Davies 2019]) enlarges the issue beyond neo-Malthusianism or natural factors revealing that the major outcome is not just ecological but also anthropological. In depredating rivers, soils, fauna, and flora, the result is the tragedy of hunger, which, in turn, is a sign of underdevelopment. In sum, a geopolitical question arises (Castro 1946, 1957). During the last three decades, Carlos Walter Porto-Gonçalves (whose fieldworks at the Acre State drove him to meet, and to learn with, the great Amazon Forest ambientalist and rubber tapper Chico Mendes in the 1980s[3]) has blended class struggles and decolonial thought to put on view the history of Amazon as a long-term process characterized by the denegation of native inhabitants and their knowledge in the name of "energivorous" procedures from the global agribusiness targeting the fall of the Rain Forest (Porto-Gonçalves 2001, 2018[2017]). At this backdrop, in times of the Anthropocene, a critical theory concerning *post-extractive societies* has been elaborated in Latin America (Dilger et al. 2016).

Similar examples could be mentioned, but they are sufficient to underline a Brazilian geopolitical tradition—which is radically different in terms of form and content from the imperial tradition represented by Ratzel, Vidal de la Blache, and Mackinder, whose critical heritage we will try to follow in the next sections.

2.3 Environmental Geopolitics and Sustainable Development

The sustainable development—a geopolitical and geoeconomic normalization that has been consolidated worldwide since the 1980s—implies a progressive and selective economic-ecological-spatial adjustment of the current order. It thus comprises a formula of the instrumental rationality of nature, belittled as raw material, resource, and fuel, and which, for this reason, presents the need for production and reproduction. Using natural resources without exhausting them; creating non-polluting technologies; and producing an environmental discipline of political, economic, and social management are unquestionable demands of the current spatial order. Sustainable development is at the heart of the productive needs of this new century, getting inserted in contemporary spatial restructuring.

Since the 1970s, the territorial-productive restructuring process has promoted changes in stocks and time for production, the flexibility of manufacturing work (Harvey 1989; Hoogvelt 1997), and the spread of industries beyond metropolitan

---

[3] Chico Mendes was murdered by the son of a land grabber in Xapuri in 1988 when he was 44 years old.

centers (Wallerstein 2003). In addition to the progressive adaptation to economic neoliberalism, an important change in the new productive model is the adoption of the concept of sustainable development, albeit in a selective way and with insufficient results. The continuing effects of neo-liberalization on social relations (Keil 2005; Heynen and Robbins 2005) create profound impacts on nature. As important as the industrialization, the formation of productive clusters or the metric of knowledge spill-overs (Massey 1995 [1984]; Scott and Storper 1986; Storper and Walker 1989; Scott 1993; Storper 1997; Storper and Venables 2005) is the protection of nature as a resource and the adoption of sustainable development. Political ecology, geopolitics, and economic geography are essential for understanding the concept of sustainable development and the political agenda on the environment in Brazil (Oliveira 2020).

Sustainable development was consolidated particularly after the United Nations Conference on Environment and Development (1992) held in the city of Rio de Janeiro (Oliveira 2019) as an important coordination mechanism between various social actors. Agents promoting sustainable development are governments at multiple scales, companies of different sizes, social activists and NGOs, residents' associations and unions, and universities.

It is necessary to remember that the environmental issue as we know it has deep roots. According to Jean (1976), the construction of castles in England was responsible for the destruction of forests; the river Seine in Paris was already polluted in the Medieval Era. The saga of Europeans in the distant Indies and incursions on the Atlantic coast riches reinforce the destruction of civilizations, social groups, and traditional communities. The exploitation of natural wealth is characteristic of the dominant mode of production.

The concept of sustainable development has its genesis in major conclaves international. Influenced by the publication of the work *The Limits of growth* (Meadows et al. 1971), the Conference United Nations Convention on the Human Environment (Stockholm-1972) had a dissent between central and peripheral countries in solutions for the destruction of nature. In the following decade, *Our Common Future* report (1987) was the document that presented sustainable development as one that meets the needs of the present without compromising the needs of generations' future.

Sustainable development was accepted at the United Nations Conference on the Environment and Development held in Rio de Janeiro by signing a prescription called Agenda 22. With Agenda 21, all central and peripheral countries would adopt and implement sustainable development in their territories. The Rio de Janeiro Conference 1992 was carried out in Brazil, a peripheral country with late industrialization, which had most of the Amazon Forest in its territory.

Twenty years later, the city of Rio de Janeiro once again hosted a major environmental conference: the United Nations Conference on Sustainable Development (Rio + 20) took place on June 13 and 22, 2012 and the document *The future we want* was signed. The Rio + 20 also addressed the legacy of Rio-92, assessing the concrete actions of the *Rio Declaration*, Agenda 21, the *United Nations Framework Convention on Climate Change*, and the *Convention on Biological Diversity*. Finally, Rio + 20 proposed the green economy, a term with an economic perspective on environmental issues. As there was no consensus on the green economy, unfortunately, the years after Rio + 20 were tragic for the environment. Inept political leaders have emerged in important countries worldwide, and there have also been attacks on nature protection.

The adoption of sustainable development led to the privatization of territories, the full financialization of nature, and the deterritorialized rationalization of economic and environmental arrangements on a global scale, from a business perspective. The construction of sustainable development has not broken with the industrial civilizing model that destroys nature, because the changes are technological; it is carried out by the same hegemonic corporations and is driven by economic interests. As all of this is hidden by the ideology of sustainable development (Oliveira 2019), all investment in industrial machinery is celebrated as a virtuous action to reverse ecological problems.

Thus, we reinforce the need for an exercise in political economy, fundamental in the interpretation of human-environmental relations in contemporary geography (Walker 2005). With respect to the political ecology of the peripheral nations of the world, urgent research agendas arise on environmental change, access conflicts, and political actions for transformation (Bryant 1992). In Brazil, where the daily environmental impacts of industrialization and urbanization in its territory have intensified, political ecology is an important field of critical thinking and political action on society-nature relations.

2.4 The Crossroads of Brazilian Territory in the Twenty-First Century: Geopolitics, Development, and Sustainability

All the debate on issues involving intersections between geopolitics, development, and sustainability gain notability in the management of Brazilian territory in the twenty-first century. In the last years, intense debates on preservation and development strategies indicate how multiple interests cross the use of territory, of economic actors, state political agents, traditional peoples, and participation of NGOs and research entities (Agnew 1994; Bryan 2012; Laing 2020).

The challenge of building an "environmentally fair" development model finds limits in the way the country fits into the world economy, as well as the historical and strategic plan that its territory has in Brazilian geopolitical thinking. The strategies taken for the country's economic development, mainly based on the dynamics of agribusiness, call into question the sustainability of large Brazilian ecosystems, such as the Amazon, the *Pantanal,* and *Cerrado* Brazilian. These ecosystems are located, respectively, in the North and Center-West Regions of the country. In traditional geopolitical thinking (Agnew 2005), they are understood as priority areas for national defense, especially because they are areas that border other countries in South America (Fig. 2.1).

In the middle of the twentieth century, the Brazilian geopolitical challenge was to seek its integration in these spaces, which was considered an "empty space" (Mattos

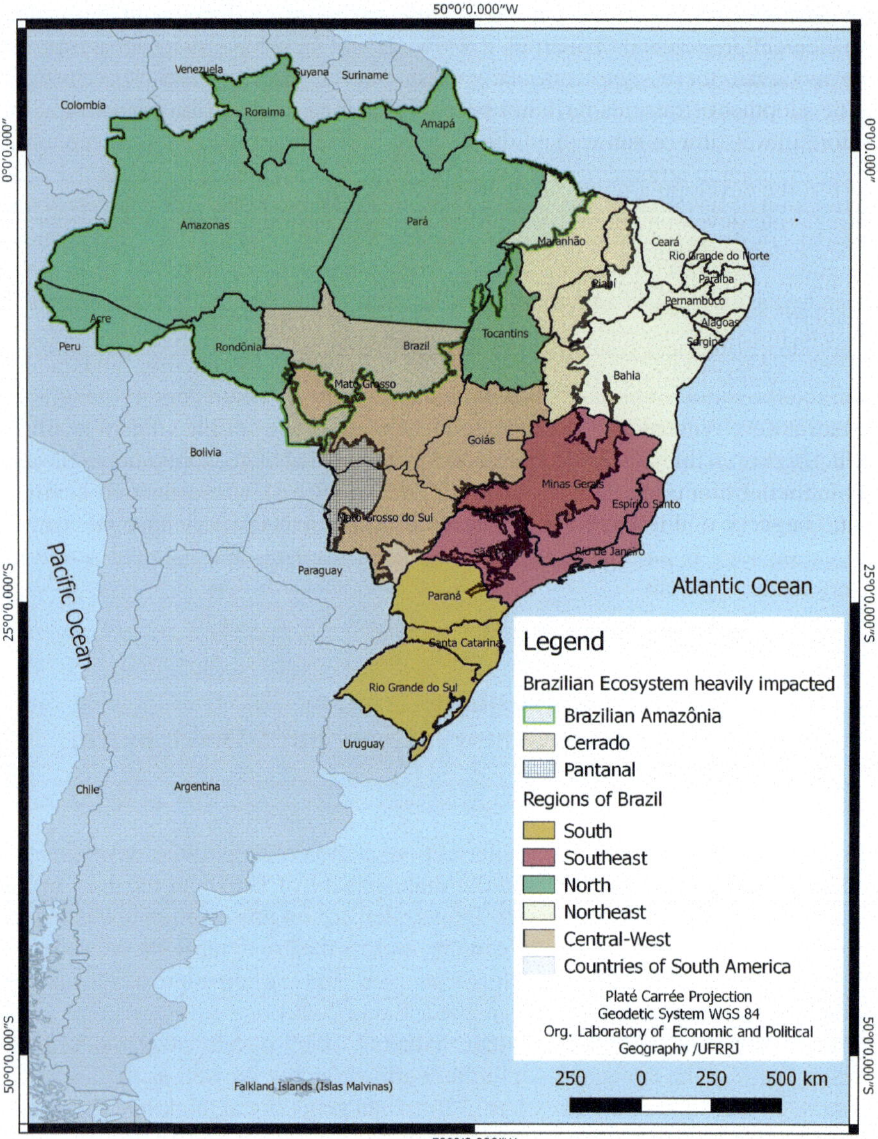

Fig. 2.1 Location of Brazilian ecosystems strongly impacted by the expansion of agriculture (*Source* Prepared by the authors)

1975). Since the greatest demographic density of Brazilian territory occurs on the coast due to the natural facilities of installation in the period of colonization (seventeenth to eighteenth century), the succession of dynamics and economic processes manifested themselves with greater density in the coastal strip. Thus, the establishment of military and economic activities for the inland areas and borders with other countries was eminently strategic (Mattos 1984). From the horizon of critical geopolitics (Cairo 2008), "empty space" geopolitical discourse serves as a state prejudice in that depriving land rights of indigenous people, as well as to legitimate a territorial occupation from above.

Since the 1970s, those empty spaces have been transformed into the new Brazilian economic frontier (Melstrom et al. 2016; Becker 1995), mainly with regard to the exportation of soy and of corn. As a result, we are facing an economic re-primarization process (which means a regression from an industrial economy to an agricultural one) which was intensified by neoliberal politics developed by Michel Temer's (2016–2018) and Jair Bolsonaro's governments. This economic strategy positions the country as the granary of the world (Lamoso 2020) and calls into question areas and strategies for the preservation of these natural spaces at the same time as it points to the suppression of the rights to the territories of traditional communities.

Here we find the paradox between development and environmental protection. In Brazil, Eco-92 (Rio de Janeiro) is a formula to think about sustainable development in the twenty-first century. In this period, from the practices of agents of the State and agents of the economic circuit, rights and economic sustainability of large natural regions are put in "check" before the need for economic development based on the aggressive expansion of agricultural areas and the exploitation of commodities. As that which occurs into the direction of the Amazonic space.

As described by Bertha Becker (the most famous Brazilian geographer on Amazon), Amazon Forest corresponds to 1/20 of the earth's surface and 2/5 of South America, besides 1/5 of potable water and 1/3 of the broadleaf forests in the world. For her, "thanks to the new value attributed to the natural resources, Amazon becomes a scientific and technological frontier for the use of nature" (Becker 2009, p. 33). In other words, Amazon represents a geopolitical trump card toward sustainable development.

However, the paths adopted by Brazil's development policy and insertion into the world economic system, as well as the scrapping of policies adopted for environmental control in the territory, have weakened the protection of these ecosystems. The most remarkable example is the national and international repercussion in the last 2 years [2019 and 2020], mainly because of the brutal advances in the burning that has affected several ecosystems, especially the Amazon, as demonstrated in the National Institute of Space Research (INPE) data.

Based on data from the National Institute of Space Research (INPE), we can see the expansion of fires in South American countries. Table 2.1 indicates the expansion of the fires in the last 5 years. Brazil appears as the country with the largest area affected by fires.

Table 2.1 Fires accumulated per country in the last 5 years

Country	2020**	Dif	2019	Dif	2018	Dif	2017	Dif	2016	Dif.*
Brazil	147.279	9%	135.496	57%	86.257	10%	131.094	10%	118.817	10%
Argentina	54.340	170%	20.127	−30%	28.694	2%	25.416	2%	24.929	17%
Bolívia	17.352	−44%	31.210	73%	18.009	−37%	16.922	−37%	26.988	193%
Chile	2.911	−5%	3.070	−6%	3.257	138%	6.381	138%	2.682	−40%
Colombia	19.985	35%	14.817	−13%	16.976	−30%	11.501	−30%	16.538	22%
Ecuador	505	-21%	641	-29%	908	-20%	450	−20%	563	−7%
Falkland Islands	–	–	–	–	–	–	–	–	8	300
French Guyana	29	61%	18	−14%	21	75%	12	−76%	50	61
Guyana	877	−8%	955	135%	407	15%	355	−56%	805	36%
Paraguay	24.453	58%	15.474	4%	14.822	−25%	19.830	22%	16.252	72%
Peru	11.489	−1%	11.639	27%	9.179	32%	6.933	−38%	11.222	29%
Suriname	195	3%	190	116%	88	19%	74	−6%	79	−40%
Uruguay	975	91%	510	33%	383	25%	306	−6%	327	−75%
Venezuela	34.792	29%	26.878	19%	22.622	76%	12.833	−31%	18.573	18%

* The difference presented in 2016 was capped compared to 2015

Source INPE (2020)-Reference Satellite (AQUA_M-T)-Accumulated data by Country of South America between 01/Jan and 21/Sep of each year, from 2016 to 2020

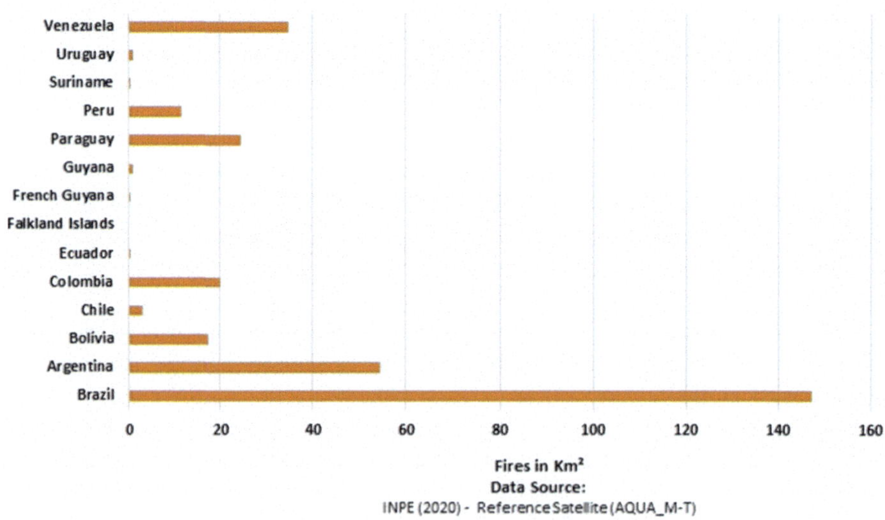

Fig. 2.2 Graph of accumulated fires by country in 2020 (January 1–September 21, 2020) (*Source* Prepared by the authors)

In Fig. 2.2, taking into account the area of fires in 2020, based on data provided by INPE, we will see that between January 1 and September 21 of this year, Brazil accumulated more than 140 km of area in outbreaks of fire.

The spatial distribution of the fires in Brazil (Fig. 2.2) is comparatively more intense in the States of the North and Center-West Regions. The Brazilian states with the largest extensions of fires are Mato Grosso (37,437 km^2), and Para (24,885 km^2), followed by Amazonas (14,512 km^2), Tocantins (8,649 km^2), Maranhão (8,806 km^2), Rondônia (7,431 km^2), and Mato Grosso do Sul (7,341 km^2). These states are in the regions which are exactly the location areas of ecosystems such as the Amazon, the Pantanal, and Cerrado Brazilian (indicated in Fig. 2.3). However, they also coincide with the area of expansion of agribusiness. In this sense, although some may indicate that the fires are natural effects in a portion of the territory, the high numbers accumulated in these regions indicate criminal actions that place the sustainability of these territories in question.

The spatiality of the largest incidence of fires in Brazil is an element that needs to be understood as a territorial strategy of the illegal advance of agricultural frontiers and severe aggression to the biomes and traditional peoples that inhabit these localities. The territorial policies of recent years marked by the flexibilization of environmental laws and the dismantling of environmental protection institutions such as Instituto Brasileiro do Meio Ambiente e dos Recursos Naturais Renováveis (IBAMA) and Instituto Chico Mendes de Conservação da Biodiversidade (ICMBIO) are the institutionalized part that cooperates for the advancement of fires.

2.5 Final Considerations

This fact has placed the Brazilian territory at a crossroads between its geopolitics, development, and environmental protection. The development as an export strategy of agricultural products serves to encourage actions to expand agricultural areas in territories of natural protection but severely attack many Brazilian biomes (Mamadova et al. 2020). This reveals an impasse since the international community gives indications of rejecting Brazilian products from the international market, in view of the increased deforestation of fires. The issues surrounding development also need to be rethought from a strategic point of view for Brazil. Once an agro exporting model is adopted, in practice, we adopt a model of exporting nature itself (Porto-Gonçalves and Leff 2015), since together with the grains and meats produced in Brazil, in its composition the soil, water, and nutrients are exported, but we could also include the spoliation of traditional peoples and communities.

A geopolitical and geoeconomics debate then emerged from the sovereignty of the territory. For example, the advance of burning over the Amazon highlights a question for national security, due to its border condition and its bioecological potential, but the precariousness of defending this natural space feeds international manifestations as to the position of the Brazilian state in fighting fires and deforestation, which brings back the debate on the internationalization of this space.

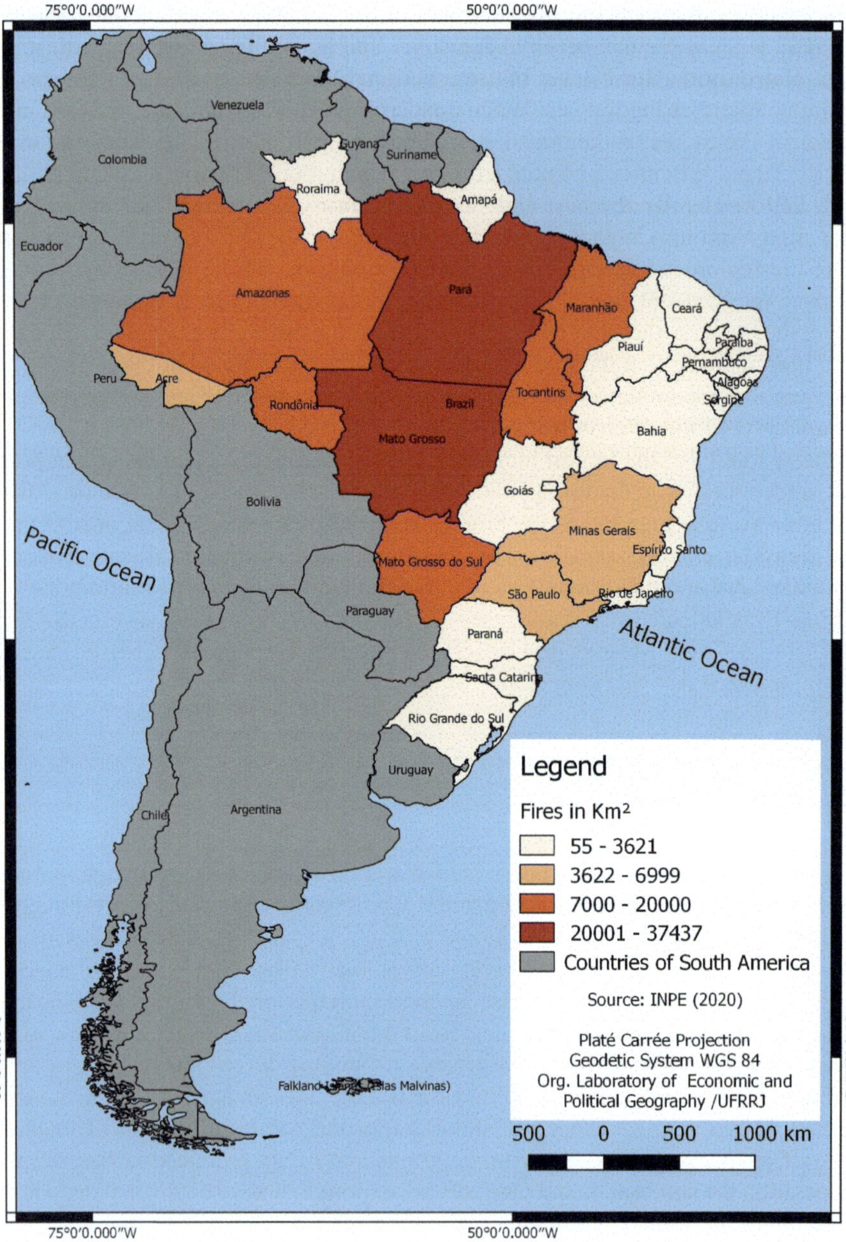

Fig. 2.3 Brazilian States—fires in Brazilian states in Km^2 in 2020 (January 1 to September 21) (*Source* Prepared by the authors)

This crossroad is, in fact, the fruit of the juxtaposition of the interests of economic and political actors and multiple scales (national and international), but also of the very dynamics of environmental geopolitics of the twenty-first century.

References

Agnew J (2005) Sovereignty regimes: territoriality and state authority in contemporary world politics. Ann Assoc Am Geogr 95(2):437–461

Agnew J (1994) The territorial trap: The geographical assumptions of international relations theory. Rev. Int. Political Economy 1(1):53–80

Anderson B (2006) Imagined Communities. Verso, London

Becker B (2015) As Amazônias de Bertha K. Becker: ensaios sobre geografia e sociedade na região amazônica. Garamound, Rio de Janeiro

Becker B (2009) Amazônia: Geopolítica na Virada do III Milênio. Garamound, Rio de Janeiro

Becker B (1995) A (des)ordem global, o desenvolvimento sustentável e a Amazônia. Becker, B. et all (ed.) Geografia e Meio ambiente no Brasil, Hucitec, São Paulo, p 46–64

Bowman I (1942) Geography vs. Geopolitics. Geogr Rev 32(4):646–658

Bryan J (2012) Rethinking territory: social justice and neoliberalism in Latin America's territorial turn. Geogr Compass 6(4):215–226

Bryant R (1992) Political ecology: An emerging research agenda in Third-World studies. Polit Geogr 11(1):12–36

Castro J (1957) Geopolítica da fome: ensaio sobre os problemas de alimentação e de população do mundo 4th edn. Editora Brasiliense, São Paulo

Castro J (1946) Geografia da fome: o dilema brasileiro: pão ou *aço*. Edições Antares, Rio de Janeiro

Cairo H (2008) A América Latina no século XXI: geopolítica crítica dos Estados e os movimentos sociais, do conhecimento e da representação. Caderno CRH 21(53):201–206. https://doi.org/10.1590/S0103-49792008000200001

Couto e Silva G (2003) Geopolítica e poder. UniverCidade, Rio de Janeiro

Cowen D, Smith N (2009) After Geopolitics? From the Geopolitical Social to Geoeconomics. Antipode 40:2–48. https://doi.org/10.1111/j.1467-8330.2008.00654.x

Davies A (2019) Josué de Castro's Geografia Combatente and the political ecology of hunger. PhD in Geography, King's College London.

Dean W (1995) With Broadax and Firebrand: The Destruction of the Brazilian Atlantic Coastal Forest. University of California Press, Los Angeles

Demangeon A (1939) Géographie politique, à propos de l'Allemagne. Annales De Géographie 48(272):113–119

Demangeon A (1932) Géographie Politique. Annales De Géographie 41(229):22–31

Dilger G, Lang M, Pereira Filho J (eds) (2016) Descolonizar o imaginário: debates sobre pós-extrativismo e alternativas ao desenvolvimento. Fundação Rosa Luxemburgo/Editora Elefante, São Paulo

Febvre L (1922) La Terre et l'évolution humaine. Introduction géographique à l'histoire. La Renaissance du Livre, Paris

Foucault M (1971) [1970]) L'ordre du discours. Gallimard, Paris

Frederico S (2015) Economia política do território e as forças de dispersão e concentração no agroengócio brasileiro. GEOgraphia, 17 (35): 68–94. https://doi.org/10.22409/GEOgraphia2015.v17i35.a13729

Harvey D (1989) The Condition of Postmodernity: An Enquiry into the Origins of Cultural Change. Basil Blackwell, Oxford

Heynen N, Robbins P (eds) (2005) The neoliberalization of nature: Governance, privatization, enclosure and valuation. Capital Nat Social 16(1):5-8

Hoogvelt A (1997) Globalization and the PostColonial World: The New Political Economy of Development. The Johns Hopkins University Press, Baltimore/ Maryland
Holdar S (1992) The ideal state and the power of geography. The life-work of Rudolf Kjellén. Political geography, 11(3): 307–232
Jean G (1976) The Medieval Machine: The Industrial Revolution of the Middle Ages. Holt, Rinehart and Winston, New York
Keil R (2005) Progress Report—Urban Political Ecology. Urban Geogr 26(7):640–651. https://doi.org/10.2747/0272-3638.26.7.640
Lamoso L (2020) Reprimarização no Território Brasileiro. Espaço e Economia. https://doi.org/10.4000/espacoeconomia.15957
Laing AF (2020) Re-producing territory: Between resource nationalism and indigenous self-determination in Bolivia. Geoforun. 108:28–38. https://doi.org/10.1016/j.geoforum.2019.11.015
Mackinder HJ (1904) The geographical pivot of history. Geogr J 23(4):421–437
Mamadova A, Behagel J, Masiero M (2020) Making deforestation risk visible. Discourses on bovine leather supply chain in Brazil. Geoforum 112:85–95. https://doi.org/10.1016/j.geoforum.2020.03.008
Massey D (1995[1984]) Spatial Divisions of Labour. Social Structures and the Geography of Production. 2rd ed. McMillan Press, London
Mattos M (1984) Geopolítica e trópicos. Biblioteca do exército, Rio de janeiro
Mattos M (1975) Brasil – Geopolitica e Destino. Biblioteca do Exercíto, Rio de Janeiro
Mbembe A (2015) Critique de la raison nègre. La Découverte, Paris
Melstrom R, Jones L, Caviglia-Harris J (2016) Patterns of frontier development: A dynamic model of resource extraction in the Brazilian Amazon. Environ Dev Econ 21(3):350–370. https://doi.org/10.1017/S1355770X15000303
Mezzadra S, Neilson B (2013) Border as method, or, the multiplication of labor. Duke University Press, Durhan and London
Mignolo W (2005) The idea of Latin America. Blackwell, London
Mignolo W (1995) The Darker Side of the Renaissance: Literacy, Territoriality, and Colonization. University of Michigan Press (1995), Ann Arbor
Oliveira LD (2019) Geopolítica ambiental: A construção ideológica do Desenvolvimento Sustentável [1945–1992]. Autografia, 2019, Rio de Janeiro
Oliveira LD (2020) Ecologia política, reestruturação territorial-produtiva e desenvolvimento sustentável no Brasil: lições do extremo oeste da Região Metropolitana do Rio de Janeiro. Espaço e Economia. https://doi.org/10.4000/espacoeconomia.16203
Ó Tuathail G (1998) Thinking critically about geopolitics. In: Ó Tuathail G, Dalby S (ed), Routledge, P. (ed.). *The geopolitics reader*. Routledge, London, pp 1–12.
Ó Tuathail G (1996) Critical geopolitics. The politics of writing global space. Routledge, London
Pádua JA (2002) Um sopro de destruição: Pensamento político e crítica ambiental no Brasil escravista, 1786–1888. Jorge Zahar, Rio de Janeiro
Porto-Gonçalves CW (2018 [2017]). Amazonía: Encrucijada Civilizatório. IPDRS-CIDES-UMSA, La Paz
Porto-Gonçalves CW (2001) Amazônia, Amazônias.Contexto, São Paulo
Porto-Gonçalves CW, Leff E (2015) Political Ecology in Latin America: the Social Re-Appropriation of Nature, the Reinvention of Territories and the Construction of an Environmental Rationality. Desenvolvimento e Meio Ambiente (UFPR) 35:65–88. https://doi.org/10.5380/dma.v35i0.44561
Quijano A (2000) Coloniality of Power and Eurocentrism in Latin America. Int Sociol 15(2):215–232
Ratzel F (1988) [1897]) Géographie politique. Economica, Paris
Ribeiro G (2015) Geografia fronteira do mundo. Ensaio sobre política, epistemologia e história da geografia. Geographia (UFF), 17 (34): 39–73. https://doi.org/10.22409/GEOgraphia2015.v17i34.a13711

Santos BS (2007) Beyond abyssal thinking: from global lines to ecologies of knowledges. Review (fernand Braudel Center) 30(1):45–89
Schmitt C (2003)[1950] The nomos of the earth. Telos Press, New York
Scott AJ (1993) Technopolis: high-tecnology industry and reginonal development in southern california. Un. of California Press, Berkeley
Scott AJ, Storper M (eds) (1986) Production, Work, Territory. The Geographical Anatomy of Industrial Capitalism. Allen and Unwin, Boston
Storper M, Venables A (2005) O Burburinho: a força econômica da cidade. In: Campolina Diniz, C, Lemos MB (eds) Economia e Território. Editora da UFMG, Belo Horizonte
Storper M, Walker R (1989) The Capitalist Imperative: Territory. Blackwell Publishing, Oxford, Technology and Industrial Growth
Travassos M (1938) Projeção continental do Brasil. São Paulo, Rio de Janeiro. Companhia Editorial Nacional, Recife e Porto Alegre
Vlach VRF (2003) Estudo preliminar acerca dos geopolíticos militares brasileiros. Terra Brasilis (nova Série). https://doi.org/10.4000/terrabrasilis.359
Vidal de la Blache P (1994[1917]). La France de l'est (Lorraine-Alsace). La Découverte, Paris
Walker PA (2005) Political Ecology: Where is the Ecology? Prog Hum Geogr 29(1):73–82
Wallerstein I (2003) "Mundialização ou Era de Transição? Uma Visão de Longo Prazo da Trajetória do Sistema-Mundo" In: Chesnais F, et al (eds). Uma Nova Fase do Capitalismo? Xamã, São Paulo

Official Documents

Agenda 21 (1992) United Nations Conference on Environment & Development Rio de Janeiro, Brazil, 3 to 14 June 1992
Convention on Biological Diversity (1992), United Nations, Rio de Janeiro, 5 June 1992
Instituto Nacional de Pesquisas Espaciais (2020). Programa Queimadas—Monitoramento por satélite— Relatório Diário 22 set/2020. www.inpe.br/queimadas. Access 30 sep 2020
Meadows DH et al (1971) The Limits to Growth: A Report for the Club of Rome's Project on the Predicament of Mankind. Universe Books, New York
Our Common Future: Report of the World Commission on Environment and Development (1987) United Nations / Oxford University Press, Oxford, England, United Kingdom
Rio Declaration (1992) United Nations Conference on Environment and Development, Rio de Janeiro, Brazil, 3–14 June 1992
The future we want (2012) Rio+20 (United Nations Conference on Sustainable Development), Rio de Janeiro, Brazil, 20–22 June 2012
United Nations Framework Convention on Climate Change (1992) Rio de Janeiro, United Nations, 5 june 1992

André Santos da Rocha received the degree of Graduate in geography for Faculty Philosophy, Science, and letters of Duque de Caxias. He has post-graduate (lato sensu) in Territorial Policies for State University of Rio de Janeiro (UERJ), Master in Geography for Federal Fluminense University (UFF) and Ph.D. in Geography for Federal University of Rio de Janeiro UFRJ. Currently carries out posdoctoral research at PPFH in UERJ. Presently, he is a Professor in undergraduate and graduate courses (master) at UFRRJ. He works in Human Geography, especially with thematic of political and economic geography. Analyses on the Baixada Fluminense and its political and economic dynamics stand out in their studies. He is also dedicated to studies of International Cooperation with a phase in Brazil-Africa-Latin America relations. He coordinates and participates in research projects with other researchers from UFRRJ and colleagues from Latin American Network Space and Economy (REELE). He serves as Editor of the Academic Journal

Continentes (UFRRJ), and Coordinator of the Laboratory of Economic Geography, Policy and planning (LAGEP).

Guilherme Ribeiro is awarded Ph.D. in Geography from Federal Fluminense University (UFF), with doctoral stage at the Paris IV Panthéon-Sorbonne and postdoctoral degree at Minas Gerais Federal University (UFMG). Guilherme Ribeiro is Associate Professor II from the Department of Geography and the Graduate Program in Geography at the Rio de Janeiro Federal Rural University (UFRRJ), as well as Coordinator of the Laboratory of Politics, Epistemology and History of Geography (LAPEHGE). Guilherme Ribeiro is Co-editor of Continentes. Revista de Geografia da UFRRJ (http://www.tiagomarino.com/continentes/index.php/continentes/issue/view/19) and Espaço e Economia. Revista Brasileira de Geografia Econômica (https://journals.openedition.org/espacoeconomia/), and Vice-leader of the Brazilian Network of History of Geography and Historical Geography (http://redebrasilis.net/).

Leandro Dias de Oliveira earned his Masters in Geography from the State University of Rio de Janeiro (UERJ), PhD in Geography from the State University of Campinas (Unicamp), and did Post-Doctoral research in Public Policy and Human Training in the UERJ. Currently, he is an Associate Professor in UFFRJ's undergraduate and graduate courses in Geography. He is dedicated to studying society-nature relationships from the perspective of EconomicGeography, with a focus on the adoption of sustainable development. He is the coordinator of the Laboratory of Economic Geography, Policy and Planning (LAGEP) and an editor of the Space and Economy: Brazilian Journal of Economic Geography. Member of the Latin American Space and Economy Researchers Network (RELAEE) and of the Political Ecology Network (POLLEN). Homepage: https://r1.ufrrj.br/geografiaeconomica.

Pablo Ibanez has graduated in Geography from the State University of Campinas (2002), and has completed Master's (2006) and Ph.D. (2012) in Human Geography from the University of São Paulo. He is currently an Adjunct Professor at the Federal Rural University of Rio de Janeiro and a researcher linked to the research groups: Welfare in Health, Development and Territory, at the Faculty of Medicine of the University of São Paulo; Research Center on Public Policies and Territory, at Fluminense Federal University; and Group of Studies and Research on Politics and Territory (GEOPPOL), from the Federal University of Rio de Janeiro. He was a Visiting Researcher at Fudan University, Shanghai. He has experience in the research area, with emphasis on Geopolitics, Political, Economic, and Regional Geography, acting mainly on the following themes: science, technology, and innovation; regional development, regionalization, and decentralization; analysis of public policies, industrial, and innovation policies; and development and territory.

Chapter 3
The Transversal and Territorial Praxis of Regional Development

Marcos Aurelio Saquet and César Adrián Ramirez-Miranda

Abstract This text is the result of a process of scientific and political cooperation carried out between Brazilian and Mexican researchers and activists, working on two regional development projects based on a concept of territorial approach and the methodology of participatory action research. The results achieved are summarized here, of course, as it is a complex and popular movement that generated many learnings, especially on the territory-region relationship, food sovereignty, agroecology, and cooperation aimed at sustainable development, as described in the text.

Keywords Territory · Region · Development · Participatory action research

3.1 Introduction

Since the 1970s, many studies and debates have been held in the academic-scientific and public policy fields at the international level about the problem of territorial and regional development. There are significant advances in countries such as Italy, France, Spain, as well as more recently in Brazil and Mexico. In the latter countries, the debate and implementation of public policies and regional and territorial development projects have generally followed the experience of the LEADER program (*Liaison Entre Actions de Développement de l'Économie Rurale*), implemented in Europe since 1991.

A criticism, therefore, to the import of European models has been made. For example, Ramírez-Miranda (2014), when carrying out a careful review of development concepts linked to neoliberalism, reproduced in Latin America precisely to favor the insertion in the market from the agrochemical production of food, as it is verified when we study the historical process of Latin American colonization.

M. A. Saquet (✉)
Integrated Laboratory of Territorial Studies (GETERR), State University of Western Paraná (UNIOESTE), Cascavel 85819-110, Brazil

C. A. Ramirez-Miranda
Laboratory of Rural Regional Development, Chapingo Autonomous University (UaCh), 56230 Texcoco, Mexico

M. Mishra et al. (eds.), *Regional Development Planning and Practice*, Advances in Geographical and Environmental Sciences,
https://doi.org/10.1007/978-981-16-5681-1_3

Unlike Europe, Latin America, since its conquest in the XV–XVI centuries, became merchandise linked to the reproduction of capital on an international level. The conquest and exploitation of Latin America integrate, by different mechanisms and strategies, the expansion of world capitalism, providing prosperity to others (Galeano 2019 [1978]).

In general, since the beginning of the sixteenth century, there have been profound impacts on nature and society in Latin America, as it was involved in the international movement of appropriation and capitalist territorial expansion, serving as a colony of exploitation of the Spanish and Portuguese people. It is a violent process of expropriation, subordination, discrimination, domination, and enjoyment of natural and cultural riches which, due to America's particular conditions, exacerbates the capitalist logic of concentration of wealth and centralization of power (Quijano 2000).

There are currently high rates of urban and rural poverty in countries like Mexico and Brazil, along with considerable infant mortality rates despite there has been substantial progress in medicine. Many urban and rural workers have low wages, precisely because of the logic of production-circulation-consumption that corresponds to dependent capitalism based on overexploitation of labor (Marini 1974). In the agrarian issue, predominate the cultivation of products destined to national and international markets, such as soy, corn, sugarcane, orange, coffee, avocado, and tomato, with a productive structure dominated by large corporations, but with an essential contribution from peasant agriculture.

The hegemonic class reproduces itself creatively, promoting dependent industrialization accompanied by urbanization, slumming, and agricultural modernization, continuing ideological and political domination based on military, geopolitical, economic, and welfare mechanisms. Moreover, it continues to generate extreme inequalities, diminishing agricultural diversity, the predominance of large-scale monocultures, the introduction of genetically modified organisms, the use of chemical inputs and machinery, rural expropriation, loss of decision-making autonomy, water contamination, soil erosion, etc. (Harguindeguy 2014). International banks have played a central role in controlling interest, prices, and financing, promoting a process of agrifood financialization.

These dependent and exploitative characteristics strongly threaten the existence of all of us (Bartra 2014), since colonialism and coloniality are reproduced as mechanisms of domination of Eurocentric capitalism that favors an economic-political minority by spoliation, extraction, and concentration of wealth (Ribeiro 2015 [1995]; Quijano 2000; Mignolo 2003 [2000]; Ramírez-miranda 2015; Saquet 2020).

Evidently, this process requires, for its explanation and reversion, an approach contrary to Eurocentric and colonial capitalism (Quijano 2000), as well as a practice of cooperation and solidarity in favor of the Latin American people, keeping nature, preserving and valuing popular culture, and producing food for the domestic market, especially craft and agroecological (Saquet 2015, 2018, 2020). We believe that working with food-producing peasants, we are investing in counter-hegemonic processes of territorial and regional development, built for and with the people, against the dominant political-economic-cultural order, which inscribes us in the broad movement toward food sovereignty (Ramírez-Miranda 2017).

Thus, our main objective in this opportunity is to describe and analyze experiences of territorial regional development carried out in Brazil and Mexico-socializing the adopted concepts, the action research methodologies, and results achieved of the projects we carry out at our public universities through various cooperation with other public institutions, associations of peasant farmers and non-governmental organizations. In the realization of our socialized projects in this text, we decided to address a participative, solidary, and dialogic research-action praxis, which we will elaborate on in the course of this text.

3.2 Territorial Transversalities and Practices in Regional Development

There are different theoretical and political perspectives of regional development, especially of Euro-American origin. However, the one that interests us most is the one that is sustained by a historical–critical and multidimensional conception of the region and development, recognizing and evidencing in theory and practice economic, political, cultural, and environmental actions. The region and development are constructed historically and geographically according to social (plural: different and unequal) and natural (diverse ecosystems: soil–water-plants-climate-animals) conditions in each space–time. The conditions of development, therefore, vary from one region to another. Although often, the regions are interconnected through communication and circulation networks.

The region, understood in this way, has a direct relationship with each territorial social formation because the territory is the result of the historical and social process of appropriation of space, transformed (with more or fewer impacts) through social works, techniques, technologies, knowledge, interests of subjects, groups, and social classes. According to the broader movements of territorialization, de-territorialization, and reterritorialization, nature is always in the territory, where social processes are regionalized and de-regionalized. So, region and territory are different concepts. However, they are not separated in theory and reality, that is, they occur at the same time, crossing each other transversely, overlapping in space and historical time. Thus, the region has territorial characteristics linked to power relations, cultural identities, natural diversity, and circulation and communication networks, fundamental elements and processes also in development (regional). According to the tradition of regional studies, this last can be carried out within an area or in different areas, that is, in different regions.

However, in a specific region, territorial development can occur in the form of an area-network or network-network without a precise delimitation, as we demonstrated in Saquet (2012). The territory is produced in a transtemporal and trans-scalar way, which is regionalized historically and geographically. It is transversal and multidimensional in a reticular perspective (of the networks of circulation and communication), and it does not have a standardized form and meaning. On the contrary, it is always singular, with specific social and natural characteristics, defined by each territory's inhabitants and other conditions.

Although local or regional development is stimulated by economic globalization, the manifestations are usually different in each territory due to the power relations that exist in it, the needs of people, interests, ecosystems, etc. In the current period, it is evident that technical, scientific, and technological innovations alter territories, often facilitating their insertion in the world market, as well as accentuating processes in the form of development axes (Sposito 2008). However, territorial development processes that are restricted to the national and/or regional level coexist, as we will demonstrate in the final part of this text.

Transversality happens precisely because of the coexistence of processes. Temporalities and territorialities overlapped in time–space in each territory and between territories. It happens because territorialities occur in the same place and/or between different places, at the same time and/or in different temporalities as relations of competition and dispute, but also as reciprocity, that is, as cooperation and solidarity, social participation and dialogue, resistance and confrontation in the face of the hegemonic class within the state, the market and the civil society (Saquet 2012, 2015, 2020).

In this way, development happens at local and/or regional level, with articulations and territorial transversality and means a continuous process of social (economic, political, and cultural) conquests manifested in our daily life by healthy eating, quality housing and education, medical assistance, access to entertainment, hygiene, political organization, and decision-making autonomy, among other aspects that are fundamental to live with dignity and quality, such as the recovery and preservation of nature, in the countryside and in the city.

Degrading the environment means degrading the territory; sustainability is thought beyond the protection of nature, incorporating the territory at the regional level, that is, its economic, natural, political, and cultural characteristics. Nature is a patrimony of the whole society and needs to be managed by the local society articulated to other social groups, with capacity for self-management, decision-making autonomy, integration of different knowledge (scientific and popular), cooperation and solidarity, management carried out by the inhabitants of each region, of each territory, by the subjects of each development process.

3.3 A Synthesis Between Brazil and Mexico: The Praxis of Transversal and Territorial Regional Development

The following projects' results, highlighting the concepts used, the methodologies, and the main results achieved, are part of international scientific cooperation initiated in 2016 by the project entitled 'Territory and food sovereignty: short networks, cooperation and management of food production and trade in peasant groups: Case studies in Mexico and Brazil' (*Territorio y soberanía alimentaria: redes cortas, cooperación y gestión de la producción y comercialización de alimentos de grupos campesinos. Casos de estudio en México y Brasil*), associated to the other project entitled 'Network for Territorial Management of Sustainable Rural Development' (*Red para la Gestión Territorial del Desarrollo Rural Sustentable*) (Red GTD, México), of which the Territorial Studies Group (GETERR/UNIOESTE, Brazil) is part. One of the main objectives of this project was to deepen the study of different peasant experiences related to food sovereignty at the regional and territorial levels, highlighting the creation of networks of cooperation and solidarity.

A fundamental premise for carrying out our projects is the recognition that dialogue is fundamental in the process of building knowledge between different subjects, in a mixture of knowledge produced together in an interdisciplinary and transversal way (Dematteis 1994). The knowledge is simple and complex, plural and singular, individual and collective, urban and rural, technical and technological, popular and territorial, local and trans-multi-scale. Using dialogue, we integrate transversally different life experiences, territorial and regional, inside and outside the university, in different spaces where knowledge and thought to happen to the people (Ribeiro 2015[1995]; Mignolo 2003 [2000]; Fals Borda 2013 [2007]).

3.3.1 The Project 'Regional Prototypes for Food Security and Sovereignty and the Fight Against Poverty (PROSOA)': A Territorial Approach

This participative action research project was financed by the National Council of Science and Technology, was carried out from April 2017 to April 2019 and resulted in the elaboration of the mentioned prototypes in thirteen states of the Mexican Republic. Its objective was to design comprehensive, multi-scale regional prototypes for food security and sovereignty and the fight against poverty, built through the dialogue of knowledge between territorial actors and universities, in order to help improve the food situation of rural families and provide elements for improving public policies under a territorial management approach.

This objective could only be approached from an interdisciplinary and transdisciplinary perspective, reflected in the concepts addressed and in the methodology of the research. Therefore, it is important to mention the characteristics and previous trajectory of the interdisciplinary team of professionals responsible for talking to the territorial actors to achieve this objective.

The research team was integrated by four agricultural engineers, two biologists, two sociologists, one anthropologist, one agro-ecologist, one economist, one agricultural development planner, one public administrator, and one graduate in international relations. This team was integrated by two different aspects and trajectories in which previous experience regarding the research problem was generated. A group of professionals was assigned to the Direction of Regional Centers of the *Universidad Autónoma Chapingo* (UACh), a public university with thirteen foreign academic units in the same number of federal entities. Ten of these federative entities were covered in this project, and three more were incorporated in which, due to their proximity to the central headquarters, do not have facilities. The second strand was made up of a group of members of the Territorial Management Network for Sustainable Rural Development (*Red de Gestión Territorial para el Desarrollo Rural Sustentable*—Red GTD) created in 2009 under the auspices of the Inter-American Institute for Cooperation on Agriculture (*Instituto Interamericano de Cooperación para la Agricultura* - IICA), to consolidate the territorial approach to development in the area of public policies aimed at the rural sector. In 2014, it received recognition and funding from the National Council for Science and Technology (*Consejo Nacional de Ciencia y Tecnología*—CONACyT).

3.3.1.1 Main Concepts

The theoretical and conceptual support of this research was based on five positions, from which specific concepts were derived. These positions are (1) criticism of productivist and modernizing developmentalism due to its negative results in terms of food and poverty; (2) recognition of the peasant rationality and the complexity of their forms of social reproduction; (3) adherence to the approach of food sovereignty due to its scientific and political relevance; (4) the need for agrifood systems based on short circuits of production and marketing; and (5) appreciation of the territorial approach as an essential element for the management of better public policies in the rural space.

Each of these elements implies a broad conceptual background and provides important theoretical and political debates. These involve the theories of rural development in Latin America and the debates about the peasantry's future and the conceptualization of short circuits and the economy of proximity. Of course, they refer to the importance of a territorial approach to development that manages to discard its neoliberal hindrance. Moreover, they incorporate the conceptualization of food Sovereignty as a multidimensional project, criticizing the food status quo, whose significance varies in time and space, but is increasingly interwoven with the agroecological paradigm. The important thing to emphasize in terms of practice is that the five elements mentioned are articulated by a general positioning of epistemological order: the centrality of the peasant subject in the process of change.

The above means that the process of knowledge generation aimed at transforming the alimentary and poverty situation should be focused on the action of the peasant subjects, from their reflection about the specific and concrete problems that affect them in their daily life, and at the same time articulating this reflection with the general elements mentioned: the criticism of the global corporate agro-food regime, the revaluation of the forms of peasant social reproduction, the relevance of the short circuits of production and commercialization, the relevance of building alternatives from the territories and, articulating all of this, self-recognition within the multidimensional and multi-scale movement for food sovereignty.

In compliance with the centrality of the peasant subject, this research was carried out using the methodological resources of Participatory Action Research. It is an approach with a long tradition and continuous renewal in Latin America, aimed at generating transformative and emancipatory knowledge in the territories through social actors' reflective action and a subject-to-subject relationship with professional researchers.

3.3.1.2 Participatory Action Research Methodology

The wide range of methodological resources related to Participatory Action Research, as well as the accumulated experience of the researchers responsible for the project in the thirteen federative entities with their prior knowledge on the general problems, constituted the starting point for the formulation of a methodological approach. It had the challenge of advancing toward a common purpose (the development of the prototypes) but recognizing effectively the regional diversity that characterizes Mexican agriculture. In addition to considering the ecological, technoproductive, and sociocultural diversity of the federative entities addressed, it was fundamental to recognize the heterogeneity of the previous processes of reflection of the subjects involved in the Participatory Action Research Groups (*Grupos de Investigación Acción Participativa*—GIAP).

The general methodological approach considered four stages, which did not necessarily follow a sequential process: (1) Participatory diagnosis and identification of the problem, (2) Dialogue of knowledge and reflection for action, (3) Training, adoption and adaptation of technology, (4) Baseline and system for evaluation of results. These stages correspond in a general way, although with important nuances, to those indicated by Soriano Niebla et al. (2012)—participatory observation, participatory research, participatory action, and evaluation. The following are the most relevant elements incorporated in each of these stages.

Participatory Diagnosis

Social participation does not guarantee processes of reflection and action of a transformative and emancipatory nature. This is clear in Latin America, where neoliberal governments created channels of participation that were fully functional to an excluding and wealth-concentrating project. For this reason, special attention was given to the formulation of a methodology for the diagnosis focused on the critical and reflexive participation of GIAP members. In addition to considering the different tools and techniques useful for participatory research, among which those systematized by Verdejo (2003), FAO (2008), and Geilfus (2002) stand out, it was also decided to carry out a specific dialogue with the experiences of Colombian colleagues from the Coffee Research Center (*Centro de Investigación de la Caficultura* - Cicaficultura). This is a transdisciplinary project based at the *Universidad del*

Cauca, focused on promoting the agroecological transition of coffee farms, as well as the Brazilian colleagues of the Territorial Studies Group (GETERR/UNIOESTE).

As a result of a scientific mission to the Cauca region in May 2017, which included community workshops, an international seminar, and a special session about methods and tools for participatory diagnosis, the project generated 16 participatory techniques to be made available to GIAP in the process of preparing the diagnoses, equally important, eight fundamental principles were agreed upon for this phase of participatory action research. Some of these principles are elementary, such as clear and honest communication with the population involved, the return of information, the multidimensional nature of the diagnosis, and attention to generating a useful inventory of the territorial assets of the communities involved. On the other hand, others hold proper keys to the viability of the research. This is the case of the orientation to join reflection-action processes already underway and incorporate previous experiences in each region. The knowledge accumulated in the field is reflected in the following alerts to researchers: (a) Avoid becoming obsessed with detailed information, as the focus is on building trusting and reflective exchange of ideas, (b) Avoid addressing issues that are conflicting in the community, (c) Do not make quick judgments about complex issues, (d) Address issues relating to costs of production and income only once relationships of trust are established. All these factors are articulated around a general statement of the highest relevance: the objective of the assessment is not to gather information but to trigger processes; consequently, collective reflection is central in the three moments of the assessment, in the descriptive moment to know and characterize, in the analytical moment to build indicators, and in the transition moment to design proposals.

With these tools, GIAP began the diagnostic phase in the thirteen federative entities. These groups, led by the aforementioned nucleus of researchers, combined more than 150 family experiences, bringing together dozens of people who reflect today's rural world's diversity. Thus, in addition to peasants, agricultural technicians, secondary and higher education teachers, undergraduate and graduate students, there were participants with long careers in the management of initiatives in the productive or educational field. In all cases, a formal protocol was made accordingly to GIAP.

Dialogue Between Knowledge and Reflection for Action

The project proposed to carry out a systematic dialogue of knowledge between the participating universities and the territorial actors via GIAP. In order to promote the dialogues in the regions and considering the generalized process of chemical and physical degradation of the agricultural soils in the country, the collection of the Center for Validation and Technology Transfer (*Centro de Validación y Transferencia de Tecnología*—CENVyTT), of the *Universidad Autónoma Chapingo*, was discussed. It consists of the bio-intensive production of polyfunctional compounds, elaborated from biomass, and other resources available in the different regions.

This stage was carried out practically simultaneously with the next one and is only presented separately for expository purposes. In fact, once the diagnostic tools were approved, the dialogue began to recover the knowledge and practices of GIAP about the components of the local agrifood systems.

Training, Adoption, and Technological Adaptation

One of the project's methodological contributions is that the assessment phase was carried out simultaneously with various training activities in low-cost agroecological practices. It was done under the premises of *dialogue by doing and learning by doing*. At this stage, agroecology showed its great potential, not only technologically but mainly dialogically, by unfolding as an inherent element of the proposal of food sovereignty. Over 50 training actions adapted to the conditions of each region and GIAP were carried out together. Among these actions, 8 Chromatography Courses for Soil Vitality were given in 6 federative entities and a plenary course with the participants of the 13 federative entities. With these courses, the project became part of a National Network of Studies on the Vitality of Peasant Lands, promoted by Sebastião Pinheiro, a Latin American agroecology reference.

Baseline, Indicators, and Results Assessment System

The project designed the planning, following-up, monitoring, and evaluation system with indicators to establish a baseline that allows the subsequent evaluation (Chávez and Ramírez-Miranda 2017). This tool, which basically constitutes an adaptation of the elements of the Framework for the Evaluation of Management Systems incorporating Sustainability Indicators (*Marco para la Evaluación de Sistemas de Manejo incorporando Indicadores de Sustentabilidad*—MESMIS) (Massera et al. 2000), consists of a matrix of attributes, diagnostic criteria and indicators proposed to GIAP, with 22 indicators, grouped according to the attributes of the MESMIS and accompanied by a set of diagnostic criteria.

In accordance with the IAP principles (participatory research and participatory action), it was decided that each GIAP would manage the type of data to be obtained and its monitoring method, according to its monitoring scales, approaches, and contexts. The challenge was for the indicators to be adjusted to each regional context to ensure relevance to each territory's management systems and strategies. The analysis attributes were conceptualized as follows:

- Productivity is understood as the efficiency of peasant production systems, expressed in the capacity to increase their food production performance.

- Stability, resilience, and reliability are grouped into a single attribute and refers to the capacity to maintain the diversity of species within the systems and avoid their degradation. Adaptability is understood as the capacity to strengthen learning and innovation processes. Equity, referring to the fair distribution of the system's costs and benefits and the improvement of gender relations. Self-dependence and self-management, seen as the capacity to respond to and control changes induced from outside.

Subsequently, as a result of the exchanges among GIAP, the measurement systems were adjusted. It was decided to observe ten indicators common to all the GIAP to build a platform of observables elements among the various local experiences. These indicators are (1) System yields, (2) Species managed and present, polycultures, rotations, (3) Degree of self-sufficiency of families (4) Soil and water quality, (5) Reappropriation, generation, and adoption of innovations, (6) Beneficiaries of the system, (7) Degree of dependence on critical external inputs, (8) Degree of use of own seeds, (9) Degree of dependence on external food products, (10) Decision-making power on critical aspects of the system's operation (Mechanisms for participation and conflict resolution).

3.3.1.3 Results

The project was concluded in April 2019 with the development of 14 prototypes in thirteen states (Fig. 3.1). The methodological process mentioned and the participatory action research carried out by the GIAP gave rise to these socio-technical arrangements called Regional Prototypes for food sovereignty and the fight against poverty. Each of the prototypes includes a particular process and characteristics that cannot be explained in this text. Their common denominator is to express a collective reflection on each of the territories and a bet on a route to follow to advance toward food sovereignty and poverty reduction. Figure 3.1 presents the name of each prototype, which reflects the diversity of the approaches taken by the GIAP to the common purpose of food sovereignty.

Finally, it should be noted that each of the prototypes is accompanied by the corresponding Regional System for Innovation Management (*Sistema Regional para la Gestión de la Innovación*—SIRGI). It is the device developed specifically for the prototypes to achieve regional dissemination (Table 3.1) via the assistance of public institutions and territorial actors in the various regions.

Fig. 3.1 Location of Regional Prototypes for Food Sovereignty *(Prototipos Regionales para la Soberanía Alimentaria, i.e., ProSoA)*

3.3.2 The Agroecological Family Agriculture Project in the Municipalities of Verê, Itapejara d'Oeste and Salto do Lontra (Southwest Paraná, Brazil), as a Strategy for Social Inclusion and Territorial Development (AGROEC/USF)

Our Brazilian project also happens as a mechanism to counteract the green revolution and its harmful territorial and regional impacts. That is why we chose to work from the perspective of research-action-participation, integrating scientific and popular knowledge. With transversal practices between different municipalities, groups, and social subjects, we consider the need to produce food without chemical inputs for the farming family and urban consumers, as well as to enhance peasant family work contributing to increase their income in agricultural practice. Thus, our main objective was to understand agroecological peasant agriculture characteristics, food producers, collaborating to qualify agroecological production, expand the consumer market, and preserve the biodiversity of rural establishments. We believe that in order to reorganize agricultural production, the way of life, and the territory, it is necessary to consider and organize power relations again, valuing and strengthening small political, cultural, and productive organizations.

Table 3.1 Denomination and location of the ProSoA

No	Denomination	Location
1	Regional prototype for the treatment and reuse of greywater for the strengthening of food production	Álamos Village, Sonora
2	Prototype for food sovereignty: revitalizing the cornfields and the land	Lacandona Forest, Chiapas
3	Peasant participation for agroecological production and food consumption	Cuextotitla, Tlaxcala
4	Inclusive social innovation networks: prototype for food sovereignty	Veracruz High Mountains
5	Food sovereignty: country-city relationship, knowledge, and balances for territorial sustainability. Zacatecas Prototype	Central Region of Zacatecas
6	Prototype for the integrated production of self-consumption in the rural areas of Zacatecas	Zacatecas bean region
7	Prototype with a gender perspective for food sovereignty	Michoacán
8	Prototype for food sovereignty: the central role of women in food culture	Istmo de Tehuantepec, Oaxaca
9	Fertile soil and bioenergy for food sovereignty	San Bartolo Tutotepec, Hidalgo
10	Energy recycling for food sovereignty and the fight against poverty	Nayarit
11	Agro-food systems: a proposal for territorial management for food sovereignty and security	Mexico State
12	Between the recovery of food sovereignty and the commercial production of fruits and vegetables	Tlahuapan, Puebla
13	School of Agroecological Management of Coffee (*Escuela de Manejo Agroecológico del Café*—EMAC)	Occidental Mountain of Jalisco
14	Peasant training, community action, and socio-environmental resilience	Low Mountain of Guerrero

Source Own elaboration, 2019–20

We start from the fundamental assumption that knowledge—from opinions and experiences, memories and traditions, speech and listening—influence the qualification of class consciousness that Fals Borda (1978, 1981) calls reflection-action, using techniques for the collective production of knowledge (Fals Borda 2013 [2007]). As we have already mentioned, this production of knowledge is interdisciplinary and transversal, fundamental concepts for the construction of the project and for carrying out actions with the agroecological subjects of each selected municipality.

The territory, as a reticular and/or regional formation, has interdisciplinary and transversal meanings since it is shaped historically and geographically based on social-nature relations, using material and immaterial mediators that cross and complement each other—such as techniques and technologies, knowledge, circulation and communication networks, the power of the state and other institutions (NGOs, associations, cooperatives, etc.) and the identities of different social groups. The territory means polysemy, transversality, multidimensionality, trans-temporality, and trans-scalarity (Saquet 2012, 2015, 2020).

Based on that, our team was composed to meet our theoretical and practical concept of regional and territorial development: three geographers, two agronomists, one economist, one sociologist, and one veterinarian—from different public universities of Brazil and Italy. A team that was even more complete with the insertion of students (with scholarships) graduated in geography, with the technicians from the NGOs of Small Farmers Support Center (*Centro de Apoio ao Pequeno Agricultor*—CAPA-Verê; currently called Center for Support and Promotion of Agroecology—*Centro de Apoio e Promoção da Agroecologia*) and the NGOs of Ecovida Network of Agroecology (*Rede Ecovida de Agroecologia*). This action research project was funded by the *Universidade Sem Fronteiras*, of the State Department of Science and Technology and Higher Education of Paraná (Brazil) (2010–2014).

To carry out cooperation activities with the selected farmers and the institutions of each municipality (Verê, Itapejara d'Oeste, and Salto do Lontra—Paraná Southwest—South Brazil) we chose the participatory methodology. It incorporates data collection via discussion of the methodology to planning and carrying out actions aimed at agroecological-based territorial development. To achieve the project objective, we work with the following procedures: bibliographic research, collection, and analysis of secondary data (Brazilian Institute of Geography and Statistics—IBGE), conducting interviews with ecological farmers based on a previously defined script, mapping of agroecological production, meetings with representatives of local institutions and agroecological farmers, writing a regional diagnosis with the main problems identified and possible solutions to be implemented in each municipality. All the actions carried out were related to agroecological practices (production and commercialization). We understand agroecology as an alternative for development linked to the achievement of peasants' decision-making autonomy, proper management of soil, plants, and water. It is also directly related to eliminating intermediate traders and the valorization of peasant knowledge and cultural heritage. The territory-region-development relationship assumed in the AGROEC/USF project characterizes this action research project with its multidimensional social dimensions. As a political-ideological, environmental-agroecological movement that is culturally and territorially fixed, the centrality remains in food production without the use of chemical inputs and its commercialization in short networks.

Based on the discussion of the participatory methodology in the UNIOESTE team, we defined the following phases and procedures of the research and actions (despite in the temporal succession of phases, some activities happened simultaneously, due to the participatory and territorial conception of research and action):

- **1st phase**: Bibliographic research about agroecology, peasant agriculture, social inclusion technologies, development, political organization, and autonomy;

planning of other project management activities; meeting with the boards of agroecological producer associations in the three selected municipalities.

- **2nd phase**: Collection, treatment, and analysis of secondary data (land-ownership structure, agricultural production, agricultural machinery, soil management techniques, instruments, and labor relations) from the municipalities Verê, Itapejara d'Oeste, and Salto do Lontra.
- **3rd phase**: Survey of complementary data at the Municipal Agriculture Departments and at the NGO ASSESOAR.
- **4th phase**: Analysis of the performance of each association or group of agroecological producers: production and commercialization strategies; number of members, production, and commercialization mechanisms used. Interviews and meetings with farmers from each association or group in order to create a place for dialogue to analyze the data collected in the field work and also to discuss and define strategies to qualify the social inclusion of agroecological producers.
- **5th phase**: Field research, considering the following variables: (i) production organization: job and income generation, labor relations, productive and post-harvest techniques, and technologies; (ii) social living conditions: schooling, basic sanitation, housing, health, roads, household appliances, and cultural characteristics; (iii) associative bonds and political organization dynamics.
- **6th phase**: Discussion and definition, based on participatory methodology, of the strategies and management policies of the agroecological production units studied, favoring improvements in family income, social inclusion, and territorial development mechanisms.
- **7th phase**: Participatory implementation of the university extension actions defined with the agroecological producers—described below.
- **8th phase**: Monitoring and evaluation of actions carried out with agroecological farmers.
- **9th phase:** Cartographic representation of the agroecological units of the municipalities and associations studied; writing of reports and texts for presentation in scientific events and publications.

In the diagnostic phase, we identified that in the municipality of Itapejara d'Oeste, in 2010, there were 9 agroecological producers (only 3 certificates). The rural establishments had an average area of 17.14 hectares, and the average area for agroecology was 3.8 hectares. The establishments are small but have diversified productions as we verified in the field work: fruits, vegetables, and/or grains (beans, soy, corn), everything produced by family work. In the municipality Salto do Lontra, we identified eight ecological farmers, with a significant variation in the size of rural establishments, between 0.5 and 40 hectares, with a diversified family production, such as the municipality of Itapejara, highlighting organic soy and corn, grown for export. In the municipality Verê, also in 2010, there were 19 agroecological producers (17 certificates): most of the rural establishments were small, from 0.5 to 28 hectares, with the predominant management and workforce of peasant families, a regional characteristic. Agroecological production is also diversified, with emphasis on vegetables, corn, soy, beans, cassava, grapes, and other fruits (Saquet et al. 2012).

Faced with this situation briefly described, we realized that the implementation of our project "Agroecological family agriculture in the municipalities Verê, Itapejara d'Oeste and Salto do Lontra [...]" was fundamental. Thus, we intensified collaborations in each municipality, trying to cooperate systematically to increase agroecological production and families' incomes. In this university-society movement, we create opportunities for dialogue and social participation, valuing political self-organization, culture and municipal governments in each territory, as well as the participation of the *Universidade Estadual do Oeste do Paraná* (UNIOESTE) in territorial development processes. During this project, we identify and analyze agroecological rural establishments, working with farmers in different activities and giving courses focused on agroecological production and labeling of transformed foods, as well as promoting the expansion of the urban consumer market in each municipality.

The dialogue opportunities in the three municipalities (Itapejara d'Oeste, Verê, and Salto do Lontra) were focused on conversations with farmers organized in associations and NGOs. We were clear about the political and cultural importance of activating the daily territorialities of the peasants in order to broaden the practice of cooperation and solidarity aimed at development through a praxis of transformation (dialogical, critical, and reflexive), also contributing to them obtaining a little more decision-making autonomy (Freire 2011[1974]).

So, based on the agroecological problem, project's objective and participative methodology, the main activities carried out were (i) meetings of the work team—interdisciplinary and inter-institutional—formed by professors and university students, to discuss and define the research and actions that would be carried out; (ii) meeting with the board of the association of agroecological producers from Verê; (iii) meetings with representatives of the different institutions linked to peasant agriculture in each municipality to present and discuss the research and cooperation project, in order to create the aforementioned dialogue opportunity; (iv) identification of ecological farmers in each municipality, presenting the project and discussing the difficulties of production-commercialization, initiating the definition of priorities for future actions, after interviews with family farmers and subsequent diagnosis; (v) delivery and discussion of the diagnosis made to municipal institutions and farmers; (vi) implementation of the actions/cooperation that were defined and planned with the ecological peasants, with our direct involvement in each action carried out; (vii) thematic mapping of agroecological rural establishments; (viii) writing, publishing, and delivery of the informative guidebook: Agroecology and conscious consumption—actions that transform, 2010 (*Agroecologia e consumo consciente—ações que transformam*, 2010); (ix) organization of some training courses in partnership with the Universidade Tecnológica Federal do Paraná (UTFPR—Dois Vizinhos): *Soil and water management; Fruit management and cultivation; Post-crop and labeling*; (x) reorganization of the Salto do Lontra and Itapejara d'Oeste producer's market in the southwest of Paraná; (xi) diffusion of agroecological production in each city of the three municipalities, including Francisco Beltrão, the largest city and consumer market in the region.

The trans-territorial actions reinforced the importance of the spatial dimension in our project, which was planned for three municipalities but was expanded to four neighboring municipalities, precisely because of the need to strengthen the consumer market through short commercialization networks. Besides this aspect of the increase of municipalities, the spatial dimension was also reinforced through the reorganization of the free fair of Salto do Lontra and the Market of the Producer of Itapejara d'Oeste, always aiming at the best presentation of the agroecological product and its direct commercialization. These actions also reinforced the regional importance and centrality of the networks of reciprocity, within each municipality and between the four municipalities, characterizing concomitant and overlapping territorialities, that is, transversality between urban and rural areas.

Thus, the analysis of the consumer market also gained prominence in the project, trying to encourage the consumption of agroecological products and help the qualification of environmental awareness and place. This survey resulted in a diagnosis that was given to the farmers of each municipality to guide changes, if it was their will, in agroecological production and commercialization, reinforcing a little more the territorial transversality of this type of regional development (Saquet 2020).

In this way, we believe that we can practice the transversal dimension of territorial and regional development, with actions (research, mobilization, information, and transformation) in the scalar scope of four municipalities, relating these municipalities with peasant producers and urban consumers (short commercialization networks), as illustrated in Fig. 3.2, forming overlaps between the singular and the regional (intermunicipal). These relationships took place through collaborations among different individuals and institutions (from the state, NGOs, associations, and

Fig. 3.2 Illustration of the territorial transversality of the project AGROEC/USF (*Source* Authors own elaboration, 2020)

unions), that is, via inter-institutional and multidimensional (economic, political, and environmental) territorialities.

These transversal relationships of collaboration are not harmonious, they are difficult to accomplish, because there are limits and difficulties that were identified in each municipality, to produce, certify and commercialize. In production, there are few people to work, specific public financing is lacking, and the existing agroecological technical orientation in the region is not enough. The certification carried out by the Ecovida Agroecology Network is participative and has no cost for the peasants, however, the transition process is slow and often farmers end up giving up this productive practice. Trading is relatively easy, however, the short networks still need to be expanded and better distributed in each city in the region. So, we also verified that there is a constant oscillation, in each municipality, of the quantity of certified agroecological peasants.

In 2017, by way of illustration (Fig. 3.3), there were only 2 certified farmers in Itapejara d'Oeste, 2 in Salto do Lontra and 21 in Verê (Ministério da Agricultura, Pecuária e Abastecimento, 2017), with a decrease in Itapejara and Salto do

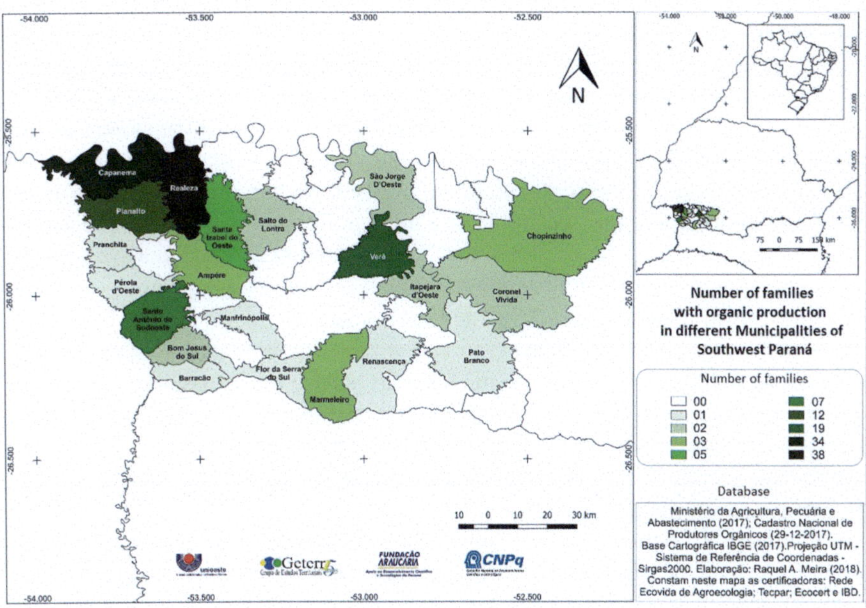

Fig. 3.3 Spatial distribution of certified agroecological peasants in southwest Paraná (*Source* Ministério da Agricultura, Pecuária e Abastecimento, December 2017. Map elaboration: Raquel Meira)

Lontra compared to 2010. The increase occurs in Verê, where there is the efficient performance of CAPA-Verê.

Currently, the number of certified agroecological producers in the three municipalities did not change so much: it went from 28 in 2010, to 25 in 2017 and to 32 in 2020, but production is much more diversified. Salto do Lontra now has 10 certified farmers, who produce cassava, bean, corn, pumpkin, sweet potato, sugar cane, and oat. Itapejara has 3 farmers producing avocado, lettuce, banana, bergamot, beet, kale, carrot, orange, strawberry, cucumber, bell pepper, radish, cabbage, and tomato. Verê has 19 certified agroecological producers and has a rich and varied agroecological production: avocado, cassava, pumpkin, banana, bergamot, kale, broccoli, guava, orange, papaya, cucumber, bell pepper, peach, radish, cabbage, tomato, peanut, sweet potato, chayote, carrot, pea, corn, rocket, garlic, onion, lemon, strawberry, etc. All produced, in the three municipalities, by peasant families, together, in the specific case of Verê, with the production of canned food (beets, broccoli, carrots, cauliflowers, cucumbers, pods, pumpkins, bananas, peaches, grapes, and papayas), jam (orange, strawberries), and tomato sauce (Ministério da Agricultura, Pecuária e Abastecimento, May 2020).

Finally, it is relevant to highlight: (i) the process of research and action revealed that cooperation between the University and local society is fundamental. By actions discussed and concretized in the chosen municipalities, integrating knowledge and techniques, urban and rural subjects, and creating opportunities for joint learning, we contribute to encourage the peasants through workshops, courses, exchanges, political mobilizations, publicity of their productions, etc. (ii) At the same time, we learned a lot from the peasants and representatives of municipal institutions about their daily practices and the rural and urban needs of each municipality. This aspect reinforces our argument that we still have a lot to do for and with the people, significantly increasing investments, public and self-organizing actions, in a synergy that is certainly contradictory, but necessary and vital to build development as a quality of life for all, in a more just and ecological society. (iii) Our project followed a process of interaction and political organization very similar to that described by Fals Borda (2012 [1986]), overcoming some tensions among the different public institutions, associations, and NGOs; establishing relationships of trust, especially between producers and consumers; integrating different knowledges, notably academic and popular from a collective coordination and social participation that happened in the different phases of the project.

In this way, we believe we can contribute a little to activate the daily sense of consciousness about place and territory, popularizing knowledge and, especially, the importance of working in solidarity and cooperation based on culture in each territory, highlighting in the popular sphere the importance of struggle, resistance, and territorial counter-hegemony.

3.4 Final Considerations

Despite the difficulties that exist when we work in a movement for food sovereignty and environmental preservation that we consider counter-hegemonic, contrary to neoliberalism and agribusiness and environmental degradation, we believe that we are able to cooperate in the transversal and trans-scale fields by activating our universities and the scientific knowledge produced, interacting with subjects from the countryside and the city, especially with peasants who reproduce their ancestral knowledge for food production.

Research-action-participation activities in Brazil and Mexico demonstrate that the territorial level is fundamental for training, provide information, research, and cooperative transformation actions. It is in the territory where people live, work, and reproduce as subjects who think, feel, have needs and dreams, producing a significant part of our daily nutrition. Our research showed that, in the historical formation of the territory, there are also regionalizations mediated by work, techniques, technologies, and different knowledge. Thus, it seems that there is not a single scale of action in development, it is essential to work, as we have done in our projects, at the level of the establishments of peasant families, public institutions, NGOs, and municipal governments. Consequently, concretizing transversal territorialities of cooperation and solidarity, anchored in each territory and in common objectives in a movement toward food sovereignty, the conquest of decision-making autonomy, the production of ecological and artisan food, the preservation of biodiversity, etc., as we summarize in Table 3.2.

For this, action-participatory research was fundamental, because we researched with the peasants of each project, taught and learned from them. In addition, we discussed aspects of their daily lives, creating relationships of trust and cooperation that were fundamental for the university to fulfill its social role in regional development and for farmers to reproduce it, being more valued in terms of culture, politics, and the environment.

Finally, we emphasize that these projects had a very well defined political content, in favor of peasants and urban inhabitants who value their food, as part of a movement to contest the historically reproduced dependence in Latin America on Eurocentric capitalism, so we work in a counter-hegemonic and decolonial praxis anchored in the territory. The latter was conceived as a fundamental concept for understanding the problem of oppression and colonization. It is also an opportunity to disseminate information, offer training, and provide a kind of cooperation that takes place by means of research-action-participation in a popular conception in which the public university still has much to offer and collaborate with the Latin American people.

Table 3.2 Synthesis of the research-action projects PROSOA and AGROEC/USF

	PROSOA	AGROEC/USF
Theoretical-methodological concepts applied	– Territorial and regional, interdisciplinary, research-action-participation, transformation, and political emancipation	– Territorial, multidimensional, reticular (cooperation networks), research-action
Scale of territorial transversality	– Regional → National (transprovincial) – With integrated, multi-scale actions, based on short networks and regional culture	– Municipal →Regional (transmunicipal) – Anchored in the territory and in the short commercialization networks
Public policies	– The objective is to subsidize the construction of regional and national public policies	– Actions linked to the national policy of acquisition of school meal
Work Teams	– Interdisciplinary and inter-institutional – Centrality for peasant subjects → dialogue of knowledges	– Interdisciplinary and inter-institutional – Centrality for non-governmental institutions and peasant associations
Phases of research-action	– Participatory diagnosis – Dialogues and reflections for participatory action – Training, adoption, and technological adaptation (learning and dialoguing by doing) – Evaluation of the results achieved in the project	– Research-participant – Discussions and planning for participatory action – Information and training for peasants – Mapping of production and dissemination of the consumer market – Evaluation of the results achieved in the project

(continued)

Table 3.2 (continued)

	PROSOA	AGROEC/USF
Main advances achieved	– Scientists: progress was made in the knowledge of the complex regional food dynamics in 13 federative entities – Methodological: the notion of prototype was formulated as an element of the participatory research processes – Organizational: local capacities related to agroecology and food sovereignty were strengthened, allowing processes of territorial appropriation – Academics: we build cooperative interactions between several universities and research centers, socializing the knowledge produced via publication of articles, books and seminars	– Scientific: we advanced in the qualification of agroecological techniques and in the knowledge of the agroecological practices in 3 municipalities – Methodology: qualification of the research-participant, of the qualitative research of peasant food production – Organizational: we contribute to the strengthening of direct commercialization of peasant agricultural production (free trade fairs) and to the political valorization of NGOs – Academics: we disseminated the importance of the public university and peasant food production via seminars, articles, and published books

(*Source* Authors' own elaboration, July 2020).

References

Bartra A (2014) Campesindios: ethos, clase, predadores, paradigma. Aproximaciones a una quimera. In: Hidalgo F, Houtart F, Lizárraga P (eds) Agriculturas campesinas en Latinoamérica—propuestas y desafíos, Quito, Editorial IAEN, pp 269–276

Chávez P, Ramírez-Miranda C (2017) Referentes metodológicos: Fase diagnóstico. Universidad Autónoma Chapingo, Material de trabajo

Dematteis G (1994) Possibilità e limiti dello sviluppo locale. Sviluppo Locale 1(1):10–30

Fals Borda O (1978) Por la praxis: el problema de cómo investigar la realidad para transformarla. Simposio Mundial De Cartagena, Bogotá, Punta De Lanza 1:209–249

Fals Borda O (1981) La ciencia y el pueblo. In: Grossi F, Gianotten V, Wit T (eds) Investigación participativa y praxis rural, Lima, Mosca Azul, pp 19–47

Fals Borda O (2012 [1986]) Primera lección: saber interactuar y organizarse. In: Herrera Farfán N, López Guzmán L (eds) Ciencia, compromiso y cambio social—Textos de Orlando Fals Borda, Buenos Aires, El Colectivo, Lanza y Letras, Extensión Libros, pp 123–141

Fals Borda O (2013 [2007]) Hacia el socialismo raizal y otros escritos. In: Orlando Fals Borda—Socialismo raizal y el ordenamiento territorial, Bogotá, Ed. Desde Abajo, pp 35–136

Fao (2008) Manual Diagnóstico Participativo de Comunicación Rural, DPCR, Dirección de Investigación y Extensión, 2nd edn. Roma

Freire P (2011[1974]) Pedagogia do oprimido. Paz e Terra, São Paulo

Galeano E (2019[1978]) As veias abertas da América Latina. L&PM, Porto Alegre

Geilfus F (2002) 80 herramientas para el desarrollo participativo. IICA, San José, Costa Rica

Harguindeguy L (2014) Economía solidaria: local y diversa. El Colegio de Tlaxcala, México

Marini R (1974) Dialéctica de la dependencia. Era, México

Massera O, Astier M, López-Ridaura S (2000) Sustentabilidad y manejo de recursos naturales. El marco de evaluación MESMIS, Gira A. C. / UNAM, México

Mignolo W (2003)[2000]) Historias locales/diseños globales. Editorial Akal, Madrid

Quijano A (2000) El fantasma del desarrollo en América Latina. Rev. Venezolana de Economía y Ciencias Sociales, 6(2):73–90

Ramírez-Miranda C (2017) Soberanía alimentaria y desarrollo rural. Implicaciones teóricas y políticas. Revista ALASRU, Nueva Época, México 11:93–117

Ramírez-Miranda C (2015) Repensar el territorio, repensar el desarrollo rural. In: Hernández Moreno M, Ramírez Miranda C, Menéndez Gámiz C (eds) Territorio y gestión del desarrollo, Plaza y Valdés, México, DF, pp 23–46

Ramírez-Miranda C (2014) Visión crítica sobre la Nueva Ruralidad y el desarrollo territorial rural en América Latina. Agronomía Colombiana Bogotá 32(1):122–129

Ribeiro D (2015[1995]) O povo brasileiro. Global, São Paulo

Saquet M (2012) Il territorio della geografía. Franco Angeli, Milano

Saquet M (2015) Por una geografía de las territorialidades y de las temporalidades: una concepción multidimensional orientada a la cooperación y el desarrollo territorial. FAHCE/UNLP, La Plata, Argentina

Saquet M (2018) A perspective of counter-hegemonic analysis and territorial transformation. Geographica Helvetica 73:347–355

Saquet M (2020) Saber popular, praxis territorial e contra-hegemonía. Editorial ITACA, Ciudad de México

Saquet M, Gaiovicz E, Meira S, Souza P (2012) Agricultura familiar agroecológica como alternativa de inclusão social e desenvolvimento territorial em Itapejara d'Oeste, Salto do Lontra e Verê—Sudoeste do Paraná. In: Saquet M, Dansero E, Candiotto L (eds) Geografia da e para a cooperação ao desenvolvimento territorial: experiências brasileiras e italianas. Outras Expressões, São Paulo, pp 35–62

Soriano Niebla J, Garcia T, Sanz I, Gonzalez J, Carrascosa M (2012) Mejora agroecológica participativa (MAP) y Biodiversidad Agrícola. Aplicación de la investigación-acción participativa al manejo de las variedades tradicionales en Andalucía. Agroecologia 2:21–30

Sposito E (2008) A noção de território: uma leitura pela noção de desenvolvimento. In: Heidrich A, Costa B, Pires C, Ueda V (eds) A emergência da multiterritorialidade, Porto Alegre, Editora da UFRGS; Editora da ULBRA, pp 63–75

Verdejo M (2003) Diagnóstico rural participativo: guía prático, Centro Cultural Poveda, Proyecto Comunicación y Didáctica

Marcos Aurelio Saquet graduated in Geography with a doctorate at UNESP—Presidente Prudente (São Paulo). He did his doctorate at Università Ca Foscari di Venezia and a postdoctoral degree at Politecnico and Università in Torino—Italy. He was a visiting professor at Universities in Italy, Colombia, Argentina, Costa Rica and Mexico. He is currently an Associate Professor at the State University of Western Paraná, where he teaches undergraduate courses in Geography (graduation) and in the Postgraduate Program in Sustainable Rural Development and Society, Culture and Borders. He is also a member of Red Territorio, Actores y Gobernanza América Latina, and of Red en Gestión Territorial para el Desarrollo Rural Sustentable, as well as a member of the International Network—Cultural Changes Laboratory, Inequalities and Sustainable Development and the Progetto di Cooperazione Territori, Sostenibilità and Università.

César Adrián Ramirez-Miranda graduated in Economics at National Autonomous University of Mexico, with a master degree at Autonomous University of Zacatecas, and a doctorate at Metropolitan Autonomous University/México. Research Professor and current Coordinator of the Postgraduate in Science in Regional Rural Development (Master's and Doctorate) taught by the Autonomous University Chapingo, Mexico. Also thesis director in the Ph.D. in Economics of the National Autonomous University of Mexico (UNAM). National Researcher, level II. Currently Technical Manager of the Thematic Network on Territorial Management for Sustainable Rural Development of the National Council of Science and Technology (CONACyT). He has coordinated several projects and interdisciplinary groups in the field of regional rural development; has promoted various publishing and outreach projects, among which the Journal of the Latin American Association of Rural Sociology (ALASRU) and the *Revista Textual*. His main lines of research are Food Sovereignty, Rural Development, and Peasant Strategies of Social Reproduction.

Part II
South-Asian Experiences

Chapter 4
Regional Development at Crossroad to Degrade the Environment: Regional Impact and Responses to Climate Change and Other Environmental Issues

L. Manawadu, M. D. K. L. Gunathilaka, and V. P. I. S. Wijeratne

Abstract The regional approach in development opens up a promising avenue for thousands of people residing in remote rural areas in Sri Lanka. This paper examines the regional development in Sri Lanka, regional development efforts and policy implications, and consequences of regional development. In the limelight, previous governments in Sri Lanka have proposed various development plans. Among them, the Accelerated Mahaweli Development Programme, IRDP, and Southern Development Plan are well known. Though these programmes succeeded to some extent, the overall success is a dilemma due to various environmental issues and climate change issues intensified with such development efforts. Now the environmental issues have become a barrier to achieve regional development due to natural disasters. It is recommended to apply proper sustainable land-use practices to avoid the impacts of this cyclic effect and provide responses by using the regional approach to an environmental perspective.

Keywords Environmental issues · Regional disparities · Regional development · Mahaweli development programme · Sustainable response

4.1 The Regional Approach in Development

Regions have always been powerful spatial metaphors and political concepts, but it is within the context of more recent paradigm shifts that the region has acquired its putative economic, political, and social salience. A focus on regions as a spatial scale of governance and economic development is, of course, not new. In most of the European countries and the new world, regionalization has been understood as a necessary process in the modernization of a country and as a means to adapt to shifting economic realities associates with globalization. Thus, regional development is a process toward development by regions or particular parts of a country.

L. Manawadu (✉) · M. D. K. L. Gunathilaka · V. P. I. S. Wijeratne
Department of Geography, University of Colombo, Colombo, Sri Lanka
e-mail: lasan@geo.cmb.ac.lk

M. Mishra et al. (eds.), *Regional Development Planning and Practice*,
Advances in Geographical and Environmental Sciences,
https://doi.org/10.1007/978-981-16-5681-1_4

Regional development has become increasingly important for national as well as local and regional governments across the world since the 1960–1970s after the founder of the regional science, inspiring social scientists in studying the relationship between such factors as geography, migration, and land use in local and regional economies, it was none other than Walter Isard (1919–2010). A series of books published by Isard; Locational and space economy (1956), Methods of regional analysis (1960), general theory (1969), and with Thomas Langford, Regional input–output study (1971) were most influential (Grimes 2010). The core concepts of regional science were developed and disseminated via the Regional Science Association and the Journal of Regional Science. Both of which were founded by professor Isard in the 1950s. As specified by Isard himself, regional science is defined as "in brief, regional science as a discipline concerns the careful and patient study of social problems with regional or spatial dimensions, employing diverse combinations of analytical and empirical research" (Hansen and Quigley 1976).

As indicated by Glasson, "territorial advancement can be viewed as the cycle of improvement in a specific region or area. It is regularly utilized in a relative setting, contrasting issue area, and the prosperous areas or with the public in an outright setting likewise the advancement inside a specific locale". Local arranging in its different structures is utilized to endeavor the advancement of a district (Glasson 1974). Friedmann (1970) additionally characterizes the term territorial advancement as the rate of monetary development. It is eventually the aftereffect of the area of monetary exercises in light of various provincial attractions. Movements in the area design have direct repercussions as pay, work, and government assistance. Since a spatial association is an element of action and connection design, territorial advancement is just a declaration of these examples. This accentuated that local advancement includes some of the interrelated cycles, including financial, social, and political cycles inside the spatial system. Likewise underscored the economy and financial advancement as the basic necessity which prompts territorial turn of events. Besides, Friedmann (1970) depends on where the monetary development is continued over extensive stretches of its frequency, runs after a reformist combination of the space economy.

"Provincial" arrangement started in many nations during the 1950–1960s, OECD nations (Organization for Economic Co-operation and Development) in particular. The principal objective of the local arrangement was more noteworthy value and adjusted improvement during the time of fast industrialization. Whatever the actions are taken to decrease provincial differences, government endeavors were not fruitful. In this way, local inconsistencies were not altogether diminished. Against a foundation of expanding globalization, decentralization, and spending strain since the 1980s enormous allotments for local projects have gotten unreasonable in a time of successive monetary downturns, summed up more significant levels of joblessness, and in-wrinkling tension on open use. As a reaction, the regional advancement instruments have gotten more extensive in scope. Territorial vital projects have filled in execution, mirroring an overall strategy shift toward help for endogenous advancement and the business climate, expanding on provincial potential and abilities, and meaning to cultivate development arranged drives. Afterward, staggered administration approaches were included. Better acknowledgement of the interdependencies of sectoral approaches and the effects on areas has worked with the co-activity of cross-sectoral arrangements simultaneously. Thorough provincial policies are progressively viewed as corresponding to public financial and structural approaches by assisting with creating development in the locales. The change in the outlook of territorial improvement strategy in this manner includes new destinations, another geographical extension.

This paper aims to discuss the regional development of Sri Lanka, regional disparities of the country, and the policy implications and the consequences of regional development, particular attention pays to climate change and environmental issues due to regional disparities seek for responses through regional development. The findings and the analysis is based on available scattered pieces of literature.

4.2 Regional Development in Sri Lanka

Regional development in the country is a consequence of two concurrent processes: the development and political processes. The development process of the country initially involved infrastructure development, irrigation development, and land settlement in less developed areas, where the political process involved devolution aimed at greater power-sharing by sub-national governments. The most influential power-sharing was Act No. 42 of 1987, which enabled the devolution of power to provincial councils (Ratnayake 2015). However, the development process was initiated in the country in various aspects.

Various research studies have illustrated that urban centers as sources of stimulating regional development and the advantages of urban centers (Wanasinghe 2003; Ratnayake 2015). The gravity of development in Sri Lanka mainly relies on the principal urban centers that evolved throughout history; pre-colonial like Kotte kingdom, colonial like Dutch and the British and continuously. It is none other than Colombo based urban center. In this respect, it is almost clear the apophthegm "Colombo got *Kiri,* and the village got *kekiri*".[1] Notwithstanding, the apophthegm seems to be changed as per the critics of various professionals, researchers, and politicians. It is still a dilemma among the scholars was it typically changed or just an illusion. Hence, it is essential to investigate regional disparities in the development process of the country.

[1] It is an apophthegm used to express the prosperity of one aspect and at the same poverty in another. Here "kiri" stands for curd which is a sign of prosperity/wealth. While "Kekiri" (Cooking melon) is a vegetable mostly grown and use in villages rather than city is sign of poverty or lowest living status. The contradictory meaning here is both curd and cooking melon are abundant in villages, but curd is transported to the cities while keeping cooking melon as a source of subsistence. So the best thing for the city less important kekiri for the village poor.

4.3 Regional Disparities in the Development Process of the Country

Being an island country, the country is in a perfect world just as topographically situated in an essential area in the Indian Sea. The advancement of Colombo city has effectively been impacted by different approaches and arranging endeavors taken before. This has prompted imbalance being developed alongside GDP (Gross domestic product) (Fig. 4.1). The populace dispersion and the land territory, and the monetary action across the territories show the disparity of advancement. The strength of the western territory is constantly featured and contended by numerous analysts (per capita Gross domestic product of 157% public normal in the western, northern, uva, north-focal and eastern areas have a Gross domestic product for every capita of 64, 69, 76 and 76% separately). Bandara and Jayasuriya (2010, p12) contend that "…western region contrasts fundamentally from different districts: it has a lot higher per capita pay and a lower pace of destitution, industry and administrations overwhelm the construction of the economy (with agribusiness' offer being practically unimportant), represents a large portion of the manufacturing industry in the nation (number of foundation, work and value-added), and has better framework offices (street thickness, correspondence offices and admittance to monetary offices)". The market-situated changes, like the advancement of exchange, liberating industry, and advancing private speculation, have profited the western territory because of its optimal area and framework. The historical development of approaches and political elements made "beginning conditions", favorable toward the Western area, and the almost thirty years of war have created conditions that worsen local inconsistencies.

Over the last decades, the distribution of provincial-wise per capita value discovered that difference with relevance this indicator and regional disparities square measure significantly high. It shows that stark regional disparities stay with the

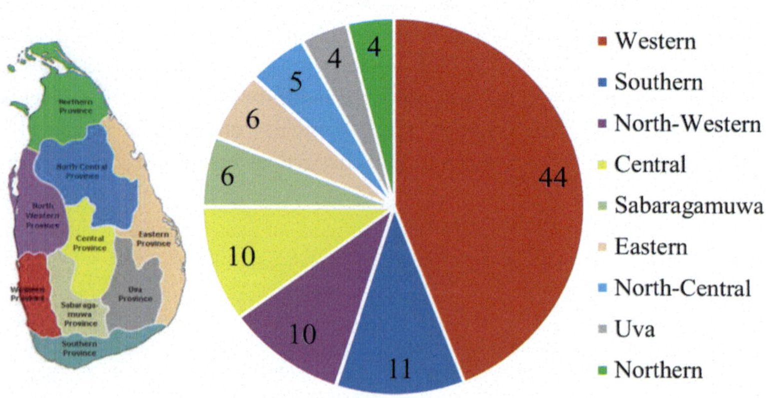

Fig. 4.1 Sri Lankan provinces and their contribution to national GDP 2011 in *percentages* (*Source* Wijerathna et al. 2014)

Western province, which accounts for 1/2 economic activity, leading by a good margin of financial gain share. Sri Lanka has the best levels of regional difference in South Asia. World Bank mentioned in 2004, the growth in the regional difference between regions has been increasing as additional prosperous regions cash in on the socio-economic and political changes experienced since independence.

The current state of affairs in the country is prejudiced to future economic development prospects of the many regions due to lack of attention by politicians, policy manufacturers, and development practitioners at native levels. Hence, the implications are going to be pricey, with irreversible damages to the socio-economic and political state of affairs within the country. First, increasing regional development disparities will cause produce economic unskillfulness due to lack of market formation method in several regions within the country. Moreover, this level of financial gain inequality is unfair, and it will be a binding reason to weaken social stability and commonness between those that board prosperous leading regions and insulant regions.

The regional difference may arise due to many reasons like endowment of natural resources to a locality, central government policies that focus on a handful of key regions, whereas others unit simply incomprehensible or missing from the foremost political agenda for receiving development aid. The country enforced a provincial council system in 1987 to supply larger autonomy to native governments to need necessary measures to increase the native share of the resources. Thereby provincial worth and employment, these efforts do not seem to possess addressed the foremost issues effectively. Still, there is a clear continuation of the quality path of development going the foremost disadvantage of distribution unaddressed. Therefore, the prevailing state of affairs over several decades of unequal financial gain accumulation implies that the regional inequality in Sri Lanka is very inopportune to the majority of different provinces in Sri Lanka. Therefore, presently within the absence of any such well-planned regionally inclusive development methods nationwide or provincially to handle this downside and thereby continuation of a better level of provincial-wise GDP disparities might take an even longer amount to cut back the event disparities in the state. This method stimulates and results in producing additional financial gain within the Western province and unequal financial gain distribution across provinces within the country. This atmosphere arises in the main because of the concentration of an oversized portion of leading economic activities and the handiness of connected facilities in the Western province compared to different provinces of the country.

In this respect, it is clear that the political and economic policies of the country are the forces that led to such regional disparities.

4.4 Policy and Planning in Regional Development

Since the 1930s different area bad habit approaches, projects, and systems have been acquainted with the advance provincial turn of events (Ratnayake 2015). After the political autonomy in 1948, the nation has made a few endeavors to advertise the socio-monetary states of the farming people that are ignored by the frontier rulers. Governments that came to control when freedom paid attention to mainly public planning for the significant instrument of advancement. Along these lines, those administrations implemented numerous public plans like the Ten years set up in 1959, the five years set up in 1970, and Public Interest in 1983–1987 that focused on all-island spatial sectoral advancement (Ratnayake 2015). Every one of these projects worked at the public level and there have been a few attempts that suggested a definite assortment of territorial turn of events. These attempts are regularly isolated basically into two assortments as decentralization of administrative capacities and improvement programs (Ratnayake 2015).

Indeed, even before freedom, the land had a limited body framework up somewhat to work with the focal government was executing its elements of the local level. Beneath this technique, Area Coordinative Councils and consequently the Divisional coordinative Boards were set up at the region level and divisional level in 1954 to work with this strategy. After 1970 the construct of provincial advancement had its spot inside the country's improvement plan. Consequently, territorial advancement planning as an assortment of formalized de-marking came when in 1970, the public authority set up a substitution administrative framework to speed up the execution of improvement exercises at the sub-public level. In 1971 a Territorial Improvement Division, in 1973 Area Political Authority Framework, in 1978 Locale Service Framework, in 1980 the Region Development Boards and 1987 the Commonplace Gatherings were set up to de-unify the administrative capacities. Albeit this institutional arrangement upheld the execution of sectoral advancement projects of the public authority, these projects weren't locale explicit.

The failure of nationwide development led governments and policymakers to introduce specially designed development programmes targeting rural areas. Strategies to develop the dry zone of the country were premiered among the policies. However, this target economically bounced back due to several constraints, especially low and varied rainfall, dry period, and poor infrastructural facilities (Ratnayake 2015). Considering the evolving issues (such as—overcrowding and poverty in the wet zone, poverty and land availability for agricultural growth in the dry zone and tourism destinations) the politicians and policymakers attempt to address these specific areas through the intervention of various programmes. Among the programmes that policies gave rise to were (Ratnayake 2015):

- Social development programmes;
- Resource development programmes;
- Decentralization of government functions;
- Regional development under the liberalized economic regime since 1978;
- Devolution of power to provincial councils.

These efforts have been contributed to regional development to some extent. However, most government policies were not initiated entirely to lessen the intra-regional inequalities. They have contributed significantly to attain several national-level development objectives of employment, food production, and power generation and population redistribution (Ratnayake 2015). Wanasinghe (2003) discussed the policy instruments that were geared to develop less developed areas in Sri Lanka. They included capital and technology transfers and transfer of public investment for education, health, and infrastructure; expansion of the area under agriculture and enhancing agricultural productivity; promotion of regional industries and the stimulation of growth of small and medium towns in rural areas. The requirement of study whether the regional development efforts taken in the country led to the development of other issues is essential when talking about numerous policies.

4.5 Regional Development Efforts of Sri Lanka

The introduction of the Integrated Rural Development Programs (IRDPs), the Mahaweli Development Project (MDP), and the programme for the South by the Southern Development Authority (SDA) are some of the popular regional development attempts in the country.

During the early 1970s administrative districts play a major role as a sub-national unit for administration and development. The government introduced a more comprehensive thrust in some districts that became popular as the Integrated Rural Development Programme (IRDP). The objectives of the project were diverse: to widen economic opportunities and enhance the general standard of living in rural areas; to focus development efforts, especially to meet local needs and to encourage local institutions; to reduce inter- and intra-district disparities and thereby promote balanced growth; encourage quick response, mutually supportive low-cost productive investments along with necessary institutional improvement; to assist in removing constraints, thereby contributing to better utilization of district resources; and, to improve the planning process in districts. The project initially started in the Kurunegala district with the World Bank's assistance and the IRDP for other districts, such as Hambantota, Matara, Nuwara-Eliya, Badulla, Moneragala, Ratnapura, Kegalle, Kalutara, and Anuradhapura, with the assistance of NORAD and SIDA. The major achievements of IRDPs have been in social mobilization and the delivery of social and economic infrastructure, as explained by Lindhal et al. (1991) and Dahlen (1992). Wanasinghe (2003) pointed out that very little consideration was given to the growth of industries or small towns or in the strengthening of rural–urban linkages.

Before the large-scale development in the Mahaweli basin, almost half of the area was covered with natural tropical wet and dry and mixed forests. Tropical montane forests were distributed in the upper reaches of the basin. Almost all of these lands were converted to various land-use types to some extent. A master plan consisted of a combination of water storage at the upper reaches of the basin and hydropower production, improved downstream agricultural facilities, and the clearing and development of new agricultural lands. The land units were divided into systems designated by letter A to H and later introduced I–J–K–L–M regions. This programme was initiated to be completed within 30 years. The Polgolla barrage and the Bowatenna reservoir and irrigation tunnel (southeast of Polonnaruwa) completed by 1976. From 1977 onwards, the project became known as the Accelerated Mahaweli Development

Programme. This accelerated project consisted of multipurpose sub-projects like Kotmale project, Polgolla diversion, Victoria, Randenigala, Rantambe, Minipe diversion, Maduru Oya, Elahara diversion, and Angamedilla diversion. Mainly Kandy: upper reaches of the basin achieved quite rapid development progress. More than 400,000 people have settled in the lower catchment area and these households contributed to cultivating approximately 127,000 hectares of land in the Mahaweli area (Tolisano and Goonesekara 1993).

Along with these developments, social infrastructure development was required for the people living and resettled in the Mahaweli basin to minimize the impact of inadequate opportunities to have a better living status. This is typically true for second especially third-generation residents who still seek access to arable land. The population is predominantly Sinhalese (Tolisano and Goonesekara 1993) and the average family size was about five and now plentiful encroaches have been distributed all over the Mahaweli basin area. More than half of families approximately earn less than 3000 rupees per year when subsiding the total cost for agricultural outputs (Tolisano and Goonesekara 1993). Though this figure increased—the average monthly per capita income is approximately 12,000 rupees without subsiding the total cost for agricultural outputs by 2018 (Mahaweli Authority 2018). Besides, the environmental issues within the basin were also accelerated with the development and natural functions, and the services were declined in line with development. Even though technical recommendations are applied to achieve sustainable regional development within the basin, various issues are emerging day by day—such as human-elephant conflict and loss of cultivation; watershed and sedimentation issues; rainfall variability and long-term dry periods; and lack of infrastructural facilities or improper maintenance of existing social infrastructure.

The southern province development programme was launched to raise the income and quality of life of the people. The project aimed to rehabilitate and construct a minor irrigation scheme and four saltwater exclusion and drainage schemes. To strengthen the private sectors of the area, the Charley Mount Industrial Estate was constructed and financed. Except for this estate, several small-scale industrial estates were proposed and road development was carried out (ADB 2004).

Successively the methodologies and techniques intended to upgrade territorial advancement issues are fairly extraordinary. The Neediness Mitigation Program or Poverty Alleviation Programme (Janasaviya) began in the arising attention to the desires of poor people and social pressures which emerge from their not being satisfied. On a more extensive perspective, Janasaviya was a group-based program that proposed to put individuals first; thought about people as an essential asset; relied basically upon independence and granular perspective in dynamic; it was socially harmonious, and it very well may be acknowledged genuinely island-wide. Janasaviya recommended improving the personal satisfaction of poor people has been the improvement of their capacities by giving food security to the family during its two-year time of activity. The Janasaviya Trust Asset, the other partner program which had a hole-spasm way to deal with neediness mitigation, included Social activation measure; training measure; Credit segment; Local area project segment (FAO 2001).

Gam Udawa, i.e., the "Village Reawakening" or "Town Stiring" lodging improvement favorable to the project, zeroed in consideration on the situation of the destitute and impeded families (Rajasingham 2012). Under this program, the public authority assembled 100,000 houses and started the 1,000,000 lodging units' venture, which was perceived worldwide by the Unified Nations. There are three sorts of constructions engaged as clarified by Rajasingham; "The primary kind of design, landmarks, and models show what performance contemplates researchers call 'surrogating'—the way toward discovering substitutes no doubt or envisioned misfortune. This sort is comprised of current and pre-frontier surrogating in small-scale size. The subsequent class comprised of houses worked in new Gam Udawa towns or re-modeled houses in re-stirred towns. The state gave re-stirred towns crude materials with the goal that occupants could revamp or redesign their homes, or guided townspeople to utilize nearby crude materials while the public authority offered some monetary help" (Rajasingham 2012). The last kind comprised of neoliberal industrial facilities presented by 1990 under the 200 Piece of Clothing Plant Program or Garment Factory Programme (GFP) as a component of the second period of neoliberalism. The entranceway to the Mahiyangana site has a plaque reporting the points of Gam Udawa as "game hadaa rata hadamu" which signifies "construct the town to fabricate the country" (Rajasingham 2012).

In 1994, the Samurdhi (or Thriving) Program was dispatched by the recently chose Sri Lankan Government as a public system to lighten destitution, and the Samurdhi Service was set up. This program covers about 1.2 million helpless families obliging 33% of Sri Lanka's whole populace (Ismail et al. n.d.). It is financed completely by the Public authority of Sri Lanka. The vital segments of the Samurdhi incorporate necessary and intentional reserve funds, human asset improvement (profitability advancement preparing, preparing in ac-tallying capacities, preparing of chief advisory groups and material asset development), foundation of Samurdhi Bank social orders (answerable for the provision of credit), a local area improvement program, work serious people groups' activities, little businesses advancement and social advancement programs (Ismail et al. n.d.).

In any case, these significant accomplishments regardless, of the past provincial development strategies have not been sufficient to advance development at the territorial levels, for instance to

- arrest the developing polarization of the CMR;
- accelerate development and versatility of districts in the outskirts and upgrade their cutthroat edge;
- achieve an all the more spatially adjusted settlement framework and development pattern;
- provide institutional systems at public, local, and neighborhood levels to embrace a more viable local turn of events.

Recent development efforts of the country, especially after the victory of armed conflict highly visible. The end of the civil war and guided by the sanguine promise of the Mahinda Chinthana aimed at redefining the country's socio-economic development. The scope of this hopeful mega-development projects seeks to provide benefits to every segment of the society in a justified manner while "promoting investments on infrastructure based on commercial and economic returns" and creating "equitable access to such infrastructure development to enable people to engage in gainful economic activities" (Mahinda Chinthana 2005). In line with the Mahinda Chinthana development framework, foreign direct investment was targeted and expansive luxury tourism developments, economic hubs were planned, including specialized economic zones for Hambanthota and Sampur and Kalpitiya (Buthpitiya 2013). Northern Spring (Uthuru Wasanthaya) and Eastern Revival (Neganahira Navodaya) are two popular state-led development projects. Both projects focused on the development of the northern province and eastern province, respectively. The projects do not merely account for development from an economic perspective but also social discourse, including political dimensions of power, control, construction of social identities, and social inclusion and exclusion (ICES 2020).

Five metro regions, including the Trincomalee–Anuradhapura–Dambulla–Polonnaruwa as the largest of all planned with railway network and expressways joining major transportation hubs, were planned to be accomplished by 2030 (National Physical Planning Department 2010). Thus the regional development is planned mostly to avoid regional disparities. However, currently, regional development distinguishes regional disparities over the regions in the country.

In this respect, it is proved the efforts taken by different governments to reduce regional disparities and focused on regional development. Fail of one regional development programme always led to another. Likewise, there are several notable development efforts in Sri Lanka.

4.6 Consequences of Regional Development Efforts in Environmental Aspect and Sustainable Responses to Mitigate Issues

The development process is a combination of clearing of forest cover and conversion to various land uses. Generally, it is well known that the bare land is exposed to severe soil degradation and soil erosion as the land cover is removed. Removing the land cover typically increases the greenhouse gas emissions and reduces atmospheric oxygen and increased carbon dioxide levels which in turn affects global warming altering the hydrological cycle (Houghton et al. 2012). It will once again damage the ecosystems. And then the sustenance of all living beings.

An environmental assessment in 1980 showed the environmental problems accurately emerged due to the massive conversion of land into agricultural lands under the Mahaweli project. Among them, degradation of watershed conditions, the decline in water quality, loss of wildlife habitat and populations, incomplete social infrastructure developments were identified as further continuous issues (Tolisano and Goonesekara 1993).

- On-site soil loss rates, particularly in the upper catchment, continue to be greater than soil replacement rates.
- Downstream water quality in the lower catchment area is being degraded as a result of nutrient loading and pesticides from farm runoff. Natural forest ecosystems are being degraded or eliminated in both the upper and lower catchments because of expanding agriculture and the unregulated harvest of forest products.
- Sediments and agrochemicals in runoff water are resulting in the eutrophication of reservoirs.
- Water management schemes have changed the natural flow regimes of the major rivers, resulting in occasional high flood flows, lower dry season flows, and degradation of riparian, and wading habitats in the lower catchment.
- Sediment inflow is beginning to fill some reservoirs and threatens to shorten the useful life of the reservoir network.
- Chemical inputs to agricultural systems in the upper and lower catchment areas are degrading local surface and groundwater quality.
- Cropping patterns are often inappropriate for the soil and microclimatic conditions of the site, resulting in reduced productivity and increased reliance on environmentally inappropriate agricultural practices.
- Poor water management practices, principally in lower catchment irrigated areas, have increased the threat of soil salinization and on-site soil losses.
- The current data on wildlife population characteristics, migration and dispersal patterns, and habitat requirements are insufficient to guide management decision making.
- Habitat conditions in many of the designated protected areas have been degraded over time, and restoration or enrichment measures have been inadequate to support existing wildlife populations.
- Land allocation schemes provide too small a land base to support more than one generation through agricultural production, resulting in declining quality of human life and an increase in illegal land uses.
- The social infrastructure in the upper and lower catchment areas cannot yet fully meet the needs for satisfactory health, education, training, or financial well-being of the region over the long term.

Several environmental problems emerged in line with the Mahaweli development programme. Due to the intensified environmental degradation regional development altered. Since most of the current development activities take place in the Western province, the costly infrastructure facilities which have been constructed in coastal areas are inundated if sea-level rise becomes an issue, the country will face asset loss of gigantic scale. If the food and water would be threatened like this it is difficult to imagine how the regions in the country would cope with such regional disparities in development. Natural disasters and compensation payments and reconstruction cost lay a heavy burden over the national budget annually, in turn, reduce the regional development activities. Identifying key threatened areas, fragile areas, and sensitive areas the plan suggests ways of managing the mentioned effects of climate change and environmental problems.

It is worth elaborating on climate change and environmental issues that have been highly influenced by regional disparities. Rapid development activities also contributed to an increase in natural disasters. The socio-economic variations within regions created impoverished villages where two-third of the settlers are under poverty or lack basic needs such as drinking water, health, and sanitation, electricity for many years, accessibility, and education. There is a lack of financial facilities to install and maintain the electric fence, lack of human resources as wildlife rangers, lack of elephant crackers derived as a side effect of financial issues, lack of prompt services, etc. Considering the human-elephant conflict, a total of 14,516 incidents were recorded in the country during 2010–2019. There were a total of 807 human deaths, 579 human injuries and 10,532 property damages caused by elephants and 2631 elephant deaths (Prakash et al. 2020). The spatial distribution of human-elephant conflict well-identified within the Mahaweli development regions (Prakash et al. 2020). A recent study has also illustrated the relationship between human-elephant conflict and climate factor, water availability factor, and river basin management practices (Zubair et al. 2010).

On the other hand, climate change and other environmental issues have highly visible due to this regional inequality. Since most of the rural areas in the country are still experiencing clean drinking water issues, this would worsen. Extreme rainfall and frequent floods, landslides, and long-term dry periods, changes in the precipitation patterns, fluctuating temperature, and humidity levels are highly visible and the risk of natural disasters also increased than ever. All of these effects, "grave effects", will affect human settlements and cities, which would likely have a large number of climate change refugees. From northwestern and south-eastern monsoons, the hill country receives sufficient rainfall, accelerating landslides every year (Wijesekara n.d.). Multipurpose reservoir areas built under the Mahaweli development are identified as landslide-dominated areas where the soil loss and siltation of tanks also higher due to various land uses (Hewawasam 2010). Anthropogenically intervened soil has resulted in many negative impacts in upper reaches as well as downstream areas. Among the impacts, a decline in soil fertility, reduction in soil depth, rapid siltation of hydropower reservoirs, sedimentation in agricultural lands, sedimentation in rivers and floods in the lowlands, increasing frequency of landslides, damage to the infrastructure, and the change of water quality of waterways and ecological systems (Hewawasam 2010). Still, there are no sustainable responses taken for such issues. If the country's population growth is considered, the uneven urban growth may be visible, which also affects the country's regional imbalance. Most of the forest areas have decreased with the rapid development activities in the recent past. Many of the protected areas are being impacted by ongoing development changes. These have been impacted by multiple infrastructure projects (e.g., construction of reservoirs and roads). Some forest reserves located in the north-central region have been impacted by road construction and encroachment along roads, while protected areas in the southeast have experienced change due to reservoir and highway construction (Rathnayake et al. 2020).

So regional development cannot provide sustainable responses for each problem to mitigate climate change's multiplier effect and other environmental issues. The challenge is to ensure that actions to address environmental issues, including climate change, will contribute to, rather than obstruct regional economic development, in turn, regional development. Thus, regional development in Sri Lanka has to be refined while reducing regional disparities and the impacts of climate change and environmental issues.

Though the priority of the country in line with sustainable development goals and country policymakers is to reduce poverty and encourage economic growth, climate change mitigation can offer an opportunity to revisit development strategies from a new angle. If food production, livestock, and fishery are considered, regional development can be used as a means to mitigate climate change impacts. Literature evidenced that sustainable regional development could mitigate climate change impacts (Wijesekara n.d.).

Sri Lanka has realized that sustainable development is the best strategy that should be adopted in the preparation of Sri Lanka for the inevitable impacts of climate change. Being a highly vulnerable country for climate change and related environmental issues, Sri Lanka is at an alarming phase. Hence, strategically well-planned resource utilization is required to mitigate climate change impacts and other environmental issues.

Among these legislations, Coast Convention Act No. 57 of 1981, Sustainable Energy Authority Act No. 35 of 2007, Environment Conservation Levy Act No. 26 of 2008, and Marine Pollution Prevention Act No. 35 of 2008 are directly relevant to the climate change issue. Emphasize the interconnections between climate change management and other environmental and social benefits. The countries in the South Asian region, including Sri Lanka, are victims of climate change. Although the Supreme Law and the Constitution of Sri Lanka are silent on endorsing the right to a clean environment as one of the key components of sustainable development, the Directive Principles in the State Policy specifically refer to protecting the environment. This statute focuses on several aspects of the protection of the environment such as environmental planning and development, preventive and corrective measures regarding environmental measures finally aiming at sustainable development. Not only the local level but international platforms are also established to mitigate climate change issues. When concerning the regional disparities and environmental issues, it is recommended to follow proper land-use practices and avoid misuse of land for various human activities, which require region-wise comprehensive study to identify the best and sustainable land-use practices for each region, ensuring a minimum impact on the environment.

References

ADB (2004) Southern development project. Sri Lanka: ADB. Bandara J, Jayasuriya S, 2010. Regional disparities: limitations in the application of new economic geography. In: Gunawardena SKaD (ed) Economic and social development under a market economy regime in Sri Lanka. Vijitha Yapa Publications, Sri Lanka

Buthpitiya V (2013) Reconciling rights, responsibilities and disjunctures: an assessment of Sri Lanka's post-war development drive. Law and Social Trust, Borella
Dahlen A (1992) Comparative evaluation study of Sri Lanka's integrated rural development projects. Ministry of Planning and Implementation, Colombo
FAO (2001) Decentralized rural development and the role of self-help organizations. FAO, Bangkok
Friedmann J (1970) Regional development policy: a case study on Venezuela. M.I.T. Press, London
Glasson J (1974) An introduction to regional development planning: concept theory and practice. Hutchinson and Sons Ltd., London
Grimes W (2010) Walter Isard, economist who studied how regions evolve, dies at 91. https://www.nytimes.com/2010/11/11/business/economy/11isard.html#:~:text=Walter%20Isard%2C%20an%20economist%20who,He%20was%2091. Accessed 11 Nov 2020
Hansen N, Quigley J (1976) An introduction to regional science by Walter Isard. Land Econ 52(3):407–413
Hewawasam T (2010) Effect of land use in the upper Mahaweli catchment area on erosion, landslides and siltation in hydropower reservoirs of Sri Lanka. J Natl Sci Found Sri Lanka 38(1)
Houghton R, House JI, Pongratz J, van der Werf GR, DeFries RS, Hansen MC, Le Quéré C, Ramankutty N (2012) Carbon emissions from land-use and land-cover change. Biogeosciences 9:5125–5142
ICES (2020) International Centre for Ethnic Studies. http://ices.lk/uncategorized/northern-spring-uthuru-wasanthaya-and-eastern-revival-negenahira-navodaya/. Accessed 11 Nov 2020
Ismail S, Immink M, Mazar I, Nantel G (n.d.) Community-based food and nutrition programmes. FAO, Sri Lanka
Lindhal CL, Softstadt L, Wanasinghe YADS, Ernfor R (1991) Mobilizing the poor: evolution of SIDA—financed rural development projects in Sri Lanka. Ministry of Planning and Planning Implementation, Colombo
Mahaweli Authority (2018) Socio-economic statistics 2018. Mahaweli Authority, Sri Lanka
National Physical Planning Department (2010) National physical planning policy and policy, Battaramulla, Sri Lanka
Prakash T, Wijeratne A, Fernando P (2020) Human-elephant conflict in Sri Lanka: patterns and extent. Gajah 51:16–25
Rajapaksa M (2005) Mahinda Chintana—victory for Sri Lanka, Presidential Election. Colombo
Rajasingham N (2012) The factory is like the paddy field: Gam Udawa performances ethnicity and neoliberalism in Sri Lanka. South Asian Rev 33(3):275–293
Rathnayake C, Simon J, Mariela SB (2020) Mapping land cover change over 25 years (1993–2018) in Sri Lanka using Landsat time-series. Land 9(1)
Ratnayake R (2015) Regional development and the role of urban centres in north-central Sri Lanka. Int J Multidiscip Res Dev 2(8):660–663
Tolisano J, Goonesekara K (1993) An environmental evaluation of the accelerated Mahaweli development programme: lessons learned and donor opportunities for improved assistance. US Agency, Maryland
Wanasinghe Y (2003) The role of small towns in the development of the north-central province in Sri Lanka. s.n, Sri Lanka
Wijerathna D, Jayatilleke CS, Narampanawa A (2014) Regional disparities in Sri Lanka: an empirical analysis. Asia Pac Dev J 21(2):77–102
Wijesekara S (n.d.) Impact of landslide of rural development of Sri Lanka. s.n, Sri Lanka
Zubair LM et al (2010) Climate influences on human-elephant conflict in Sri Lanka. International Research Institute for Climate and Society, Sri Lanka

Lasantha Manawadu is a senior professor and the Chair Professor in geography in the Department of Geography, University of Colombo, Sri Lanka. He was the former Head of the Department and presently, the Dean of Faculty of Arts. He received his BA (Hons) in Geography and Ph.D. from the University of Colombo while his M.Sc received from the Asian Institute of

Technology, Thailand. Lasantha Manawadu has contributed to several development projects lead by different organizations in the country concerning national development. His contribution to research and development much leaned to climate change, regional planning, urban studies, and geoinformatics. He has published several research papers at the international and national level while presented several abstracts at both international and national level. He is an author of several books published at an international level and authored textbooks related to geography at the national level as well. He has supervised numerous undergraduate and postgraduate dissertations while worked as a coordinator of postgraduate courses. He is an experienced senior professor in academic and administrative sectors while awarded the Vice-chancellor awards and senate award for research Excellency in 2016 and 2017.

M. D. K. L. Gunathilaka is an academician attached to the Department of Geography, University of Colombo, having a BA (sp) in Geography (2017), M.Sc in Environmental Science (2021), teaches Geography, Environmental pollution, and Management, GIS, etc. at the University. Research interests are leaned toward Geo-biotechnology, Environmental science, Bio-geography and ecology, Geoinformatics, Tourism Studies, and Sustainable Development. Gunathilaka has been published several research studies in both International and National research conferences as well as journal articles related to Environmental science, Ecology and Biodiversity, Tourism, GIS and Remote Sensing, Social science studies, and Sustainable development. Several research articles were published in peer-reviewed/refereed/indexed journals and several abstracts and extended abstracts were presented and published as proceedings. Recently published a book chapter and two textbooks, in which the books are translated into seven foreign languages as well. Currently, three research studies are carried out 1) Changing the behavior of waterbird species concerning land use and land cover changes; 2) Above ground and below-ground biomass assessment in wetlands; 3) Seasonal variations in the ecology of microbes and water chemistry across the land-freshwater gradient in geographically isolated wetlands. Currently, she is reading for the Ph.D. degree, holds organizational memberships, with administrative experiences, and extracurricular activities awarded for the best paper presenter as well.

V. P. I. S. Wijeratne is a lecturer in the Department of Geography, University of Colombo, Sri Lanka, and presently, she is a Doctoral candidate in Physical Geography at the College of Urban and Environmental Sciences, Northwest University, China. She received her M.Sc in GIS and Remote Sensing from the Postgraduate Institute of Science, University of Peradeniya, Sri Lanka in 2014 and BA (Hons.) First Class in Geography, University of Colombo, Sri Lanka in 2011. Her teaching areas are hydrology, applied hydrology, GIS, disaster management at the undergraduate level. She is broadly interested in Eco-Hydrology, Social hydrology, GIS and Remote Sensing, Climate Change, Disaster Management, and Environmental Management. She has published several research papers related to hydrology and GIS and Remote Sensing in indexed/refereed journals. Also, she has authored several books and book chapters at the international and national level. The national development-related contribution and administrative and academic contribution of Wijerathne are also significant.

Chapter 5
Regional Planning Strategy for Sustainable Development in Nepal: A Critical Analysis

Pushkar K. Pradhan and Puspa Sharma

Abstract The regional development planning policy implemented in the early 1970s in Nepal was essential to integrate the country's three geographic or ecoregions, viz. plain, hills and mountains, as well as to bring on a balanced development sustainably across the three regions and between rural and urban areas by utilizing their abundant natural resources—water, forest and biodiversity. In doing so, the implementation strategies adopted by the policy included the identification of polyfunctional settlements at three levels like regional centres, growth centres and growth points for the hill, Tarai, and mountain regions, respectively, and north–south growth corridor roads linking those regional growth centres for production and circulation. Five development regions were identified from east to west across the country, containing polyfunctional settlements in each of the three ecoregions. Accordingly, investments have been made by the government to provide essential facilities in each of the polyfunctional settlements as per their hierarchical levels and to build north–south road corridors. As a result, the private sector also found pouring investment in the growth centres. Later, two additional policies such as the service centres approach and the district decentralization system were formulated to strengthen this strategy. Despite all these efforts, the strategy attained its intended objectives only to some extent due to several difficulties it faced and now, it is virtually dysfunctional. Not only this, more than a dozen of other development policies implemented over the past six and half decades remained to discontinue, not because they were irrelevant but mainly disliked by the successive governments. The sector-based economic planning that often received high priority since the very beginning also has shown little impact on improving economic conditions in the country. For instance, the agriculture sector has not improved to the desired level, nor has the industrial sector enhanced it and only a fraction of those potential resources has been utilized. The country's rugged mountains are often blamed as the main obstacle to this shortfall achievement. Other weaknesses included the slow pace of development of road networks due to limited investment funds available compared to the desperately needed amount, ad-hoc decisions in providing facilities and infrastructure in the regions vis-à-vis their centres,

P. K. Pradhan (✉) · P. Sharma
Central Department of Geography, Tribhuvan University, Kathmandu, Kirtipur 44618, Nepal
e-mail: pushkar@reachpuba.org

M. Mishra et al. (eds.), *Regional Development Planning and Practice*,
Advances in Geographical and Environmental Sciences,
https://doi.org/10.1007/978-981-16-5681-1_5

lack of seriousness and commitment of government, among many others. If these weaknesses were corrected, we could see the impacts of the regional planning strategy on people and places. However, the strategy is still pertinent to address the unique development problems in the country, it is argued.

Keywords Growth corridors · Polyfunctional settlements · Growth centres · Balanced development

5.1 Introduction

Regional planning as a tool is being widely used in different countries across the world to integrate regions having diverse resources and to utilize these resources for development. However, one of the significant challenges the developing countries facing is how to develop their various regions by providing facilities or infrastructure in a rational manner so that people living in the different regions can get access to them comfortably. It is even more challenging and difficult in Nepal, as four-fifths of its area consists of rugged mountains, where the elevation of topography rises from 60 metres above sea level (masl) in the south to 8,848 masl in the north over an average of 193 km span.

In regional planning, a defined region consists of a development foci referring to a growth centre or a service centre and its service area or hinterland region. The centre and hinterland act as a complement to each other for reciprocal activities or transactions for the benefits to both of them. Development interventions of production-oriented infrastructure such as roads, electricity, bank, etc. and service-oriented facilities including market, health, education, etc. are needed to accelerate the growth of their reciprocal activities, keeping intact of the existing state of environmental resources.

In Nepal, regional planning is essential particularly to integrate its diverse geographic regions such as plains, hills and mountains. Each of these regions having unique environmental and socioeconomic features is ecologically bounded and offers enormous development potentials. Given this situation, efforts to be initiated to bring on a balanced development across these diverse regions in Nepal seem not only a huge challenge but an opportunity as well (Gurung 2005). However, this requires that the national planning and policy strategy be adopted towards achieving a balanced development consistently for a long period of time.

Here in this chapter, our attempt is to portray critically the regional planning efforts being undertaken by the government for a balanced and sustainable development in the country during different planning periods.

5.2 Database and Methods

The paper draws on data and information in both analog and digital forms from various public sources, which are available for use. They include topography and other physical features, population growth and distribution, spatial distribution of urban centers and rural settlements by size and growth, road network, resource potentials, and economic activities, viz. agriculture, industries, tourism and trade flows. Population and economic activities data are available by decennial census years regularly, while others on a yearly basis, though not updated regularly. Mapping data have been acquired from the public archives. As administrative units such as district, municipality and rural area and their boundaries often change in Nepal, data being adjusted to the changed boundaries at the national level are not yet available. So, the data computation has been carried out for the analysis purpose of this paper. There are also studies carried out by research organizations and individual scholars, which here have been used for verification.

In Nepal, database building and updating is a crucial issue, which is not carried out on a regular basis by the public agencies in most cases, except the decennial population census. The last population census was carried out in 2011, and the next will be undertaken in 2021, but there has been changed a lot in the social and economic features and administrative units and boundaries since the last census. So in many instances, data computation and manipulation have to be performed to fulfill the purposes of this paper. The toposheet of Nepal at the base scales of 1: 25,000 (Tarai districts and hill urbanized districts) and 50,000 (hill and mountain districts with thin population) resolutions were prepared in 1996–97, but no updating of the toposheet contents such as administrative boundaries, drainage, contour, transport and communications networks, settlements and urban built-ups, land use, schools, health, religious sites, etc. since then has been carried out. Other information about the government policies, programs and acts are available for use through both online access and hard copies.

In this paper, basic facilities or services are defined to comprise both productive infrastructure (roads and transportation, bridge, electricity, water supply and the like) and community-related facilities (health, education, market, extension services and so on). Ease of access to the given facilities is defined in terms of their availability at a convenient location for as well as in the time of need of the people living in the surrounding settlements. Some of the essential conditions for convenient location include the connection between facility or service centres and service users of the surrounding settlements by means of road access and communication media, efficient delivery and availability of given services (good governance system) and geographical centrality, i.e. convenient to the majority of population clusters. The definition of scattered and agglomerated forms of settlements has been derived from the available studies in Nepal.

The Geographic Information System tool-based thematic layers have been prepared to portray Nepal's geographic, administrative and regional divisions, drainage, resources potential, road network, population density and urban centres.

5.3 Nepal—Geographical and Socioeconomic Features

Nepal is preponderantly a mountainous and landlocked nation bordering Tibet, China to the north and India to three sides: south, east and west. The country covers 147,516 square kilometers (km^2) with 885 km (km) east–west (MoLMCPA 2020). It is divided into seven provinces, 77 districts, 293 urban municipalities, and 460 rural municipalities. Kathmandu, the capital and largest city, is situated in the central hill region (Fig. 5.1). With a population of approximately 26.5 millions (CBS 2011), the country has a density of 180 persons per km^2. The population comprises diverse (59) ethnic groups, inhabiting distinctive pockets of territory across the country. Their population makes up about 42 percent of the country's total population (CBS 2011).

In terms of physiographic settings, the most obvious regional differentiation across the country are the plains, the hills and the mountains, which run parallel through east–west and are arranged systematically—the plain called *Tarai* in the south, the mountains (Himalayas) in the north and the hills in between (Fig. 5.2). These natural regions have distinctive environments and resource endowments and offer various development potentials. They are drained by three major river systems—the Koshi, the Gandaki and the Karnali originating from the Himalayas and flow down to the south, dividing the country into four longitudinal river basins east–west (Fig. 5.3). These rivers have deep gorges and vast banks, and therefore, the inhabitants of each of the river watersheds feel easier to move within the river basin itself than to other basins. The basins are yet more accessible to both Tibet (north) and India (south) than to each other, given their socioeconomic links. Thus, the Tarai has easier access to the Indian northern plain than to the interior areas of the hills and mountains of the

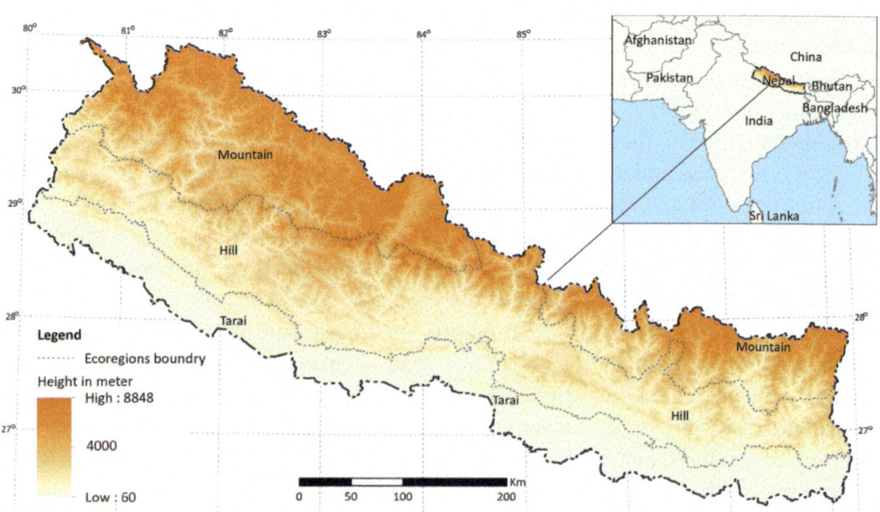

Fig. 5.1 Location of Nepal in South Asia, showing elevations (*Sources* EROS 2015; Survey Department 2020)

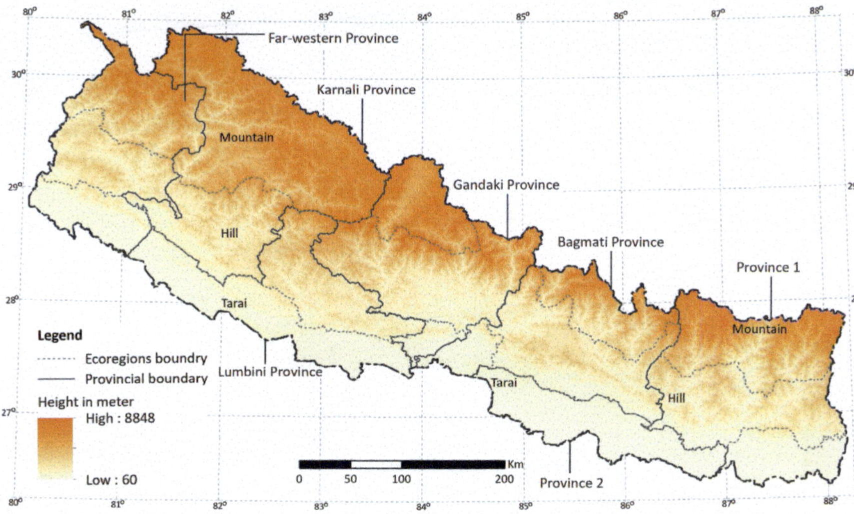

Fig. 5.2 Ecoregions and provincial boundaries, Nepal (*Source* Survey Department 2020)

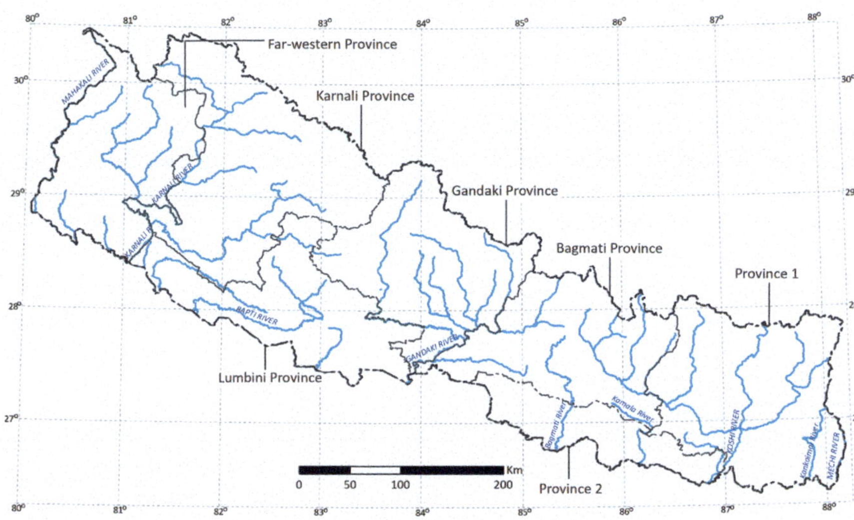

Fig. 5.3 Major river systems, Nepal (*Source* Survey Department 2020)

country. In this context, the regional planning approach is useful to provide a sound basis for integrating the regions having diverse features for sustainable and balance development.

The Tarai region comprising generally flat plain, making up less than one-third of the country's area is a major agricultural region and home to over half of its population (Fig. 5.4). The humid subtropical climate prevailing here supports rich vegetation, i.e.

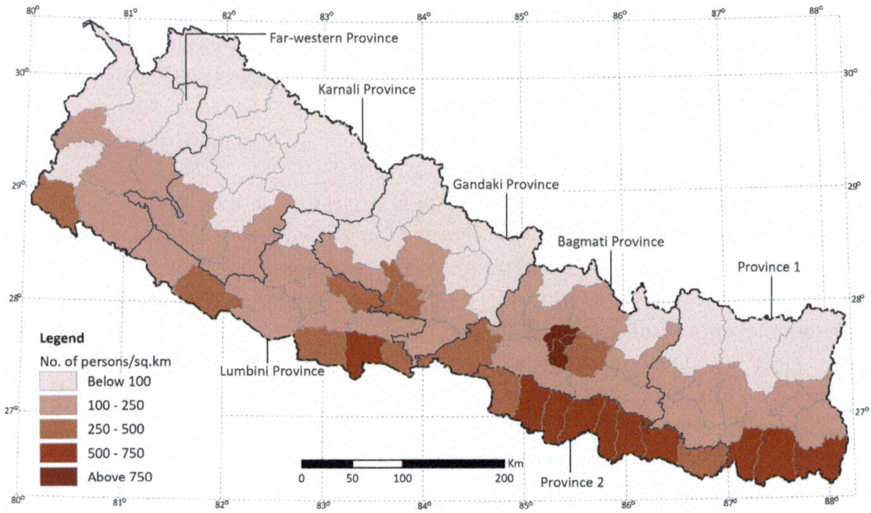

Fig. 5.4 Population density by district, Nepal (*Source* CBS 2012)

the forests of commercial importance, and the fertile soils are favorable for large-scale agricultural operation. It contributes about two-fifths of GDP and over 45 percent of the commodity export (MOF 2000). Major agricultural crops grown in this region are paddy, wheat, oilseeds, jute, sugarcane, tobacco and potato, and seasonal vegetables and fruits. In the past, the production of some of these crops appears to have been large enough not only to meet local needs but also to contribute to the export trade, particularly to neighbouring Indian states (Regmi 1971). But the quantity of food grains supplied from the Tarai to meet the needs of the hills and mountains was not so large, mainly because of transport difficulties. Notwithstanding the importance of the Tarai from the viewpoint of agriculture, considerable areas of arable land appear to have been waste or under forest or remain barren, particularly during the winter season due to lack of irrigation facilities.

The highlands, including both the hills and the mountains (Himalayas), constitute two-thirds of the country's total area but share less than half of the total population (Fig. 5.4). The mountains having lofty peaks contain eight of the world's ten highest mountains, including Mount Everest. Here, pastoralism is the main basis of the economy, while agricultural activity is limited due to inclement weather and poor soils. Livestock rearing (yak, sheep and mountain goats) is widely practiced in a traditional manner with adopting a transhumance system (movement of herds between cold and warm locations with seasonal change in search of grass fodder). However, since the recent past, it has begun to produce and export apples, herbal medicines, vegetables and potato seeds. Thus, pastoralism and trade are still the main bases of the economy, and tourism is the most potent activity for development. The population is sparse, and there is considerable seasonal out-migration during the harsh winter. The hills, on the other hand, consist of a maze of spurs, intermontane valleys and upland plain (locally known as *Tar*). The economy is supported by subsistence farming, and there is considerable pressure on the population on land resources. Due to fragile geology, degradation has severely affected the hill landscape. However, the hills are a homeland of traditional Nepalese origins. Settlements over the hill slopes are generally scattered and tiny in size, with predominant settlement units having below ten houses, while in the river valleys, settlements are agglomerated and large (CDG 2002).

Owing to climate variability, diverse ecosystems and abundant natural resources across the north–south and east–west, the highlands of Nepal have huge potentials for development. But the highlands having naturally built ecological sensitive features are facing a constant threat to hazard and tremendous pressure due to the continued expansion of cultivated land and the building of non-engineering roads. Here there often occur landslides by runoffs and streams every year during the Monsoon rainfall season, devastating hundreds of buildings and complex infrastructure and facilities. Moreover, the fertility of the soil has decreased tremendously due to erosion of topsoil over the hill slope that has led to accelerated erosion of productive soils, undermining the productivity of farmland and increased sedimentation in downstream areas. For instance, the annual soil loss by the major rivers is estimated at 5 tonnes per hectare, which in turn replenishes the Tarai flat plain, as well as the neighbouring Gangetic plain of India, by sediments of soils (Carson 1992).

Each of the three ecoregions has unique production potentials and resource flows (both natural and fabricated) take place between the ecoregions (Table 5.1). It is argued that the strategy for the development of the ecoregions should not be competitive but rather complementary, providing scope for specialization. The advantages of such a wider regional complementarity seem realistic if one notes that the Tarai is a densely populated area and has potential developments of extensive agriculture and industries, while the Hill and the Mountains specialize in sub-tropical and temperate products (Jodha 1989–1992) that would find favourable markets in the south.

Table 5.1 Production potential items and resource flows (both natural and fabricated) by ecoregions

Ecoregions	Natural resources and human-made product flows
Mountain	• Herbs, timber, water, and topsoil • Livestock, its products and fruits • Migration of people and tourists
Hill	• Fuelwood, timber, topsoil, water and electricity • Livestock and its products • Agro products— vegetables, fruits, spices, cut flowers • Manufacturing—carpet weaving, traditional jewelry making • Migration of population and tourism
Tarai	• Fuelwood, timber • Livestock, fruits and grains • Industrial products

Source ADB (2010)

In this context, the concept of landscape (ecosystem) integrity planning seems to be an important and useful tool for Nepal's local and regional development. The concept is fundamental not only for maintaining the conservation of significant ecologies and environmental resources but also for supporting the livelihood systems of the majority of people who live on the use of resources locally available. The country's three ecoregions, also referred to as landscape units, are interlinked and interdependent with each other. It is asserted that the development initiatives of national or regional importance by isolating any one of these ecoregions or by basing on a single ecoregion or landscape unit may break natural integrity that is closely tied, on which the social and economic features have been relied since the long. So, the landscape integrity of the ecoregions can be viewed as a 'chain development system', and according to this concept, isolated development efforts of national or regional importance focusing on a mere single ecoregion may have a dissentient effect on natural, social and economic links of other two ecoregions in the long term.

Along this line, three agro-ecological zones identified by the country's agriculture perspective plan (APROSC 1995) for agriculture development also follow exactly to what is described in Table 5.1. The plan has identified potential high-value agricultural productions for each of the three agro-ecological zones, based on the physical environments, social, and provision of infrastructure and facilities. Further details of this plan are given in the subsequent unit.

5.4 Development Policies in Nepal

This section consists of two sub-units. The first unit deals with a brief historical account of development planning in the country. In the second unit, a description of national plans and policies undertaken for the development of the country and its critical analysis including changing state vis-à-vis political change is provided.

5.4.1 Brief Historical Account of Development Activities

Nepal witnessed many political changes but without having any specific development efforts during the historical time. Until the early-mid-eighteenth century, there was not a single nation such as Nepal, and during that time, there were forty-six mini-states (22–24 states in the eastern and western parts respectively) ruled by petty rulers. These mini-states were no more than the small geographical and economic units bounded by innumerable mountain barriers. Elsewhere in the country, there were only minor concentrations of population in small settlement agglomerations on the hill slopes, which were sufficiently contained and self-sufficient primitive agricultural societies and also had carried out minimum commercial activities (Stiller and Ludwing 1975; Regmi 1971). The warm river valleys and the foothills were not habitable for living mainly on accounts of their weaker strategic location and prevalence of malaria but used for subsistence production of food crops (Gurung 1976).

During that period, development activities were very much limited, and whatever existed were based on the agriculture economy, cottage and craft productions and limited trading activities. Mostly the settlements were tiny and scattered and they had a closed economic unit producing the basic requirements of life (Regmi 1971; Hagen 1971). With the lack of road, the rugged hill terrain imposed severe limits on the movement of goods and people. However, exceptionally the Kathmandu valley's settlements such as Bhaktapur, Kathmandu and Patan, which were also the tiny states lying in the hill region, could properly be described as towns. They were strong enough in terms of economic condition since they were based on handicraft and trade and commerce and acted as entrepôt centres of goods originated in India and Tibet (Regmi 1988). Items exported from these towns to Tibet were cotton goods, brass and wooden products, but the trade was minimal and whatever trade existed tended to be transient, with goods such as musk, gold dust and yak tails from Tibet being carried through the hill towns for sale or barter in the lower districts of Tarai, and goods manufactured in India being carried to the north for exchange in Tibet (Gurung 1973). The trade between Tibet and the Hills of Nepal was created due to their specialized pattern of production. The inhabitants of Tibet could not grow sufficient food because of inclement weather and poor soil, but had numerous salt lakes. On the other hand, the hills of Nepal raised large quantities of food grains, but lacked sources of salt supply and so had to exchange their food grains with Tibetan salt (Regmi 1988). India was the market for a wide variety of products like musk, hawks and horses from the Himalayan or mountains and cotton, cardamom, medical herbs and metal goods from the hills of Nepal. Likewise, agricultural and forest products, including food grains, pulses, oilseeds and timber of the Tarai were exported to India. In turn, kerosene oil and manufactured goods from India were imported into Nepal.

Since the beginning of the second half of the eighteenth century, Nepal saw a gradual change in the social and economic activities throughout the country. After the unification of the tiny states into a single nation Nepal by Gorkha king in 1769 and subsequently with the decline of entrepôt trade role of the Kathmandu Valley, movement of the Newar ethnic, the native of the valley and were traders and manufacturers by tradition, took place to several strategic locations or break of bulk points along the traditional trans-Himalayan trade routes across the country and set up long-distance trading posts there (Blaikie et al. 1976). The very creation of small towns was indeed an outgrowth of the military and administrative needs of an expanding Gorkha empire but not of any urbanizing tendencies within the local peasantry (Caplan 1975; Giuseppe 1799). Another major breakthrough that took place during the early nineteenth century was an improvement in communications and transport based on the administration of the postal system of locally organized porterage and the establishment of a series of staging posts manned by locals and operating under the supervision of administrative offices situated at strategic points along the trails. Some of the offices of this kind were located in important towns along the crucial routes across the country.

During the beginning of the nineteenth century, agriculture development efforts were undertaken by the successive rulers in the Tarai, as it was the most valuable region for its abundance of cultivable lands. The main crops grown in this region were paddy, oilseeds, cotton, jute, tobacco and sugarcane. The production of these crops appears to have been large enough not only to meet local needs but also to contribute to the export trade, mainly to India (Regmi 1971). Notwithstanding the importance of the Tarai from the agriculture viewpoint, considerable areas of land appear to have been waste or under forest (Kirkpatrick 1969). During the first half of the nineteenth century, the rulers encouraged people from neighbouring India for land reclamation and settlement with the main objective of increasing the area available for tax-free land grants, as well as of contributing to the security of border areas (Regmi 1988). They imposed a crop-tax system as an integral part of their economic policy in order to provide landholding rights to a specific plot of land and linked revenue with both the extent and the productivity of agriculture (Hamilton 1971). This policy remained to continue during the latter part of the nineteenth century as well, during which the successive rulers encouraged immigrants from Bihar of India into the Tarai, and as a result, the cultivated area in the Tarai increased sharply during the end of the nineteenth century. That trend inevitably had a positive impact on the role of Tarai as the major source of state revenue (Regmi 1988). According to the historical evidence, around the turn of the nineteenth century, there was an increase in the volume of trade between Nepal and India due mainly to the extension of the Indian railway network north to the border of Nepal that facilitated the development of trading centres near the border in the Tarai. Those centres acted as important entrepôt for the long-distance salt and grain trade, as well as for the export of local grain surpluses using the railheads at border towns of India. The improved facilities for marketing provided by the Indian railway system encouraged merchants from the hill towns to invest in land in the Tarai and to move into the sphere of exporting Tarai-grown crops like rice and mustard to India. But the growing importance of the Tarai centres had not inevitably led to their physical growth as towns, for the Tarai remained inhospitable, and as a result, most of the trading took place on a seasonal basis with maximum activity during the winter when the dangers of infection were least. At the same time, the import trade from India affected badly the once active small-scale industrial production of the hill towns to decline from the competition by imported cheap Indian manufactured goods and their importance as centres of petty industrial commodity production was seriously threatened (Blaikie et al. 1976).

Agricultural development efforts continued to remain as the main focus in the Tarai even during the twentieth century. For instance, at the beginning of this century, the successive ruler made the earliest effort of resettlement in the Tarai and announced that any Nepali citizen could acquire as many land area as he wished in the Tarai and could cultivate it (NRC 1980). Notwithstanding this encouragement, no hill people (*Pahadis*) moved to the Tarai for permanent settlement. The reasons for this reluctance were fear of malaria, dense forests, the prevalence of poisonous snakes, hot climate and wild animals. In the early 1950s, the subsequent government had also made efforts to encourage hill people for permanent settlement in the Tarai through launching a malaria eradication program. In the latter 1950s, a resettlement project, as well as a malaria eradication program, was initiated for the first time in the Chitwan valley of the central Tarai to encourage *Pahadis* to permanently settle there (Dusenberry 1957; Tuladhar 1972; NRC 1980). Since then, major streams of population movement from the hills and mountains and neighbouring India took place in the Tarai (Kansakar 1973–74; Gurung 1989; KC 2003).

5.4.2 *National Development Plans*

Planned development efforts in Nepal began in the mid-1950s when the first national periodic plan was initiated. Till now, the country has implemented 14 national periodic plans and the current one is the Fifteenth National Periodic Plan (2018–2023). The plans so far undertaken have been focusing on building physical infrastructure for setting up the basis of development. The plans also accorded priority to developing production-related infrastructures like irrigation for agriculture growth and others such as roads, bridges, communication and power. However, the planning exercise has set out every time priority sectors among those infrastructure types for development across the country. Most national plans have generally envisioned improving the living standard of the people by reducing the number of people under the poverty line and raise the per-capita income of the people by focusing on production base activities and generating employment.

The national plans initiated several strategies for developing the rural areas and thereafter, several measures were initiated to support rural development through various programs,[1] focusing mainly on agricultural development in order to alleviate poverty (Pradhan 1981). However, the sheer variety of these institutional attempts showed poor performance in the rural development sector due to several factors such as lack of coordination and cooperation among the line agencies, the unclear concept of people's participation in plan formulation, implementation, evaluation and benefit-sharing, ineffective monitoring mechanisms, administrative constraints and communication gaps (Pradhan 1981). The major weakness was that most of the benefits of these development activities accrued to the few rich and landlord groups who had vested interests, while the poor mass remained away from the process of development.

Indeed most national plans have never attained the intended key goals, which are yet reflected through the under-developed economic condition of mass people in the country. Economic growth has been lackluster and slow, and even the fast-growing urban areas have failed to encourage growth in adjacent rural areas and spread the benefits of growth equitably (ADB 2010). The pace of development works appears to be slow with respect to the rising aspiration of the people, which seems

[1] They included Back to Village Development, Land Reform, Cooperative Organization, Decentralisation, Local Development and Integrated Rural Development Program (IRDP), Small Area Development Program (SADP), Small Farmer Development Program (SFDP), Remote Area Development Program and Hill Transport Development Program.

to be undoubtedly normal as compared to the average rising levels of development elsewhere in neighbouring developing countries. The country has also passed through a series of political transitions; the most crucial one being the Maoist insurgency that lasted for a decade (1996–2006) hampering the implementation of development plans, as well as regular works. On top of them, efforts remained to be futile due to largely inadequate budget, as well as lack of commitment, integrity, and seriousness on the part of successive governments.

Even the agriculture sector, the backbone of the Nepalese economy with receiving the highest priority ever since the very beginning of an undertaking of the national plans, has not adequately improved and its performance has always been below expectation (Pant and Thapa 1981). Such failure to achieve significant growth in agriculture included a number of factors such as the lack of irrigation and road transport, inefficiency of agricultural service delivery system, inadequacy of farm inputs supply, paucity of agricultural research and inefficiency of marketing system (FAO 1987). It is evident from the fact that the country once was food sufficiency and exporter of some of the agricultural crops until around the 1970s, became an importer of most of the food crops afterward.

In Nepal, the agricultural perspective plan (APP) formulated during the mid-1990s for comprehensive agriculture development has received little or no attention (Pradhan and Pradhan 2006). The plan has already identified high-value agricultural activities and products best suited to the three ecoregions, based on physical environments (terrain, soil, and climate), social conditions (poverty), and provision of infrastructure (roads, irrigation, electricity, fertilizers and extension services), and marketing facilities (APROSC 1995). It has identified the Tarai as the prime region for commercial farming of grains (rice, wheat and maize), vegetables, subtropical fruits and livestock. While husbandry of yaks, sheep and goats; growing apples, off-farm vegetables and potato seeds; and herbs and non-timber forest products were potential agricultural activities in the mountains, dairy products, citrus fruits, off-season vegetables, vegetable seeds, apiculture, sericulture, tea and spices were significantly potentials for the hills. This plan of primary sector-based products has intended first for the domestic market and for generating employment opportunities for the youths locally and then the surplus exportable items to foreign markets (ADB 2010).

There has been exploited barely development potentials of many natural resources in the country (Figs. 5.5, 5.6 and 5.7). For instance, Fig. 5.8 depicts a massive potential in using hydropower to produce energy from the numerous snow-fed rivers in steep mountain territory (ADB 2010). However, the proper use of these abundant natural resources is being limited due to factors like inadequate physical infrastructure and poor transport networks, weak capacity of local institutions for managing development, lack of private sector investment, institutional constraints to joining the global economy, etc. (Blaikie et al. 1980; Pradhan 1981). The country, on the other hand, has some of the labour-intensive manufactured products competitive in the international market, such as hand-knit woolen carpets, pashmina mufflers, shawls, metalware, leather and leather products, jewelry and textiles, but they also have suffered to continue production due to a lack of favourable government policies (ADB 2010).

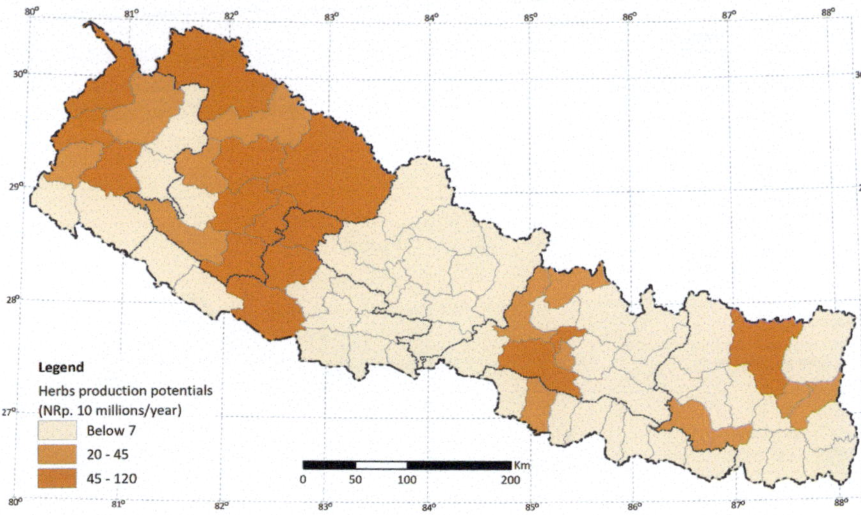

Fig. 5.5 Herbs production potentials, Nepal (*Source* Pradhan and Pradhan 2006)

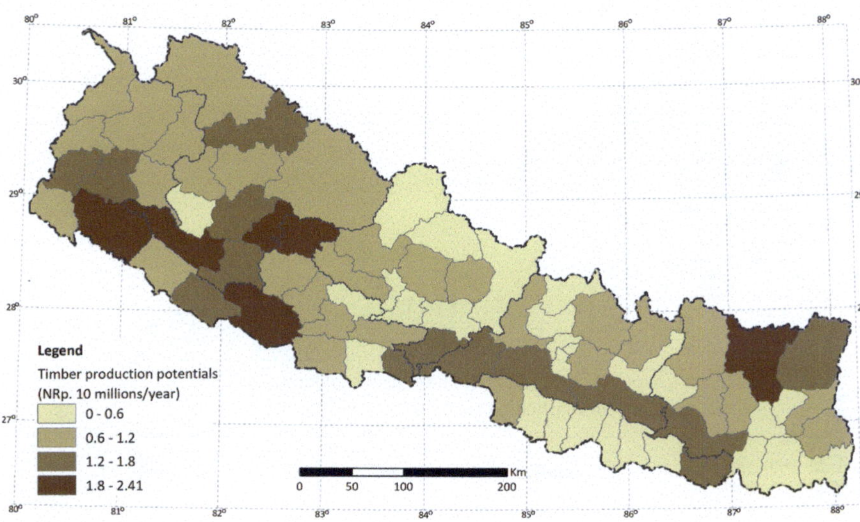

Fig. 5.6 Timber production potentials, Nepal (*Source* Pradhan and Pradhan 2006)

The industrial base is still rudimentary and there are no major industrial outputs available for international trade. About two-thirds of the country's economically active population is engaged in agriculture, contributing to mere one-third of gross domestic production (GDP). The non-agricultural productivity—with only one-third of the labor force engaged in the manufacturing and service industries, on the other

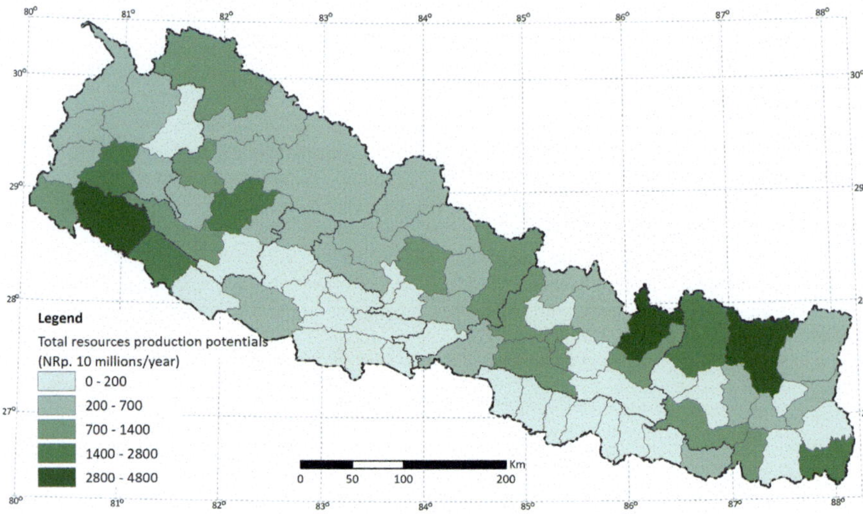

Fig. 5.7 Total resources production potentials, Nepal (*Source* Pradhan and Pradhan 2006)

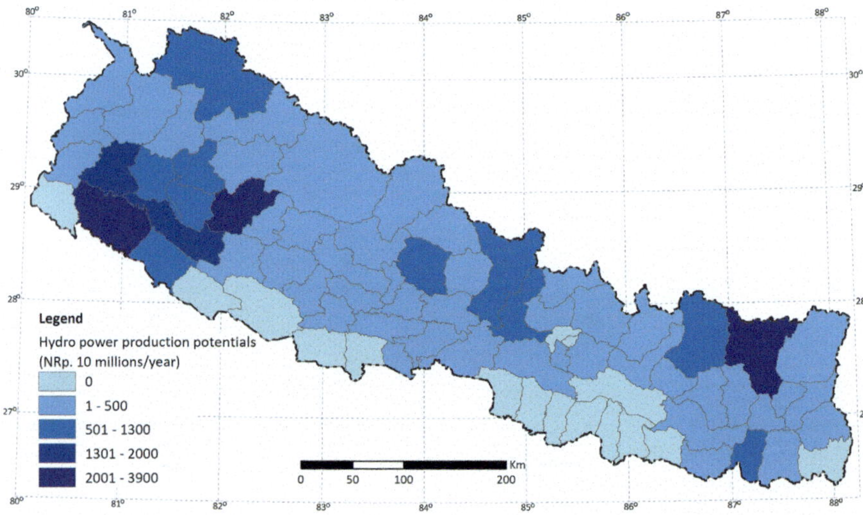

Fig. 5.8 Hydropower production potentials, Nepal (*Source* Pradhan and Pradhan 2006)

hand, contributes two-thirds of GDP (MOF 2019). The contribution of the non-agricultural sector to GDP appears to be gradually increased. In 2010, slightly over 60 percent of Nepal's GDP came from the non-agricultural sector, combining service (44.4%) and industry (14.2%). Of the two non-agricultural sectors, the share of the service sector increased to 49.5% while that of the industrial sector declined to 13.7% in 2015. The share of the agriculture sector decreased from 33.2% in 2010 to 29.4%

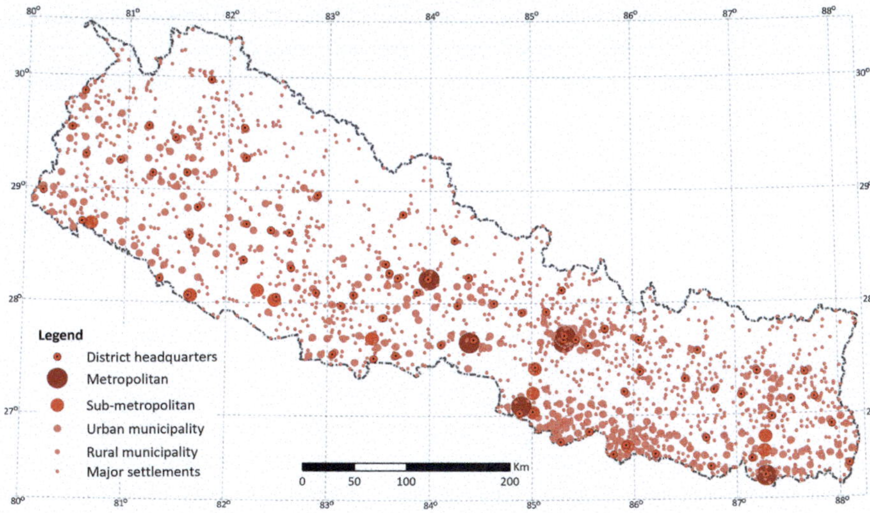

Fig. 5.9 Spatial distribution of settlements by type, Nepal (*Sources* MOFAGA 2017; CBS 2012)

in 2015 (MOF 2019). The GDP per capita was $1,034 in 2018. Yet, overall, 31% of Nepalese still live below the poverty line.[2]

There is enormous potential for tourism across different parts of the country such as sights seeing, trekking and mountaineering in the high mountain region, unique cultural and religious features in the hills and sub-tropical biodiversity conservation areas and internationally important religious elements in the Tarai. These all give the country strong advantages in tourism. Tourism shares 25% of the national GDP (ADB 2010). However, the flourishing of the tourism industry is constrained by the inadequacy of tourism infrastructure, such as poor roads to connect villages, towns and tourism places, limited efforts for tourism dissemination, lack of efficient arrangement for tourists' on-arrival visas at the international airports, poor delivery of tourism governance and so on. The investment resources are indeed severely limited compared to the needs for tourism development, and investment decisions have mostly been ad-hoc undermining intended impacts, despite the supports from bilateral cooperation and international organizations to the development initiatives.

The settlements in the hilly areas are mostly tiny and scattered (Fig. 5.9), which are not viable for sustainable and rational use of development facilities, nor are they feasible for providing a consolidated force to rural and regional development. There is no way but encouraging a compact or agglomerated form of settlement for the sustainable development of urban and rural areas particularly in the hills and mountains (CDG 2002). In 1986, the Land Resource Mapping Project (HMG/KESL 1986) indicated that over 30 percent of settlements together with properties and infrastructure located above 15-degree slope were technically not feasible, but were vulnerable to destroyable by intense rainfall-prone landslides during the monsoon

[2] Data is retrieved from www.statista.com

season. A study of accessibility to nine rural development facilities (banks, credit, fertilizer, health services, high schools, improved seeds, markets, offices, post offices and veterinary clinics) in a rural hill district of Lalitpur that lies in the Kathmandu Valley has shown the travel time ranging from just over 1 hour for post offices to more than 3 hours for high schools (Pradhan and Pradhan 2006). That study also indicated that travel time to locations with basic facilities in rural areas having agglomerated settlements and road accessibility has improved, yet the mean travel times such as 32 minutes to primary schools and 86 minutes to health clinics were far above the national parameters of 15 minutes for primary schools and 60 minutes for clinics. The case is severe in remote districts. The study by CDG (2002) showed that the average walking distance between two dispersed settlements in the hills was found to be more than 6 hours, based on 20 minuntes walking per km. The study has also provided a policy framework for developing agglomerated and compact settlements. Another study on school accessibility in Mugu of western Nepal, a remote and isolated mountain district without any road links, showed that access to school was very poor and drop out among the school going students were remarkably very high due to far distance and unsafe travel to the school locations from the villages (Ranabhat 2013). Implications of the results of these studies have however been of no use in planning and policy formulation at the government level. There has also been relatively weak investment in developing infrastructure against the mitigation of environmental problems such as land degradation and natural disasters in the country. The highlands are less accessible and have poor infrastructure and poor connection between villages and towns, making access to markets difficult for local producers despite having diverse agriculture and livestock products. Therefore, many people have a tendency to migrate from the highlands to the Tarai plain and from rural to urban areas. The government does not seem serious concerning the planning of scattered settlements, though they are being unfeasible and unsustainable for rational provision of basic development facilities. Sector-based planners argue that such dispersed settlements will be transformed automatically into agglomerated forms once the economic growth rate reaches over 6%. However, one should be cautious to the fact that mere growth rate wouldn't speed up, if no spatial and social planning has taken place for transforming the dispersed settlements into agglomerated form through intervening for provision of physical infrastructure.

5.5 Regional Planning Strategy for Development in Nepal

The regional growth model assumes that the development of a region is a function of national economic development. Its primary notion is that economic progress does not appear everywhere at the same time and that once it has appeared, powerful forces make for a spatial concentration of economic growth around the initial starting points. The model suggests that by concentrating investment and innovation in a few carefully selected potential centres, the development would either spontaneously or by inducement 'trickle down' throughout the region (Smith 1976). This is an induced growth centre or top-down planning approach. There are two sets of spatial forces, such as spread effects and backwash effects, which are set in motion by a propulsive growth centre. Spread effects refer to the mechanisms by which economic prosperity generated at a growth centre is transmitted to the surrounding hinterland or the 'growth space'. Conversely, the negative effect of the process of development is the backwash effect.

This region-based development approach appears to be essential in Nepal in order to attain a 'balanced and integrated growth' throughout its three diverse ecoregions—mountain, hill and Tarai, as well as between rural and urban areas in a sustainable manner. Further, the approach is also essential to include 'inclusive development' so as to ameliorate the existing unfair conditions, so that the risks of social vulnerability can be reduced.

Against this backdrop, the regional development policy, so-called 'growth pole model' was initiated as the first national development strategy in Nepal in the early 1970s (HMG 1970; Gurung 1969). This strategy was based on a direct bearing on the importance of growth centres in the development process (HMG 1970).

The main objective of this strategy was to eliminate social and economic disparities among the ecoregions, as well as between rural and urban areas, and the policies adopted in the strategy included resource allocation and development decentralization through its defined growth centres and growth axes. The broad objectives were to:

- reduce the interregional disparity in terms of resource use, flow, infrastructure, activities;
- integrate the national economy;
- break the vicious circle through transport network;
- eliminate the imbalances among projects;
- analyse the regional economic opportunities.

For implementation, the strategy identified five macro planning development regions such as eastern, central, western, mid-western and far-western (Fig. 5.10), which were defined by the north–south flowing major rivers such as the Koshi, the Gandaki, the Rapti[3] and the Karnali. Each development region consisted of latitudinally arranged all three ecoregions—mountain, hill and Tarai, and in each development region, polyfunctional settlements with three hierarchical levels, such as regional centres, growth centres and growth points for each of those three ecoregions, respectively, were identified (Table 5.2).

The polyfunctional settlements comprising regional centres, growth centres and growth points have been identified strategically in such a way that they would assume to integrate the Tarai and Mountain regions through the Hill region in terms of resources flows, and they would also act as centres for catering diverse needs of their hinterlands including agricultural transformation, marketing, decentralization of development activities, transmission and implementation of national programs and

[3] Except the Rapti River, all three rivers are tributaries to the Ganga River flowing through the northern India while the Rapti is a tributary to the Karnali River.

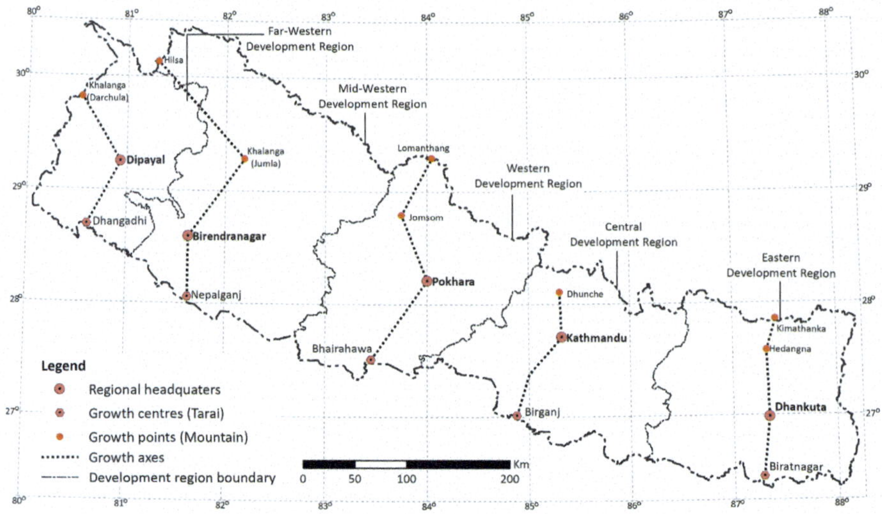

Fig. 5.10 Development regions, growth corridor axes and growth centres by level, Nepal (*Source* Modified after Gurung 1969)

Table 5.2 Development regions, growth centres and growth corridors in Nepal

Growth axis	Development region	Hill regional centre	Tarai growth centre	Mountain growth points
Koshi	Eastern	Dhankuta	Biratnagar	Hedangna
Kathmandu	Central	Kathmandu	Birganj	Dhunche
Gandaki	Western	Pokhara	Bhairahawa	Jomsom
Karnali	Mid-Western	Birendranagar (inner Tarai)	Nepalganj	Khalanga (Jumla)
Dipayal	Far-Western	Dipayal	Dhangadhi	Khalanga (Darchula)

Source Gurung (1969), modified

locations for eventual industrial developments wherever were feasible. Consequently, a planned clustering of development activities such as health, education, security, judiciary, communication and the like has been made in each of the centres according to their levels, with the expectation that the effects would spread to their surrounding hinterlands. As a result, private sector investment was also increased in the economic and social activities at the strategic centres.

Further, development corridors/axis roads were designed to linking the Tarai's growth centres with the Mountain's growth points via the Hill's regional centres with the belief that all these lines of development would hinge on the exposure of all ecoregions through transport linkages (Fig. 5.11; Table 5.3). Accordingly, axis roads began to build. Till today, i.e. over four and half decades since the inception of the

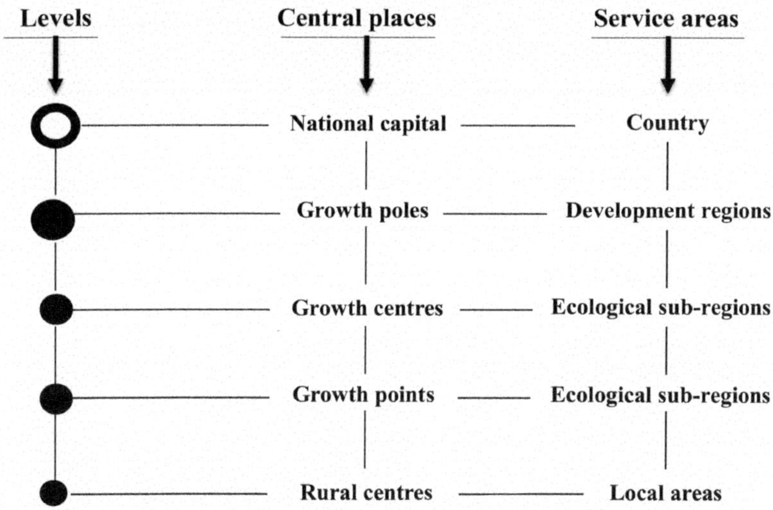

Fig. 5.11 Nepal's regional development strategy: schematic links of different hierarchical levels and their service areas (*Source* Authors' own construct)

Table 5.3 Nepal's regional development strategy: proposed road corridor linkages and their current status

Development regions	Regional headquarters	Growth centres	Growth points	Growth corridors	
				Proposed	Progress
Eastern	Dhankuta	Biratnagar	Hedangna	Biratnagar-Dhankuta-Hedangna	Incomplete
Central	Kathmandu	Birganj	Dhunche	Birganj-Kathmandu-Dhunche	Completed
Western	Pokhara	Bhairahawa	Jomsom	Bhairahawa-Pokhara-Jomsom	Completed
Mid-Western	Birendranagar	Nepalganj	Khalanga (Jumla)	Nepalganj-Birendranagar-Khalanga	Incomplete
Far-Western	Dipayal	Dhangadhi	Darchula	Dhangadhi-Dipayal-Darchula	Completed

Source ADB (2010)

policy, the construction of north–south aligned corridor roads linking those growth centres has only been completed in the central, western and far western regions and that in other two regions is still under construction. Anyway, the road links with the centres located in the north bordering Tibet (China) are indeed a need to be built for strengthening corridor strategy.

Upon implementation of the regional planning policy, efforts were made for the rational distribution of development investments in accordance with the needs of the population and areas across the country to some extent, as well as for retaining people in their own places and engaging them for the exploitation of the resources locally available. Indeed, this spatial development policy was an attempt to induce growth from below and thus a marked departure from the top-down thinking, i.e. departure from national capital planning decision to regional development centres. The strategy in fact provided a major impulse for rural regional development. The policy also realized the need for a more articulated and integrated system of urban centres to reduce disparities between urban and rural areas and ecoregions (HMG 1970).

Two parallel development policies, such as decentralization and the service centre approach, were also pursued simultaneously as important instruments of planning and regional development. These two policies were considered as complements to the regional development policy. In both policies, the district was taken as a general spatial framework for formulation and implementation of district-level projects and programs, as well as for the facilitation of national development plans and policies. The decentralization policy, though promulgated for the first time in 1963, was revitalized in 1982 under the 'Decentralisation Act 1982' to comply with the regional development policy (MLD 1982).

The service centre approach adopted in the late 1980s was considered as the main basis of a spatial framework for strengthening the regional development policy (HMG 1985). Initially, nine service centres were identified for each district and equipped those centres with basic functions such as health, education, agricultural extension, communication, post office, police post, technologies, cooperatives and others. The strategy aimed at building the capacity of the service centres to stimulate rural economies and providing an increased access for the rural population to those basic facilities, as well as for the execution of basic rural development plans. In the early 1990s, some modification of the service centre approach was made by increasing the number of service centres to a maximum of 17 in the districts with large population size, while nine minimum number of service centres remained unchanged.

However, both decentralization and service centre policies remained unsuccessful in achieving their intended goals. The failure of the former policy was primarily due to the lack of proper coordination mechanism among the concerned institutions located at various levels, and poor and unfair distribution of the services and facilities in the districts. In the latter case, the policy ignored to consider the existing market towns as service centers in most cases, though they have been playing roles in providing economic, social and other services to the surrounding rural population according to their functional size since the long (Shrestha 1987).

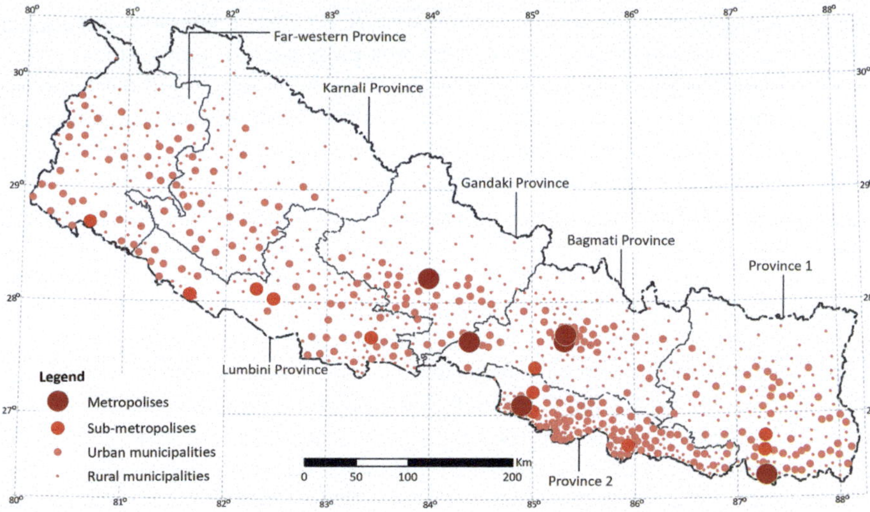

Fig. 5.12 Spatial distribution of municipalities by levels, Nepal (*Source* CBS 2014; MOFAGA 2017)

Moreover, as most of the market towns have been spontaneously developed occupying strategic locations in terms of business accessible viewpoint, they would have been suitable service centres for effective delivery of public basic services. Studies indicate that rural people in Nepal visit the market towns for multiple purposes such as marketing of goods, as well as for other services (Pradhan 1988; Pradhan and Routray 1992). Furthermore, because of resource constraints, neither was it possible to make locational decisions for the establishments of flatly nine service centres in all districts, irrespective of their area and population, nor was it practical for all seventeen service centres with the same level in larger districts in terms of population size.

In both policies, the vital idea that the economic development of urban areas as a necessary condition to encourage the economic development of rural regions has received marginal attention. It is evident from the fact that Nepal is one of the least urbanized countries in South Asia. By 2011, the population of 58 urban areas[4] accounted for slightly over 17% of its total population that increased abruptly to 293 urban areas in 2014 (Fig. 5.12), the population of which constituted nearly 63% of the total population (CBS 2014). This gives an urban density of 376 persons per km^2. Even if the employment structure of those 58 principal urban areas considered, 38% of the economically active population was employed by the primary production

[4] The urban areas in Nepal are defined mainly by population size. For instance, the minimum population size for any settlement locality to be designated as incorporated municipality has been set at least 10,000 for the hill and mountain and 20,000 for the Tarai (CBS 2012). But, the country's politicians and administrators usually treat the urban areas as political, administrative and census enumeration units.

sector, including agriculture and livestock works (CBS 2012), which is considered a weak urban economic base. Undoubtedly, this proportion of employment structure of 293 urban areas engaged in the primary production sector will definitely be increased as most of those inducted 235 municipalities have predominantly of rural character, as they have been designated by amalgamating surrounding rural areas for having a minimum population of 10,000 in the hills and mountains. Indeed, they have been designated primarily for administrative purposes rather than economic centres. However, there is no available of data on the economic and social features of all those newly inducted municipalities. In such circumstances, those recently inducted urban areas having primarily ru-urban character (officially 'urban' but with dominant rural features) cannot generate economic production and integrate surrounding rural areas because such municipalities need to focus more on developing within their own areas than to act roles for supporting to develop in surrounding rural hinterlands.

It is undoubtedly true that the road network construction will make the development regions more accessible to diverse locations, resources and activities. The regional centres connecting by feeder roads with the potential market towns in the rural region and equip them with a wide range of social and economic facilities will act as service centres and generate growth, and as a result, the regional circulation between regional centres and dispersed rural centres can be articulated (Hinderlink and Titus 1988; Pradhan and Routray 1992). However, the pace of road network building across all regions is very slow due mainly to the lack of financial resources and sincere commitment of the government. Till now, neither all north–south aligned roads as the main corridors of development in all five development regions have been completed, nor yet the feeder roads from the main corridor roads to connect major market settlements in the vast rural areas heading to the east and the west have been built. In 1956, the first road—Tribhuvan highway with 190 km long was built, linking the Kathmandu Valley to the southern Tarai, and then the construction of other roads followed but with very slow speed (Pradhan 1982). For instance, until now, not all 77 districts are connected with the road network. In the hills and mountains, the roads being built are of two types; one is the fair-weather road where vehicles are operated seasonally and another is all-weather road, where vehicles can operate all year round (Fig. 5.13). The latter type accounted for only 30% of the total rural road network, while the fair weather and non-operation roads combined represented 70% (DOLIDAR 2011). In addition, the trail networks comprising east–west tracts as the main artery and their tributary trails branched off to the south and north directions across the hills and mountains for delivering government supplies and trade and commerce were built mostly during the nineteenth century (Regmi 1988), which are still important and working in the absence of motor roads, though villagers need longer time and energy to travel on foot along such trails.

Thus, the regional development strategy has been unsuccessful to a larger extent to attain its intended balance growth objective within the development regions, but it has rather brought polarized development around the relatively large growth centres (Pradhan 1998). This has inevitably led to a certain degree of unequal distribution of benefits during the course of development. In most cases, the spread effects or trickle-down effects did not take place due to the lack of integration of such centers with

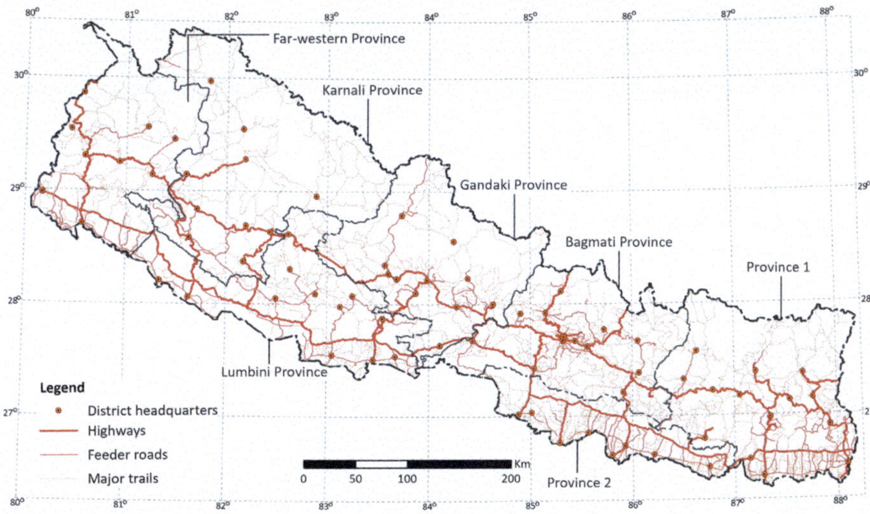

Fig. 5.13 Roads network, Nepal (*Source* DoR 2018)

the well-dispersed market towns in the rural hinterlands. For instance, the regional centres of Kathmandu and Pokhara lying respectively in the Hill valleys of central and western development regions, and Birendranagar lying in the Inner Tarai Valley of the mid-western development region have been relatively better developed than those of Dhankuta and Dipayal lying respectively in the eastern and far-western development regions. In the case of the latter two regional centres, physical constraints were the main reasons for limiting further expansion of development activities, due to their location along the hill ridge and hillslope. Regional planners believe that the trickle-down effect does not happen automatically; it requires a certain process or mechanism to act, or it requires the provision of transport and communication and setting up of different levels of hierarchical centres for public investment accordingly (Smith 1976).

This spatial development strategy still persists in Nepal, but in practice, not any of its prescribed implementation processes have been followed. The strategy has been made dysfunctional due to the fact that it never attained the status of a national commitment (Sharma 1989), and there was also a lack of coordination among the concerned government institutions that also made it ineffective. Besides, the formulation of regional planning policies has been based on a poor database, which is still a crucial, but an undeniable issue in Nepal. The database building and data upgrading often receive the least priority and their quality, consistency, availability and accessibility are often neglected, which can be considered as a key factor for the failure of most of the development plans and policies, including the regional strategy. If this practice continues to remain, any strategies to be formulated even in the future would be futile in Nepal.

5.6 Discussions

The concept of development is a process generally concerning with making comfort living of the people or improving social life entirely. In Nepal, it can be linked to the availability of essential infrastructure and facilities, proper utilization and mobilization of resources, creating opportunities for employment generation, agriculture commercialization, delivery of basic services, social inclusion, and modernization of overall society in the entire regions.

Implementation of planned development efforts in Nepal is already over six decades, but the economic growth has been lackluster and slow, and achievements of the development efforts are not satisfactory because of imposed development, unstable political situation, ad hoc investment decisions of governments, absence of people's participation and extremely lack of research and development.

Within a short span of duration, Nepal has witnessed tremendous political changes, such as the multiparty system from 1951 to 1960, party-less panchayat system from 1961 to 1990, multiparty democracy with constitutional monarchy system from 1991 to 2005, and the running one is the federal democratic republic since 2006. Development policies, including integrated rural development program, service centre approach, decentralized district planning, agriculture perspective plan, local self-governance act, to mention a few, have been discontinued like the regional development policy. Yet, some of them are useful and relevant.

The regional planning policy introduced during the panchayat political system has undergone ups and downs and since 2002, it has virtually been dysfunctional. Instead of blaming this strategy as such for its side effects such as poor trickle-down effect, polarized development of few selected cities, limited resources, poor urban–rural links, etc. the question should be posed how these disruptive side effects have happened and can be neutralized, for instance, by increasing the investment opportunities elsewhere in the potential areas (ADB 2007). Because the investment resources indeed were severely limited compared to the needs and also the investment decisions have been ad hoc undermining intended impacts. Secondly, this strategy of induced growth needs to be rethought in favour of integrated urban-based region development in terms of population, resource base, flows of goods and services and road network links. This, however, requires careful assessment of potential production sectors for pouring investment (Pradhan and Routray 1992).

However, most development strategies in Nepal have faced a great challenge. They often change with the change of government or political systems and do not remain to act even up to their own stipulated period of time. For instance, the growth corridor strategy linking all three ecoregions—mountain, hill, and Tarai, one of the integral parts of the regional development policy, is reasonably essential not only to exploit its diverse resources for development and but also to acquire benefits from transit trade with two giant neighbouring and emerging industrial countries—China and India. In these contexts, this strategy still appears to be important and essential, but unfortunately, it has received least or virtually no priority from the successive governments in recent years, but instead, they have given more focus on sectoral planning. Studies elsewhere in several countries indicate that spatial planning has acted as a complement to sectoral economic planning for integrated development and economic and social growth.

On the other hand, there is the spreading of budgets across different parts of Nepal, ignoring priority areas for investment. This is the main reason for slow growth and opposition to the spirit of development policies, including the regional development planning. The political decision is found playing a key role in budgetary allocations to invest in development works particularly in electoral constituencies.[5] There are budgetary allocation systems for two parliamentary development programs—one is grant allocation to each parliamentary member and another is parliamentarian development fund. In 2017, each of these two received NRp. 5 million and 30 million, respectively. Electoral constituencies are not local government units, and allocating fund for constituencies thus contradicts the principle of decentralization. In addition, what happened in practice is that most parliamentary members are found investing the funds in their constituencies in favour of their own voters, party workers, and supporters rather than to the benefit of 'those most in need' (Dahal 2017).

Pieces of evidence are there. For instance, the road has been a vital factor for development throughout the country. The opening of new roads appears to be the first stage in the development of intensified agriculture production, marketing activities and integration of urban and rural areas (Shrestha 1973–74; Blaikie et al. 1976; Schroeder and Sisler 1980; Pradhan 1981, 1988; Pradhan and Routray 1992). The traditional subsistence cereals crops have been replaced not only with commercial vegetables, but their cultivated areas have also increased substantially. The study by NPC/DFID (2013) shows that after the opening of the road in the eastern hills of Nepal in 1985, arable land in the hill slopes, which were previously little cultivated for substance production, or abandoned, has been intensified. The farmers there began to grow off-season high-value vegetable crops and to export them to the Tarai and even to the Indian cities. This has also increased opportunities for off-farm employment and income. In other vast areas, however poor road connectivity and road quality are major constraints hindering to enhance production and productivity and access to communities and markets. There are two types of road. One is the strategic road network, comprising national highways and feeder roads linking major centres across the country. The road density of this type is 8.76 km/100 km^2. Another is the local road network, which has a density of 39 km/100 km^2 or 2.18 km/1,000 people (DOR 2015). There are only two trunk highways traversing from east to west across the country. The East–West highway traversing all through the Tarai districts is the main thoroughfare in terms of connectivity and traffic flows, while the construction of the mid-hill highway is not yet completed. The district headquarters of Humla, Mugu, and Dolpa in Karnali Province and the district headquarters of Manang District in Gandaki Province, western Nepal are not yet connected with any road network. The railway is available only in two districts of central Tarai. Air service is available in

[5] The Constituency Development Grant was introduced in 1995 with initiating of NRp. 300,000 and was included as Electoral Constituency Development Programme in the Ninth Five Year Plan (1997–2002) for local development (NPC 1996). In 2018, the exchange rate was US$ 1 ≈110 NRp.

most of the mountain and hill districts, which is also not operational in the rainy season. The latter two types of transport service are not yet crucial for the economic link.

Scattered rural settlements are clearly a crucial problem for development in the hills and mountains. Under such condition, it is difficult to provide adequate and economically feasible basic infrastructure and services. At the same time, creating viable communities in terms of threshold demand size there is indispensable but challenging as well. Together with this, the urban network system in the hills and mountains is also fragmented by the lack of road access and by the scattered rural settlements. So, efforts to transform the scattered settlements to agglomerated forms are essential to provide economically feasible facilities in reasonable proximity to the users. Also, even the large cities are characterized by haphazard urban growth. These settlement systems—rural and urban—are to be rethought rigorously for inclusive growth, as well as for playing roles in providing linkage services to the development of surrounding hinterlands (Pradhan 2004).

At the end, it does not matter whatever good strategies or policies for development are devised and adopted, but rather it is a matter of sincerity and commitment to implement them by the government in the multiparty democracy system. This is true in the case of Nepal. For instance, there were some of the best and appropriate policies such as district decentralized planning, service centre approach, local self-governance, people participation, urban-based rural, regional development and so on undertaken in the country. However, they all remained futile and discontinue to operate, due simply to just disliked by the successive governments, without correcting their errors, if they existed, to changing scenarios. This is a misfortune to Nepalese people, and there is no way but to wait for true, genuine and pragmatic political leadership to drive the nation without adhering to the benefits of their own people, communities and parties.

5.7 Conclusions

Nepal's development strategies should aim at wider sharing of benefits through fuller utilization of diverse resources of the ecoregions with speeding up the growth itself. The strategies should be viewed in two ways: *first*, development planning is no more a closed system of hard choice among the ecoregions within the country, and *secondly*, planned development should be directed not towards accentuating inherent disparity but should be a conscious effort towards minimizing the regional disparity and between urban and rural areas.

Against this backdrop, over the past six and half decades, more than a dozen of development policies were implemented in Nepal, but most of them have been discontinued, not because they were irrelevant but mostly due to limited investment funds available compared to desperately needed amount, lack of seriousness and commitment of the successive governments, etc. Further, the rugged mountains posing obstacles to providing the infrastructure and facilities required for development appear to be undoubtedly the most crucial challenge. The regional planning policy also likewise became dysfunctional. This policy has not been able to reduce the disparities, and likewise, the backwash effects remained to be seen to some extent. These are indeed not the side effects of the policy itself, but are due to the slow pace of investment in road networks, ad hoc decisions in providing facilities and infrastructure in the regions vis-à-vis their centres, together with other reasons as mentioned above. Indeed, Nepal is by no means a resource-poor country, and it is one of the wealthiest countries in the world in terms of per capita natural resources. The country's abundant natural resources include water, forest and biodiversity, but only a fraction of these resources have been utilized.

Utilizing those resources sustainably for balanced growth between rural and urban areas and throughout the three ecoregions—mountain, hill and Tarai, is indeed a need now and in the future as well. Nepal's development lies indeed in the utilization of those diverse resources, and developing potential growth centres connected by north–south aligned corridor roads are one of the best ways to exploit those potential resources of the regions across the country, to acquire benefits from the transit trade between India and China, and to export local produce into the international market. What would be the appropriate strategy to address these issues? Nepal indeed has experienced economic-sector-based development planning since the very beginning, but neither the agriculture sector has improved nor has the enhanced industrial sector, and in fact, it alone has not been able to bring out the integration of diverse regions and a balanced growth. Yet, regional planning strategy appears to be relevant to address those issues. In this context, it is argued that a strategy blending of urban-based regional planning with economic-sector-based planning may be appropriate for Nepal to attain the balanced growth and development sustainably by integrating the regions having diverse production potentials. This strategy, however, calls for a thoughtful discussion among the development practitioners towards stable growth and development in the country.

References

ADB (2007) Regional development strategy report. National Planning Commission/Asian Development Bank (ADB), Kathmandu

ADB (2010) Unleashing economic growth: region-based urban strategy for Nepal. Asian Development Bank, Manila

APROSC (1995) Nepal agriculture perspective plan. Agricultural Project Service Centre, John Mellor Associates, Inc., Kathmandu

Baker J (ed) (1990) Small town Africa: Studies in rural-urban interaction. Seminar Proceedings No. 23, The Scandinavian Institute of African Studies, Uppsala

Barrow CJ (1991) Land degradation: development and breakdown of terrestrial environments. Cambridge University Press

Blaikie P, Cameron J, Feldman D, Fournier A, Seddon D (1976) The effects of roads in west-central Nepal, vols 3. Report to the Ministry of Overseas Development (U.K.), Overseas Development Group, University of East Anglia, Norwich

Blaikie P, Cameron J, Seddon D (1980) Nepal in crisis: growth and stagnation at the periphery. Clarendon Press, Oxford
Caplan L (1975) Administration and politics in a Nepalese town. Oxford University Press, London
Carson B (1992) The land, the farmer and the future–a land fertility management strategy for Nepal. ICIMOD Occasional Paper No. 21. International Centre for Integrated Mountain Development, Kathmandu
CBS (2012) National population and housing census 2011. Central Bureau of Statistics, Government of Nepal, Kathmandu
CBS (2014) Population monograph of Nepal, vol. 1. Central Bureau of Statistics, Government of Nepal, Kathmandu
CDG (2002) Strategy for agglomerated settlement development in Nepal. Project Report. Central Department of Geography/Department of Urban Development and Housing Construction, His Majesty's Government of Nepal, Kathmandu
Dahal G (2017) Democratic practice and good governance in Nepal. J Political Sci 17:18–35
DOLIDAR (2011) National plan for rural road maintenance. Department of Local Infrastructure Development and Agricultural Roads, Government of Nepal, Kathmandu
DOR (2015) Road statistics, 2015/16. Department of Roads, Government of Nepal, Kathmandu
DOR (2018) Road statistics, 2017/18. Department of Roads, Government of Nepal, Kathmandu
DUDBC (2008) National urbanization policy. Department of Urban Development & Building Construction, Government of Nepal, Kathmandu
Dusenberry HL (1957) Reports on Rapti valley development project. USAID/Nepal, Kathmandu
EROS (2013) Shuttle radar topography mission (SRTM) global 30m ellipsoidal. Earth Resources Observation and Science Center, (DOI) number: /https://doi.org/10.5066/F7PR7TFT)
FAO (1987) An agricultural development strategy for Nepal. Consultant Report. TCP/ NEP/4504, FAO, Rome
Giuseppe F (1799) Account of the kingdom of Nepal. Asiatic Researches, Vernor and Hood, London
Gurung HB (2005) Nepal regional strategy for development. Working Paper Series No. 3, Nepal Resident Mission (June), Asian Development Bank, Kathmandu
Gurung HB (1989) Regional patterns of migration in Nepal. Papers of the East-West Population Institute (113) September, East-West Center, Honolulu, Hawaii
Gurung HB (1973) Regional planning development for Nepal. In: Pradhan P (ed) A new dimension in Nepal's development: the regional approach planning and people oriented development strategy, CEDA, Tribhuvan University, Occasional Paper (3), p 1–16
Gurung HB (1971) Rationale for hill areas development. Nepal Ind Digest: 17–24
Gurung HB (1969) Regional development plan for Nepal. National Planning Commission, His Majesty's Government, Kathmandu
Gurung U (1976) Rapti valley project with special reference to Chitwan valley. Master's dissertation, Department of Economics, Tribhuvan University, Kathmandu
Hagen T (1971) Nepal: the kingdom in the Himalayas. Oxford and IBH Publishing Co, New Delhi
Hamilton FB (1971) An account of the kingdom of Nepal. Manjusri Publishing House, New Delhi
Hinderlink J, Titus MJ (1988) Paradigms of regional development and the role of small centers. Develop Change 19(3):401–423
HMG (1970) The fourth plan (1970–75). National Planning Commission, HMG, Kathmandu
HMG (1985) The seventh plan (1985–1990). National Planning Commission, HMG, Kathmandu
HMG (1999) Local self-governance act 1999. Law Books Management Board, Ministry of Law and Justice, Kathmandu
HMG/KESL (1986) Land resource mapping project: land utilization report. Kathmandu
Jodha NS (1989–1992) Mountain perspective and its utility: a framework for development strategies. The Himalayan Review, Vol. XX-XXIII:11–24
Kansakar VBS (1973–74) History of population migration in Nepal. Himalayan Rev VI(5-6)
KC BK (2003) Internal migration in Nepal. Population Monographs of Nepal. Vol. II, Central Bureau of Statistics, Government of Nepal, Kathmandu
Kirkpatrick C (1969) An account of the kingdom of Nepaul. Manjusri Publishing House, New Delhi

Mabogunje AL (1980) The development process: a spatial perspective. Hutchinson & Co Publishers Ltd
MLD Decentralisation act (1982) Ministry of Local Development. His Majesty's Government of Nepal, Kathmandu
MLD (1999) Local self governance acts and regulations 1999. Ministry of Local Development, Kathmandu
MOF (2000) Economic survey 2000. Ministry of Finance, Government of Nepal, Kathmandu
MOF (2019) Economic survey 2019. Ministry of Finance, Government of Nepal, Kathmandu
MOFAGA (2017) Urban municipalities and rural municipalities of Nepal. Ministry of Federal Affairs and General Administration, Kathmandu : https://sthaniya.gov.np/gis/
MoLMCPA (2020) Political map of Nepal. Ministry of Land Management, Cooperatives and Poverty Alleviation, Government of Nepal, Kathmandu.
MOPE, ICIMOD, SACEP, NORAD, UNEP (2001) State of the environment Nepal. United Nations Environment Programme, Bangkok
MOUD (2017) National urban development strategy 2017. Ministry of Urban Development, Government of Nepal, Kathmandu
NPC (1996) Ninth five year plan 1997–2002. National Planning Commission, Government of Nepal, Kathmandu
NPC/DFID (2013) Research into the long term impact of development interventions in the Koshi hills of Nepal: geographic information systems (GIS) in land use change and its impacts. National Planning Commission, Government of Nepal and Department for International Development, United Kingdom, Kathmandu
NRC (1980) Nepal resettlement company: an introduction. Nepal Resettlement Company, Kathmandu
Pant TN, Thapa GB (1981) Development potentials of Nepal's hill agriculture. Nepal's Experience in Hill Agricultural Development. Ministry of Food and Agriculture, Kathmandu
Pradhan BB (1981) Rural development in Nepal: problems and prospects a project report. Ministry of Panchayat and Local Development/UNDP, Kathmandu
Pradhan PK (1982) A study of traffic flow on Siddhartha and Prithvi highways. Himalayan Rev XIV:38–51
Pradhan PK (1988) Development of market centres and their functional base in Chitwan district. Nepal. J Develop Admin Stud 10(1 & 2):87–103
Pradhan PK (1998) Market centre and hinterland relation in Lalitpur district. Number 17, Institute of Geography & Regional Science, University of Klagenfurt, Austria
Pradhan PK (2004) Rural urban relations: with particular reference to Nepal. Rural Urban Partnership Programme, Ministry of Local Development and United Nations Development Programme, Kathmandu
Pradhan PK, Pradhan B (2006) Environment and natural resources: concepts, methods, planning and management. Quest Publication, Kathmandu
Pradhan PK, Routray JK (1992) Market centre and rural development: a study in Chitwan District, Nepal. HSD/Asian Institute of Technology, Bangkok
Ranabhat S (2013) Delivering school education in Mugu district, western Nepal. In: Pradhan PK, Buček J, Razin E (eds) Geography of governance: dynamics for development. International Geographical Union Commission on Geography of Governance, pp 11–21
Regmi MC (1988) An economic history of Nepal, 1846–1901. Nath Publishing House, Varanasi
Regmi MC (1971) A Study in Nepali economic history (1768–1846). Manjusri Publishing House, New Delhi
Schroeder MCW, Sisler DG (1980) The impact of the Sunauli-Pokhara highway on the regional income and agricultural production of Pokhara valley, Nepal. Department of Agricultural Economics, Occasional Papers No. 32. Cornell University, Ithaca
Sharma P (1989) Urbanisation in Nepal. Papers of the East-West Population Institute, No.110 (May) East-West Center, Hawaii
Shrestha CB (1969–70) Problems of regionalization in Nepal. Himalayan Rev 2–3:13–21

Shrestha CB (1973–74) The system of central place in the Arniko Rajmarg area of Nepal. Himalayan Rev 5–6:19–39

Shrestha CB (1987) Structure of rural market centres in the Kathmandu Valley. The Journal of Development and Administrative Studies. 9(1):1–12

Smith CA (ed) (1976) Regional analysis. Academic Press, New York, Volumes I & II

Stiller SJ, Ludwing F (1975) The rise of the house of Gorkha: A study in the unification of Nepal 1768–1816. The Patna Jesuit Society, Patna, India

Survey Department (2020) Political boundaries of Nepal. Government of Nepal, Kathmandu

TPC (1998) Nepal overseas trade statistics. Trade Promotion Centre, Kathmandu

Tuladhar BR (1972) Notes on Rapti valley development project. Food and Agricultural Division/USAID, Kathmandu

Pushkar K. Pradhan has earned his M.Sc. in Rural and Regional Development Planning from the Asian Institute of Technology, Thailand; an M.A. in Geography from the Tribhuvan University, Nepal; and a Ph.D. degree from the Institute of Geography and Regional Science, Klagenfurt University, Austria. Affiliating to the Tribhuvan University, Prof. Pradhan has gained over three and half decades of professional teaching and research experience. During this time period, he has served the Tribhuvan University with several academic and administrative positions. He served as an executive member of the Steering Committees of International Geographical Union (IGU) Commissions on Marginality, Public Policy, and Local Governance and successfully convened four IGU Conferences in Nepal in 2004, 2007, 2011 and 2019. He has served as a member of the Editorial Boards of several prestigious journals and IGU publications. Professor Pradhan has published over 40 research papers and books nationally and internationally, participated in over three dozens of international and national seminars, and has been leading several national and international research works.

Dr. Puspa Sharma is serving as an Associate Professor at the Central Department of Geography, Tribhuvan University, Kathmandu, Nepal, since 1996. Puspa teaches geographic subjects to the Masters' and Ph.D. students at the department, including Geographic Information Systems, Urban Planning and Management, Regional Planning, Development Planning, and Human Ecology and Field Studies. In addition, she has served as a lead trainer in the GIS training to professionals working in various public service departments, as well as private organizations at the Department. Puspa has contributed research articles and reference books published nationally and internationally and also contributed reference materials on GIS and in the newsletters. Several students have completed their Masters' thesis, and some are underway under her supervision. Dr. Sharma has been involved in various research works especially related to GIS, carried out by the university, research organizations, and international agencies.

Chapter 6
Dislocation and Involuntary Migration: Lessons from the Teesta River Bank Erosion in Bangladesh

Mst. Rebeka Sultana

Abstract Dislocation and involuntary migration is a recurrent issue in the floodplains of the Teesta river. The present research aims to identify the outcome of displacement and forced migration influenced by the active bank erosion of the Teesta river. Displacement and resettlement in the study area embrace temporary shifting, stepwise shifting and permanent shifting of the homestead. Prevalence of dislocation causes a socio-cultural and economic change in the community of the Teesta river bank. Likewise, the health and environmental state of the affected community deteriorates due to sudden changes in living conditions. The psychological consequences of dislocation and migration are also widespread among individuals. The methodology of the study includes household-based questionnaire survey, FGD, and case studies. Loss of income sources, frequent homestead shifting, isolation from society, squatter development, and mental stresses are remarkable in the study area. The study result finds that at present 39.6% house is in the Teesta riverbank, and 82.4% of respondents shifted their homestead permanently after several times displacement. Moreover, 48.6% of respondents have been displaced 7–9 times owing to bank erosion incidents. Land ownership conflict is a growing concern among the bank erosion displaces, and about 84% of respondents handled different types of estate proprietorship clashes. The study also depicts 53.1%seasonal migration and 46.9% temporary migration for livelihoods due to damage of occupation after riverbank erosion. The government should take initiatives regarding the issues of resettlement of the dislocated population. The study suggests preparing a national guideline for the rehabilitation of the river displaces. Besides, the provision of employment opportunities through establishing industries in the Teesta floodplains will be a barrier to the river victims migrating to the cities for occupation after erosion disaster and thus reducing the consequences of dislocation and involuntary migration for bank erosion.

Keywords Displacement · Relocation · Bank failure · Corrosion · Affected Community

---

Mst. R. Sultana (✉)
Department of Geography and Environmental Studies, University of Rajshahi, Rajshahi, Bangladesh
e-mail: rebeka.sultana@ru.ac.bd

M. Mishra et al. (eds.), *Regional Development Planning and Practice*,
Advances in Geographical and Environmental Sciences,
https://doi.org/10.1007/978-981-16-5681-1_6

6.1 Introduction

Dislocation and involuntary migration occur alongside the alluvial rivers of Bangladesh. The rivers of Bangladesh are susceptible to bank erosion due to recurrent floods and unconsolidated soil characteristics. Bangladesh is a country of alluvial soils accumulated by rivers during floods (Siddique and Chowdhury 2000). The country is known for its widespread river bank erosion every year along with population displacement and disaster vulnerability. In Bangladesh, riverbank erosion is a silent and endemic disaster (Islam and Khatun 2019). Displacement is the foremost effect of riverbank erosion, and it is the perennial issue of riverbank dwellers of the country. The Teesta is one of the most erosive rivers of the northern territory of the country, which prompts the bank erosion victims to displace their houses and migrate to another location for survival. The dwellers of the Teesta floodplains are prone to riverbank erosion-induced disaster in each monsoon, and involuntary migration is the aftereffect of erosion. Floodplains of Bangladesh are vulnerable to disaster, which instigates force migration (Shamsuddoha et al. 2012). However, people suffer much for shelter, food, and drinking water during riverbank erosion strikes. Dislocation and migration have a long-term effect on the community both in socio-economic and cultural backgrounds. As a result of short and long-term effects of dislocation and migration, the displaced people suffer in a volatile environment for their existence. Displacement due to bank erosion causes socio-economic disruption and also increases uncertainties of the affected community (Iva et al. 2017).

Dislocation is synonymous with displacement, which refers to shifting a house to another location owing to disruption of one's living place or living environment for different factors. In broader aspects, dislocation occurs due to natural factors and anthropogenic influences. The present research deals with riverbank erosion-induced dislocation, which is propagated with climate change and natural factors of dislocation. The person who is the victim of displacement for riverbank erosion is known as river displacee in Bangladesh. The term migration is difficult to understand. Migration is a holistic term used for population movement from one place to another place, either temporarily or permanently. The factors affecting involuntary or forced migration have varied dimensions, i.e. economic, social, cultural, political and environmental. Involuntary or forced migration occurs due to geographical location, socio-political conflicts and economic deprivations (Das et al. 2014). Migration from rural to rural and rural to urban areas of Bangladesh occurs mostly for livelihood destruction owing to a natural disaster such as flood, riverbank erosion, cyclone, etc. Migration occurs because of unemployment opportunities influenced by natural disasters, and loss of assets prompts people on permanent migration (Mallick 2011). The magnitude of flood and riverbank erosion exaggerates the severity of displacement and migration of river bank dwellers. In particular, displacement and migration have significant effects on agricultural production, occupation, economic condition, mental state and social status.

On the other hand, displacement depends mainly upon the economic circumstances of a family. Moreover, resettlement in a new location after the disruption of the homestead raises a lot of issues regarding basic needs and social requirements. The displaced people seek the favour of local people and relatives to rebuild the house in a river island or mainland of a new locality. The convenience of land for relocation and linkage with relatives and local leaders are the key factors for the resettlement of the displaced population (Elahi 1991a, b).

Due to perennial bank erosion in the study area, dislocation and migration frequently occur among the residing communities along the river banks, and they become challenged during the incident and after resettlement. Correspondingly, they have to endure the condition in their entire lives through economic insolvency, mental stress and obstacles to meet relatives in a new living place. Considering all these issues, the present study seeks to explore the consequences of dislocation and involuntary migration. The present study reviewed different literature to develop the conceptual framework and determine the research gap of the study. The reviewed literature shows the factors of migration in the studies of Haider 2010, Parkins 2010, Ishtiaque and Ullah 2013. Similarly, several studies have been conducted regarding climate change and migration, including Perch-Nielsen et al. 2008, Warner et al. 2010, Chindarkar 2012, Bernzen et al. 2019. Ahmed 1991, Elahi 1991a, b, Iqbal 2010, Das et al. 2014, Mollah and Ferdaush 2015 investigated flood and erosion induced displacement. The impact of erosion on displacement and migration was also studied in different studies by Rabbi et al. 2013, Islam et al. 2017, Rana and Nessa 2017. Similarly, Zachariah et al. 2001 studied the impacts of migration, considering population structure and socio-economic factors. Saleheen 1991 explored the features of the mobility of river displaces of massive erosion-prone areas of Jamuna river bank of Bangladesh. Haque 1991, Mahabub and Islam 1991, Sultana et al. 2018 evaluated the adjustment process of the displaced population of river erosion. The previous studies discussed diversified issues on displacement and migration due to flood and bank erosion of different river catchments from regional to international scales. Although several studies explored the factors behind displacement and migration, the consequences are mostly unexplored yet. The present study aims to investigate the consequences of the dislocated community of riverbank erosion. There is a shortage of comprehensive studies regarding the consequences of dislocation and forced migration due to riverbank erosion. In this regard, the current study made an effort to focus on this aspect and recommended some necessary measures as well.

6.2 About the Present Study

The community of the Teesta riverbank is economically backward due to riverine disasters and recurrent shifting of houses. Likewise, different studies confirmed that migration for income-earning is also a distinguishing feature of the community (Sarker et al. 2015, 2020; Mondal et al. 2020, Sultana et al. 2020). Therefore, the present study portrays dislocation and involuntary migration as a response to riverbank erosion. Indeed, the present study tries to explore the socio-cultural, economic, health-environmental and psychological effects of dislocation and migration amplified by the Teesta riverbank erosion. More specifically, the study aims to address.

- the consequences of dislocation;
- migration and its impacts.

The present study examines the effect of dislocation and migration of riverbank erosion. To fulfill the objectives of the study, a pilot survey was conducted in the study area. Moreover, the study collected information of repeated flood and bank erosion in the Teesta river and human displacement from the national newspapers. Likewise, a review of the literature of Elahi 1991a, b, Elahi and Ara 2008, Rahman et al. 2011, Mondal and Islam 2017, Karim et al. 2017 also indicates the ambience of flood, bank erosion and migration of Teesta riparian community. Therefore, the investigation focuses particularly on the communities residing adjacent to the Teesta riverbank in Bangladesh. The field survey was carried out during and after the 2017 flood in the Teesta basin, considering the erosion-induced dislocation and migration. Seven unions[1] of seven erosion-prone upazillas[2] of the Teesta river course were selected to conduct the study. The study unions are also highly erosion affected. The study upazillas and unions have been chosen purposively to evaluate the consequences of dislocation and involuntary migration among the erosion-affected communities. The study upazillas are Dimla, Jaldhaka, Hatibandha, Lalmonirhat Sadar, Gangachara, Rajarhat, and Sundarganj. Data has been collected from the unions of Purbachatnai (Dimla), Saulmari (Jaldhaka), Sindurna (Hatibandha), Khuniagacch (Lalmonirhat Sadar), Kolkanda (Gangachara), Bidyananda (Rajarhat) and Belka (Sundarganj).

The study is based on primary data gathered from direct field investigation. These data and information include the distance of the house from the riverbank, time of staying at the present house, frequency of homestead migration for erosion, type of migration, etc. The field survey for the present research comprises several methods such as questionnaire surveys, case studies and Focus Group Discussion (FGD). The total number of respondents of the study were 381. A simple random sampling technique has been applied to collect desired data from oppressed households by bank erosion. Data recruitment has been completed through a well-developed questionnaire. At first, the respondents were motivated to respond to the questionnaire, as some of them disagreed with giving any kind of information. Moreover, some respondents gave false economic information to get monetary profit. Therefore, the respondents were assured that the data will be used only for research purposes, and they will not get any financial benefit for their response.

Furthermore, FGDs have been carried out from most consequential respondents. In addition, case studies were conducted to understand the severity of homestead displacement and forced migration. After the field survey, the data and information were assembled, standardized and coded to analyse. The study used SPSS and Excel software to analyse the statistics.

---

[1] Union is the lowest administrative unit in Bangladesh.

[2] Upazilla is the lower administrative unit after district.

6.3 Dislocation and Involuntary Migration in the Light of River Bank Erosion

The banks of the Teesta river are densely populated. Therefore, the residents of the Teesta floodplain are highly vulnerable to riverbank erosion-induced displacement. Destruction of the homestead is determined by the population density of an area located in and around any river bank (Das et al. 2017). Moreover, the weak alluvial soil of the banks of the Teesta is a dominant factor for erosion after each successive flood in the study area. The community adjacent to the Teesta riverbank experienced a horrible flood along with catastrophic riverbank erosion in 2017. The study observed agricultural land, homestead land and infrastructures have undergone into the river, and people become shelterless and worried about their resettlement. Besides, the displaced people salvage their housing goods and utensils for survival and migrate to a new location to rebuild another house. Migration not only occurs for shelter management, but it also has another dimension, which is related to economic disruption resultant from bank erosion. Most of the respondents of the study area directly or indirectly earn their living with agricultural activities. Intense damage in the agricultural sector, such as land loss, crop production loss, farming commodity loss after bank erosion, impacts the occupation of the community. Disruption in occupation motivates the vulnerable people to migrate to the nearby towns and big cities. Therefore, the present study intends to explore the fact that bank-erosion victims migrate for two fundamental reasons: resettlement and searching for a new job. The respondents argued that they try heart and soul to stay in their inherited land, but when river bank erosion ruins their homestead and occupation, they have no means to live in the source region. Thus, displacement is the ultimate action in response to riverbank erosion to save life and existing household goods. Figure 6.1 illustrates the driving forces of dislocation and involuntary migration and people's response to riverbank erosion.

The consequence of displacement and resettlement in an unknown location brings a new aspect of socio-economic deprivations such as income inequalities before and

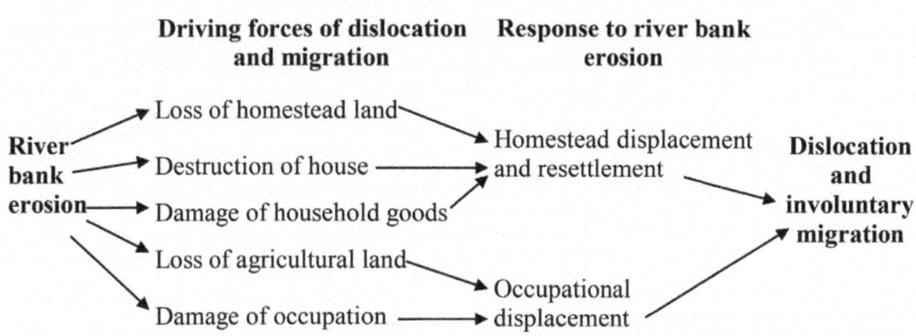

Fig. 6.1 Dislocation and involuntary migration in response to riverbank erosion (*Source* By the author)

Table 6.1 Issues of resettlement after dislocation

Resettlement after displacement	Issues
Resettlement in newly accreted *char* land	Land grabbing, Social conflict
Resettlement in *khas* land	Panic of government land reclamation activity
Resettlement in embankment	Squatter development, unhygienic environment
Resettlement near to source destination	Lack of income-earning opportunities, fear of further erosion
Resettlement in urban squatter	Unavailability of job, lack of urban facilities, drug business, drug addiction, woman and child trafficking, anti-social activities

Source FGD 2017

after displacement, shelter management, social adjustments and so on. At the onset of riverbank erosion, people become insecure with their broken homestead, and sometimes the displaced community settles under the open sky. The worst condition arises when people lost their only earning sources and are forced to migrate to the urban squatters. Squatter development, land grabbing and resettlement are growing concerns regarding bank erosion-induced displacement (Elahi 1991a, b). The respondents of the study replied about their sufferings regarding resettlement and job searching in a migrated location. Most often, the displaced community resettles houses in *char* land (reemerged land from the river), *khas* land (land requisitioned by the government) and embankments. The issues of resettlement after dislocation have been illustrated in Table 6.1.

6.4 Consequences of Dislocation and Involuntary Migration

It is evident from the study that the victim of riverbank erosion suffers from fulfilling basic needs throughout their entire life. Likewise, poverty increases for livelihood damage after erosion. The unemployed people sell productive assets such as livestock, poultry, tree, etc., for survival after the disaster. In critical conditions, the displaced people engage in low-cost day labouring instead of receiving the actual price of the labour force. Moreover, they reduce family expenditure by stopping their children from going to schools, skipping meals and starving. As an effect of erosion-induced frequent dislocation, people suffer for food and nutrition. Sometimes, the children are also engaged in day labouring to manage hunger and family expenditure. The adolescent girls migrate to cities and work in readymade garment factory to accomplish the family expenses for several times displacement. In broader aspects, the dislocated people face difficulties in their socio-cultural and economic state. Apart from that, the displaced people also deteriorate environmental conditions through squatter development in embankments, towns and cities. The condition of health of the displaced population is very vulnerable for lack of pure drinking water

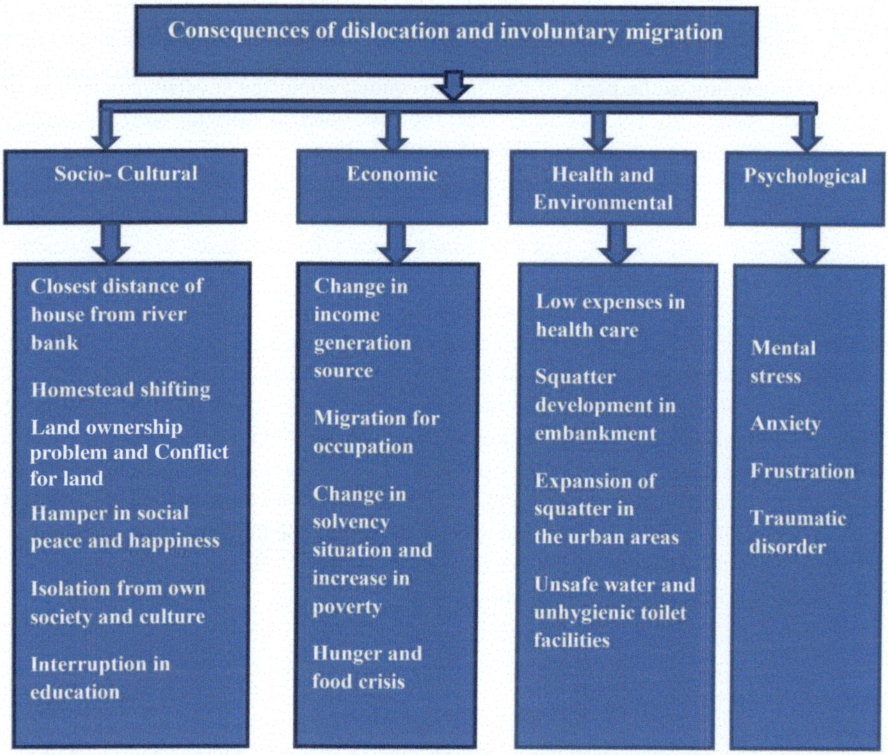

Fig. 6.2 Consequences of dislocation and involuntary migration (*Source* By the author)

and unhygienic latrine. Social deterioration also increases as a consequence of dislocation and migration through the unstable economic condition, broken family and isolation from society. The impacts of dislocation and involuntary migration have been depicted in Fig. 6.2.

6.4.1 Unstable Socio-Cultural Settings

The socio-cultural environment of the community of Teesta riverbank is unstable due to migration and frequent shifting of house, conflict for land, hamper in education, etc. Field study result unveils some of the surveyed households dislocated thrice in a year after three sequential floods and erosion. The respondents stated that riverbank erosion is like a chronic disease that impacts society and culture through displacement, isolation and conflict. Thus, instability arises in the community.

6.4.1.1 Closest Distance of the House From Riverbank

The people residing at the bank of the Teesta opined about the severity of erosion in recent years. When the river bank erodes, the dislocated people try to rebuild their houses in proximity to their eroded land with the hope of regaining the land after accretion. Hence, the study finds out that consequent cycles of resettlement along the vulnerable river bank increase the susceptibility for further phases of displacement. The location of the house is a dominant factor that increases or decreases the vulnerability of displacement.

Interestingly, the present study finds that the victimized communities of the study area live near the river. The dislocated people shift their houses on the embankments and natural levees along the river banks, which builts a linear settlement pattern of the victims. Table 6.2 represents the distance of the Teesta river from the residence of the respondents. According to the study, about 39.6 percent of respondents live on the bank of the river. Moreover, 30.7 percent of respondents have been recognized to settle within a 100-m distance, and 22.0 percent displaced population has been recorded to live within a 200-m distance from the river. The table also depicts the chi-square result of the distance of the Teesta river from the residence of the respondents in the study unions. The test result shows that the p-value is significant.

6.4.1.2 Homestead Shifting

Field survey result points out that homestead shifting is the ultimate outcome of dislocation of the Teesta riverbank erosion victim. The respondents of the study shifts their homestead permanently, owing to numerous displacement, which is a great barrier for livelihoods and economy. Stepwise homestead shifting is also followed in the study area due to bank erosion. According to the study, about 82.4% respondents shifted their homestead permanently after several times dislocation. When bank erosion occurs, the people shift their homestead to a short distance. They are recurrently affected by these events. And afterward, when they are unable to manage their livings, then they migrate to a longer distance. At first, the displaced people relocated houses near the eroded land or another location of his/her village or union location. Next, people choose to resettle a house in another village of their own union or char after erosion. Subsequently, the displaced people relocate houses in another union or nearby town and search for a new occupation for survival. Finally, the displaced people decide to leave the source region due to the fear of further erosion and destruction of livelihood and migrate to urban areas.

Similarly, people try to live near to source region due to their social relations and mental settings. Therefore, a recurrent erosion attack makes them resettle houses in the nearby village, *char*, union, and town. The study result regarding the place of displacement displays that about 51.7 percent of respondents resettled in the new location from another place of their own union. Hence, the study result reveals that most of the people moved house into another location of their own union. The displaced people shift from one union to another union, one village to another village,

Table 6.2 Distance of the Teesta river from the residence of the displaced population

Union	Distance of the Teesta river from residence									
	House is in the bank		100-m distance		200-m distance		More than 200-m distance		Total	
	N	%	N	%	N	%	N	%	N	%
Purbachatnai	12	35.3	12	35.3	6	17.6	4	11.8	34	100
Saulmari	26	40.6	24	37.5	8	12.5	6	9.4	64	100
Bidyananda	28	68.3	7	17.0	4	9.7	2	4.8	41	100
Khuniagachh	23	31.9	13	18.0	32	44.4	4	5.6	72	100
Kolkonda	25	39.1	29	45.3	10	15.6	0	0	64	100
Sindurna	8	26.7	13	43.3	7	23.3	2	6.7	30	100
Belka	29	38.2	19	25.0	17	22.3	11	14.4	76	100
All	151	39.6	117	30.7	84	22.0	29	6.6	381	100
Chi Square	$\chi 2$ value = 75.893, df = 30, p = 0.000									

Source Field Survey 2017

Table 6.3 Place of displacement after the dislocation

Union	Place of displacement							
	Another location of this union		*Char*land		Another union		Total	
	N	%	N	%	N	%	N	%
Purbachatnai	13	38.2	15	44.1	6	17.6	34	100
Saulmari	22	34.4	29	45.3	13	20.3	64	100
Bidyananda	22	53.7	16	39.0	3	7.3	41	100
Khuniagachh	32	44.4	28	38.9	12	7.7	72	100
Kolkonda	36	56.2	14	21.9	14	21.9	64	100
Sindurna	15	50.0	9	30.0	6	20.0	30	100
Belka	57	75.0	9	11.8	10	13.2	76	100
All	197	51.7	120	31.5	64	7.8	381	100
Chi-square	χ^2 value = 37.177, df = 12, p = 0.000							

Source Field Survey 2017

from char to the mainland, and from the mainland to char land. However, they do not leave their source region and live near the damaged homestead to hope the eroded land must be raised again. Ancient maps and earlier literature display new settlement advances in the bank of the river for the vested land property (Mallick 2016). Hence, the dislocated people resettle houses close to their past locality due to socio-economic intentions. People choose relocation of the house in char and within upazilla for socio-cultural adaptation after riverbank erosion (Elahi and Rogge 1990). Table 6.3 signifies the place of displacement of the Teesta river displaces. The chi-square test result indicates significant differences ($p = 0.000$) in terms of the place of displacement of the respondents.

On the other hand, the present study investigated the respondent's duration of living at the present house. The respondents opined that the period of dislocation and resettlement is the monsoon and post-monsoon season. Hence, the study result indicates that 12.6 percent of the respondents have been staying at the current location for less than one year. Field data represents that among the 12.6% respondents 1.8% people moved into the new location today (0 yearss) from their previous location. About 29.1% of respondents staying for 1–4 yearss, 36.2% staying for 5–8 yearss in the present location. Hence, the study denotes the internal displacement of the river victim from one location to another location. Table 6.4 clarifies the percentage of the duration of staying in the present house/ location.

Displacement and several times homestead shifting followed by riverbank erosion is the reality that the community of the Teesta bank has to face frequently. The study observed more than 9 times homestead displacement in the study area, and some of the respondents described experiencing 15 times displacement after homestead damage by the Teesta river. In Bangladesh, more than 40 percent of people were displaced 3–4 times, and 36 percent displaced 5–10 times due to bank erosion (Hutton and Haque

Table 6.4 Time of staying at present house/location

Time of staying in present house	N	%	Time of stay and %
Less than one year	48	12.6	0 year (Today) 1.8%, 0.02 year 0.8%, 0.03 year 0.8%, 0.04 year 1.0%, 0.05 year 2.1%, 0.06 year 1.0%, 0.07 year 0.8%, 0.08 year 0.3%, 0.1 year 0.5%, 0.12 year 1.3%, 0.2 year 0.3%, 0.22 year 0.3%,0.25 year 1.3%, 0.3 year 1%
1–4 year	111	29.1	1 y 3.1%, 2 y 11.3%, 3y 8.7, 4y 6.0%
5–8 year	138	36.2	5 y 21.0%, 6 y 3.1%, 7 y 6.8%, 8 y 5.2%,
9–12 year	66	17.3	9 y 1.0%, 10y 11.8%, 12y 4.5%
13–16 year	18	4.8	13 y 1.6%, 15y 2.9%, 16y 0.3%
Total	381	100.0	

Source Field survey 2017

2004). Sometimes the displaced people build a temporary shelter and afterward seek a suitable place for resettlement and then shift house building materials and household goods (Fig. 6.3). The temporary shelters are built on the roadsides and earthen mounds. If the distressed family may not salvage the housing materials, then temporary shelters are made with polythene.

Surprisingly, the community lives through repeated bank erosion, but they have the mental strength to combat the disaster and build a new shelter in another location with a unique hope of survival. The study result indicates 48.6 percent of respondents have been displaced 7–9 times owing to bank erosion incidents. About 34.1 percent of respondents answered regarding their displacement of more than 9 times. Figure 6.4 illustrates the percentage of respondent's displacement for river bank erosion. Case study 1 illustrates the scenario of several times homesteads displacement by the Teesta riverbank erosion.

Fig. 6.3 Temporary shelter after displacement and shifting of housing materials (*Source* By the author)

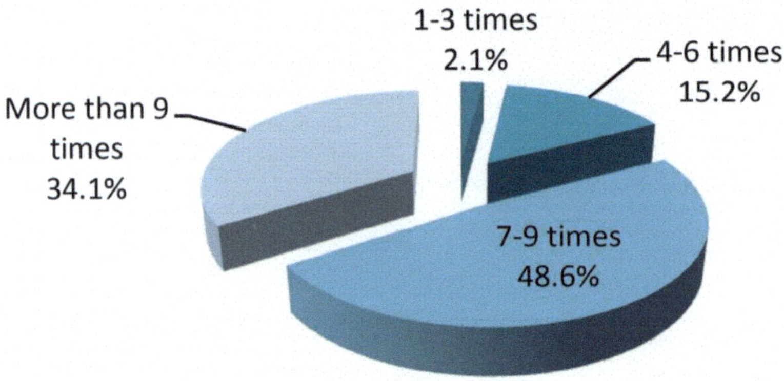

Fig. 6.4 Displacement for riverbank erosion (*Source* Field survey by the Author, 2017)

Case study 1

Marjina begum is from Khuniagacch union of Lalmonirhat Sadar upazilla, and after displacement, she lives at Saulmari union of Jaldhaka upazilla of Nilphamari district. She is 48 years old. She is displaced from her own union in the 2017 flood and riverbank erosion of the Teesta. Margina begum is a maidservant, and she has five children. She was married to a farmer who owned 3 acres of land. But recurrent river bank erosion made her family economically insolvent. After losing the arable lands, her husband becomes a day labourer, and they lived hand to mouth. They stopped their children from going to school and engaged them in day labouring. She was forced to marry her two daughters at the age of 12–13 years. Her two sons migrated to Nilphamari town and pull rickshaw. Her husband died six years ago. Now she is landless and lives in a house of two rooms in the embankment of the Teesta with one of her sons. Two months ago, she shifted her house in the embankment. It is the eighth time displacement of her homestead. She described homestead shifting brings untold sufferings to the victim's house owner like herself. Marjina begum described that once upon a time, her house was full of different fruit trees, cattle and domestic fowl. She lost all her property within a few minutes into the river gorge. Riverbank erosion destroyed her homestead, and she became landless and shelterless.

6.4.1.3 Land Ownership and Conflict for Land

The people who live in the floodplains lose homestead land as well as agricultural land after bank erosion. It needs a few years or decades to emerge the eroded land, and social conflict arises after accretion of eroded land. For example, before erosion, a farmer possesses land in one village or river island. However, the problem arises after land erosion and further accretion of the same land in different reemerged river islands (*char*). In several cases, it has been observed that many people claim the land as their own land. Accordingly, ownership clash ascends in society. During the field surveys, the displaced people claimed about the conflicts of accreted land. Among the respondents, 84 percent faced different types of ownership conflict. Hence, eroded land accretion becomes a dilemma for the people who lost their land once. When accretion occurs after thirty years of erosion, then it is considered as government land or *khas* land. Therefore, the owner has no legal authority to achieve the land after accretion. On the other hand, the local musclemen and land grabbers capture the accreted land. As a result, there is no hope for the helpless landowners to regain the accreted land. If the landowner claims the land, they are violently tortured by the 'powerful' land grabbers. Pieces of evidence are there that the landowners are hurt severely to injure to restrain them fighting against the local 'power'. In maximum cases, land grabbers and musclemen are local leaders, and they are politically empowered. They keep *lathyial bahini* (a group with physical strength and weapon to fight with weak landowners) to spread fearsome situation in the society and capture the reemerged land. The local leader controls riverine Bangladesh's socio-political system to get the reemerged land by *lathyial* organization (Lein 2009). Thus, social conflict arises with land grabbing in the *char* lands along with the riparian environments. Though erosion made the people destitute and displaced once, they have the confidence to get back the eroded land after accretion. But the landowner ruins through losing heart for the most hopeful accreted property due to the ill-treatment by land grabbing activities. The case study regarding ownership conflict has been illustrated in Case study 2.

Case Study 2

Md. Moslemuddin lives at Kolkonda union of Gangchara upazilla of Rangpur district. He is 42 years old agricultural day labour and several times displaced by the Teesta riverbank erosion. At present, he is landless, and his two brothers migrated to Dhaka city forever. He described the death of his father for conflict of emerged land from the river bed. Once his father had 2 acres of land, and afterward, the land was eroded. After few years, 0.9 acres of land emerged from the river, but the land grabbers captured the land. His father claimed for the land and was frequently injured by the *lathyials* of the land grabbers and passed away. At that time, Md. Moslemuddin was a little boy, and his brothers were also brutally beaten by the musclemen. After the shocking situation, his brothers migrated to the city, and they never think about coming back to claim the land. Even Md. Moslemuddin also does not claim for the accreted land ever.

6.4.1.4 Hamper in Social Peace and Happiness

The study examined that social peace and happiness declined due to living standard fall, broken family after economic disruption, and a decrease in social bonding and relationship after displacement. The displaced families move from one place to another place and have no feelings of peace and happiness in their uncertain life. Hence, they have to live away from relatives and have no one to share their stress and anxiety. Riverbank erosion prompts people to involuntary migrate and moves away from kinfolks (Saleheen 1991). The displaced people who cross a shorter distance and rebuild the houses in nearby villages, char and union have bonded with relatives. But once displaced people migrate to urban areas, they have no connection with relatives. The respondents argued that many of them have no communication with cousins and other relatives for a long time, who migrated to big cities once. Therefore, displacement causes a decrease in social bonding along with a hamper in social peace and happiness. The social structure breaks down, followed by riverbank erosion displacement. The multiple response analysis of hamper in social peace and happiness indicates that 39.1 percent of respondents opined about the collapse of the standard of living. The economy of the study area is predominantly agriculture-based. Land loss due to bank erosion causes a decreasing trend of employment opportunities. Destruction of occupation in a sudden by riverbank erosion brings sub-standard life to the affected people. The respondents argued that before displacement and migration, their living standard is much better than the present. Living standard falls after the loss of income, and it never becomes equivalent to the past earnings (Islam and Rashid 2011). Join family breaks down into single-family after erosion occurrences and shifts in rural areas, towns and cities for income generation and seeking a new shelter. Consequently, the declining trend of social bonding and relationship creates an impasse in society. The study finds out that 32.7 percent of respondents answered about ruined families and 28 percent replied for a decrease in social bonding and relationship impacts on social peace and happiness (Fig. 6.5).

6.4.1.5 Isolation from Own Society and Culture

Shreds of evidence collected from the field investigation denote that homestead damage after erosion makes people take decisions instantly regarding resettlement. In the case of homestead shifting in an unknown location, people have no alternative option without shifting of family. Sometimes mass displacement follows when massive erosion attacks a village or union. While devastating erosion occurs in a locality, people immediately decide to migrate anywhere to take shelter. Thus, the displaced people separate from their locality, relatives, and society. On the other hand, individual migration is also responsible for isolation from their own society and culture. The field investigation also implies that many male family heads migrated to cities and have no connection with their own family and resettled there with another marriage. Dislocated populations face difficulties for isolation through social network destruction and separation from their own families (Arobi et al. 2019). The study

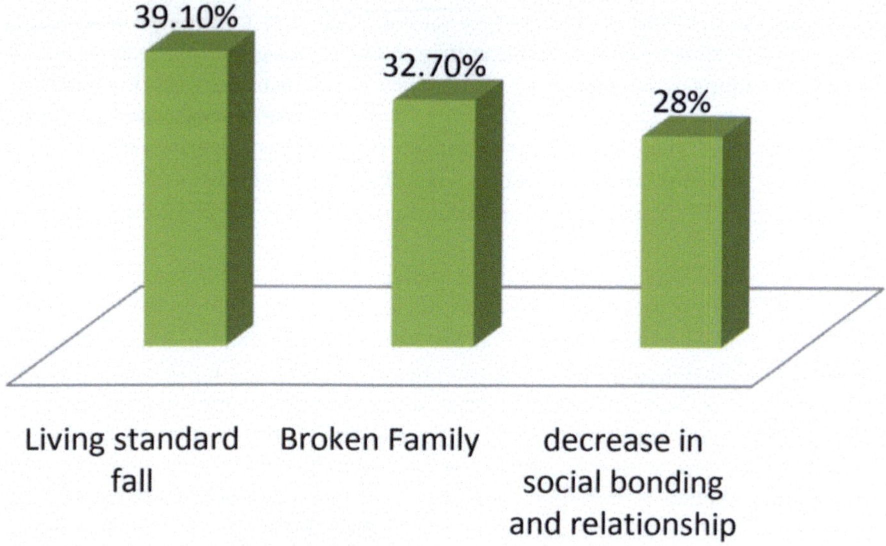

Fig. 6.5 Hamper in social peace and happiness (*Source* Field survey 2017)

unveils that people resettle houses close to their previous homestead for regaining the eroded land. However, another fact is that the displaced try to live with relatives and their own society. Furthermore, most of the people who are uneducated and unskilled, are unable to do jobs with higher skills rather than farming and day labouring. Considering all the factors, people try to live to the source of land where they were born or close to the inherited land. Despite the severity of bank erosion, the people inhabited in the banks disagree to shift their house for social networking and to continue the river-based livelihoods (Dun 2011). But when several cycles of displacement occur in a family, they become isolated from their own society and culture and moves towards an uncertain life in an unknown area. The displaced people feel anxiety without having their own society to live in a fearsome condition at a new shelter. The study result depicts 89 percent of respondents replied feeling isolated apart from their own society and culture after displacement. The respondents of the study mentioned the breakdown of the strength of unity after dislocation and resettlement.

6.4.1.6 Interruption in Education

The occurrence of erosion brings immense damage to the sector of education. Children become deprived of education after a long break in the study due to erosion and flood in their locality. Furthermore, every year a lot of educational institutions break down into the river womb due to erosion. Interruption in education after erosion through destruction of the transportation network is also a frequent event in the study area. During the field survey, a temporary school, made with corrugated iron sheets, has been observed in Bidyananda union (Fig. 6.6a) because the two-storied school building was plunged into the river. In the newly reconstructed school building, there are shortages of accessories, furniture, and educational tools such as desks, chairs, boards, etc. Therefore, the environment of the school deprives the students of the school, and thus, erosion causes severe obstacles in education. Alternatively, the environments of displaced households are not favorable for children to study at home.

Moreover, the affected family becomes unable to send their children to school due to homestead damage and economic insolvency. Hence, the children stop education at the primary level and never go to school after displacement. Another problem arises in educational institutions because, in the flood and erosion-affected areas, the schools are used as flood shelters during and aftermath of flood and erosion to resolve the temporary shelter management of river displaces. Therefore, schools are closed for weeks to months, which interrupts the educational environment. Likewise, the displaced take shelter in the school building with domestic fowl and livestock that causes bad smell through stool and urine. Figure 6.6b illustrates the school building used as a flood and erosion shelter both for displaced populations and livestock. When the school reopens after the disaster, it becomes a great issue for the school committee to regain the past healthy environment of education due to dirt in the school rooms, lawns and toilets. The study result unveils that about 49 percent of enrolled students' education hampers for the economic destruction after erosion, whereas, 27 percent of students' education is hampered due to the infrastructural loss, and another 24 percent gets interrupted in education for unhealthy educational environment in school after erosion.

Fig. 6.6 **a** Reconstructed new school after erosion and **b** school building used as flood and erosion shelter (*Source* By the author)

6.4.2 Fragile Economic Condition and Migration

The economy of the study area is dominated by agriculture. Lack of education and any professional skill, the community of the study unions are mostly dependent on agricultural activities, day labouring and livestock and cattle rearing. However, when river bank erosion occurs, it destroys the agricultural sector and most people's economic activities. In fact, these incidents motivate people to displace and migrate in the search for alternative jobs. Provoked by fragile economic conditions, people change occupation, and that impact on solvency situation, increases hunger and poverty within the victimized communities.

6.4.2.1 Dislocation and Change in Earning Sources

The earning source changes after agricultural land loss and dislocation. Sometimes people become unemployed for several months. Damage of occupation after erosion and displacement brings forward many issues, including unemployment, change in income generation source, alteration in economic condition, etc. Individual family shifts house in different location according to their economic condition in that brittle environment. Besides, the choice of a new destination also depends upon available income earning sources. After losing agricultural land, farmers resettle houses where lands are available for sharecropping. The study identified that before erosion and dislocation, most of the study area people engaged in their own farming, but after the disastrous impact, the large farmers initiate cultivation through sharecropping and leasing others' land. However, the small or marginal farmers try to cope with cattle rearing, petty business and wage labouring. Though farmers of large land-holding size have been affected by riverbank erosion, the small landowners have been much affected by erosion-induced hazard (Hutton and Haque 2004). The study also observed a seasonal unemployment crisis. The families with unemployed household-head face a serious food crisis. Seasonal unemployment and joblessness impacts on household food security and nutritional status (Elahi and Ara 2008). In response to bank erosion-induced displacement and occupational change, the wage labour of farmlands engage themselves in both farming and non-farming wage labouring such as Rickshaw/van pulling, luggage pulling, dragging agricultural commodity, labouring in the brickfield, working in garment factory, etc. The change in occupation has been illustrated in Table 6.5.

6.4.2.2 Migration for Occupation

In the study area, migration is a well-known strategy after the loss of livelihood. Usually, the working population ranging from the age group of 15–50 migrates for job search. However, when there is no means to survive in a brutal environment

Table 6.5 Change in income generation source/ occupation

Occupation (before displacement)	Occupation (After displacement)
Farmer (Farming in own land)	Unemployed, sharecropping, farming in leased land, cattle rearing, petty business, wage labouring
Wage labour (in farm)	Unemployed, wage labouring in both farm and non-farm activities
Business	Unemployed, petty business, grocery, wage labouring

Source FGD 2017

due to bank erosion, then people migrate with family and engages wives and children in work. Age, gender and structure of family play a significant role to control the decision of migration for occupation (De Jong et al. 2002). The study finds out step migration in response to riverbank erosion and adjustment with occupation. At first, the displaced cross a shorter distance, then relatively longer, and finally longest distance from the source region for different socio-economic factors. Displaced population moves firstly in rural areas, then in embankments, afterwards in nearby *char* and lastly in the cities (Hossain 1991).

Joblessness due to erosion pushes the displaced people in cities and towns to search for earning sources. Both temporary and seasonal migration occurs in the study unions for job-seeking after displacement. The multiple response analysis of migration for occupation represents that 53.1 percent respondent follows seasonal migration, and 46.9 percent answered for temporary migration (Table 6.6). Seasonal migration shows the highest percentage because most of the respondents migrate to the countryside for agricultural labour during the harvesting period. Apart from that, the respondent also temporarily migrates to the town or big cities for earning. The respondents replied that the people of the study area temporarily migrate to earn their living through rickshaws or van pulling in the urban areas. However, several times,

Table 6.6 Migration for earning/job seeking

Union	Migration for earning/job seeking					
	Temporary migration		Seasonal migration		Total	
	f	%	*f*	%	*f*	%
Purbachatnai	25	43.9	32	56.1	57	100
Saulmari	56	47.4	62	52.5	118	100
Bidyananda	37	48.0	40	52.0	77	100
Khuniagachh	68	48.5	72	51.5	140	100
Kolkonda	60	48.3	64	51.6	124	100
Sindurna	22	44.8	27	55.2	49	100
Belka	58	45.0	71	55.0	129	100
All	326	46.9	368	53.1	694	100

Source Field Survey 2017 (Percentages and totals are based on responses)

displacement and frequent loss of occupation for erosion make people permanently displaced and resettle in urban areas. People choose temporary migration for survival, but extreme erosion leads them to make decisions towards permanent migration (Islam et al. 2017). The permanently migrated family or individuals work as garment workers in a readymade garment factory, maidservants at house or earn a living through rickshaw pulling. On the other hand, the people who have no permanent job, engage themselves in drug smuggling, woman and child trafficking and different anti-social activities.

However, in the case of seasonal migration, people choose agricultural day labouring and working as labour in brickfields. Though day labouring in agriculture and the brickfield is ill-paid, the displaced people choose these jobs due to seasonal joblessness and skillfulness. During summer, autumn and late-autumn season, the folk of unemployed male household heads and young male family members of the Teesta river displaces, migrate in rural areas of Dinajpur, Bogra, Noagaon and Comilla district for paddy harvesting. The study result depicts 33.4 percent of respondents migrate in summer, and 66.6 percent migrates in autumn and late-autumn season in the study area for paddy harvesting. The percentage of migration in autumn and late autumn is high due to food shortage, or *Monga* syndrome becomes critical in that season after erosion. Likewise, migration for day labouring in brickfields has been observed in the winter season. The villages of the Teesta river banks seasonally turns into the locality of women when the male migrates for the job in different places. Case study 3 depicts the condition of dislocation and migration in the study area, which is almost the same history as most of the affected families of riverbank erosion.

Case Study 3

Tohubar Hossen was the owner of 5 acres of land and lived at the Kolkonda union of Gangachara upazilla of the Rangpur district. He is 52 years old. At present, he is landless and lives in a squatter in Dhaka city. He migrated to the city thirteen years ago. After the land was lost for riverbank erosion, he resettled his house into the Gajghanta union of the same upazilla. Then, he has no available job opportunity in his surroundings and no own land to take shelter. He built a house on government acquired land of the Teesta river accretion. Just then, he earned his living through seasonal migration in the areas where the labour of paddy harvest was needed. In the Gajghanta union, he shifted house two times for erosion. He suffered much for livelihood and was unemployed for a long time. He faced extreme poverty after homestead shifting and loss of income opportunities. At that time, every morning, he went to Rangpur city and worked as construction labour. He returned home in the evening. He became a temporary migrant for livelihood. Tohubar Hossen mentioned that to reduce the transportation cost, he decided to resettle in Rangpur city with his three children and wife. He lived for 3 years in Rangpur and worked as a rickshaw puller. But his income was not satisfactory to fulfill the basic needs of the family, and then he shifted to Dhaka along with his family and lives at Kuril slum. Now he works as a night guard of a shop. At first, Tohubar Hossen followed seasonal migration, then temporary migration and step migration, and finally permanent migration for his job destruction and house damage owing to riverbank erosion.

6.4.2.3 Worsening Poverty Situation

The study area is one of the significant poverty pockets of Bangladesh. Loss of land and occupation impoverishes the displaced people and intensifies poverty. Likewise, poverty upsurges in cyclic order after each erosion and displacement. The growing trend of unemployment, lack of household assets and insufficient food consumption are the signs of aggravation of poverty of the displaced community of the Teesta riverbank erosion. The study displays the multiple response analysis of the poverty situation of the displaced population of the study unions (Table 6.7). The highest responses regarding the situation of poverty have been pointed out by the respondents are the growing trend of unemployment 37.4 percent (N = 373). The second-highest poverty situations experienced by the respondents are insufficient food consumption 34.8 percent (N = 347) and lack of household assets has been opined by 27.7 percent (N = 276) of respondents. As unemployment increases due to livelihood damage, it is worthwhile to recognize that the solvency situation of the displaced community deteriorates, which indicates the acceleration of the poverty situation. The study result unveils that the solvency situation of the respondents decreases after erosion and displacement. Before erosion and displacement, 45.1 percent of respondents

Table 6.7 Situation of poverty

Union	Situation of poverty							
	Growing trend of unemployment		Lack of household assets		Insufficient food consumption		Total	
	N	%	N	%	N	%	N	%
Purbachatnai	33	39.3	19	22.7	32	38.0	84	100
Saulmari	62	37.3	44	26.5	60	36.1	166	100
Bidyananda	41	39.4	26	25.0	37	35.6	104	100
Khuniagachh	72	36.4	57	28.8	69	34.8	198	100
Kolkonda	64	37.4	50	29.2	57	33.3	171	100
Sindurna	30	37.5	26	32.5	24	30.0	80	100
Belka	71	36.8	54	28.0	68	35.2	193	100
All	373	37.4	276	27.7	347	34.8	996	100

Source Field survey, 2017(Percentages and totals are based on responses)

were solvent, but at present, only 2.6 percent are solvent. In contrast, at present, 97.4 percent of the victims responded about economic insolvency.

6.4.2.4 Hunger and Food Crisis

The community of the Teesta floodplain battle against hunger and malnutrition and they become needy by repeated displacement. The study reported that with few exceptions (the large farmers or service holders), most of the displaced families encounter with hunger in their entire life. A serious food crisis arises during the post-monsoon months, September–November when people become unemployed after erosion and have no money to buy food. The condition of the food crisis in the study area is locally known as '*Monga*'. The community of the Teesta- Brahmaputra floodplains meets acute food crises annually (Elahi and Ara 2008). Scarcity of food due to joblessness is a sign of *Monga*. The study unions are *Monga* prone, and in the condition of the severe food crisis, people have no option but starvation or skipping a meal in a day. Consuming food once daily, avoiding meals and sharing food by turn with family members are the strategies followed by the *Monga*-prone community of Northern Bangladesh (Baqee and Azim 2019). During FGDs, people mentioned most of them starve for one or two days in an unemployed situation or take a meal by turn with the household associates. The study result shows that about 25.3 percent of respondents consume food two or three times daily. Among the respondents, 74.7 percent consume food once daily, which pictures the real condition of hunger and poverty in the study area. Moreover, to meet up hunger, people collect any kind of edible leaf from the wilderness. Alleviation of hunger with wild vegetables minimizes the expenses of buying food (Sultana et al. 2018). Likewise, during and after resettlement, people face acute food crises due to a shortage of cash and unemployment. Little savings of the victimized households is finished in the reconstruction of the house. Therefore, they have to face a food crisis until having a new income-earning source. The situation exists year after year, and malnutrition becomes a long-lasting crisis among destitute people.

6.4.3 Impact on Health and Environment

The food crisis and hunger of the displaced population affect the condition of nutrition and health of the community. The study further discloses that a community with no capacity to buy food is also unable to take care of health. Likewise, dislocation and involuntary migration adversely impact the environment, including squatter development in the embankments and enlargement of a squatter in cities and towns. In Bangladesh, the primary reason for the floating population in the urban centres is migration. The displaced community increases the population pressure of cities. Thus, the enlargement and growth of multiple squatters in the cities occur. Similarly, various criminal activities increase by the river displaces owing to unavailable job opportunities.

6.4.3.1 Low Expenses in Health Care

The devastating situation of flood, erosion and different natural disasters cause disease as well as ill-health conditions. But the affected people are not conscious enough about health because the respondents replied that they do not go even to the village doctor if they are not seriously affected by the disease. Common diseases such as fever, cough and diarrhoea, which require serious treatment from physicians, are often neglected by the community. The respondents mostly depend on quacks to spend a minimal amount of money on health care due to flood and erosion-induced illness and disease. About 51.2 percent of respondents expend 100–300 taka (Currency of Bangladesh), 42 percent pay 301–600 taka and 6.8 percent spend more than 600 taka for treatment of disease of their families. Sometimes the people of the study area trust *kobiraj* (rural unskilled people dealing with treatment who have no institutional training in medical education) to cure disease. Self-treatment with different medicinal plants and herbs is also observed in the study area, which is recognized as a cost reduction technique. The field survey result also indicates that people avoid upazilla health complex for treatment to diminish the transportation cost. To quote one of the respondents, ‘we starve due to shortage of money then how can we manage the expenditure to receive better treatment facilities’?

6.4.3.2 Squatter Development in Embankments

The displaced population considers the embankment as the land of rehabilitation. Thus, settlement has been recognized in the Teesta river embankments. The study finds out that about 33.1 percent of the displaced population lives in the embankments of the Teesta river. Similarly, squatter development in the embankment causes damage to the flood defense technique and increases the vulnerability of further flood and erosion. Likewise, the concrete blocks of the embankment have been stolen by the local people and by the displaced community, which is a great threat to riverbanks. Due to resettlement in the earthen embankment emerges the issue of degradation of the river protection initiatives as well. The mud of the earthen embankment also has been used by the resettled people for different purposes. Therefore, squatter development in embankments increases the susceptibility to breach in the protective measure and intensifies further erosion and displacement of the destitute community. Moreover, the environment of the embankment deteriorates with domestic garbage and hanging toilets.

6.4.3.3 Expansion of Squatter in the Urban Areas

A large portion of riverbank erosion victims of the Teesta resettles their houses in Dhaka, Chittagong, Khulna, Rangpur and other urban areas of the country. The squatters in the urban areas are expanding due to the riverbank erosion victims of the Padma, Jamuna, Meghna, the Teesta and other erosive rivers of Bangladesh. Squatter development in cities and towns of Bangladesh largely depends upon the involuntary migrants of riverbank erosion, and 10–25 percent of erosion victims move in the urban areas (Hutton and Haque 2004). The ultimate result of squatter development in the city deteriorates the urban environment of Bangladesh. The growth of squatter in the urban area increases population density, pollution and pressurizes employment opportunities. The unemployed dwellers of squatter often get involved in the drug business, drug addiction, woman, child trafficking and different similar illegal offenses that exaggerate violence in urban society. Therefore, displacement and migration enforce urban crime, poverty, environmental degradation as well as enlargement of squatter, but the resettled community deprives from urban amenities. Though, the migrants pushed into the cities with the hope of employment, a dwelling place and a better living environment. No doubt that a healthy living environment is beyond the means of the resettled river bank erosion victim in urban squatters. River displaces get poor access to social facilities such as safe drinking water, toilet and treatment services (Elahi and Rogge 1990).

6.4.4 Psychological Consequences

The existing study investigates the psychological consequences of dislocation and involuntary migration. Though the migration practitioners have not paid the psychological consequences of dislocation much attention. The disrupted community also neglects the psychological consequences of erosion-induced dislocation. The destruction of property and resettlement makes people mentally shocked. They often talk about their past wealth and homestead, which they achieved from the previous generation. Displacement for riverbank erosion and resettlement has short-term as well as long-term effects on human psychology (Arobi et al. 2019). Psychologically a displaced family and its member get a great shock after the dramatic fall down of the economy. The household head absorbs tremendous pressure and psychological stress for damaged households, loss of agricultural land, loss of crops, and spoiling of most desired employment opportunity. Besides, the woman of the study area gets immense psychological stress when she comes to know about the permanent migration and second marriage of her husband in an urban area. Hearing the misfortunate news, the displaced woman gets mentally distressed and, along with her children, falls into a misfortunate condition. The respondents mentioned their anxiety regarding resettlement, migration and search for a new job. Sometimes, they become frustrated after migration due to unavailable employment opportunities. The study investigates about 63 percent of displaced people replied regarding mental stress, 31 percent mentioned anxiety and 6 percent respond to traumatic disorder after erosion and displacement. Continuous mental stress, anxiety and frustration bring traumatic disorder among the affected individuals. During FGDs, the respondents replied that they have no soil under their feet and no roof above their heads after erosion. They do not know where to go, where to resettle, what to do and what to eat the next day? Thus, the economic condition, as well as the mental condition of the river displaces breaks down all of a sudden. The displaced population becomes mentally depressed after the loss of belongings and anxiety for the future makes them traumatic patients. However, mental trauma and disorder is such a condition that cannot be seen in normal views rather than great physiological disorder appears.

6.5 Recommendations and Concluding Remarks

The above-mentioned research findings demonstrate the appalling condition of the river displaces, which impacts both rural and urban society in all aspects. The recommendations of the present study are.

- Set up of industry in the study area to ensure job facilities for the involuntary migrants.
- Encourage the river displaces in income intensification in their locality to reduce occupational migration.
- Skill development and training for non-farm activities to discourage occupational migration.
- Boost the affected farmers for further agriculture through agricultural relief facilities such as seed, fertilizer and modern equipment.
- Farmers should motivate to grow short-duration crop varieties to reduce economic loss.
- Proper planning and implementation of government rehabilitation programs for the displaced population of riverbank.

The present study identifies homestead shifting, change in income-generating source, isolation from own society, squatter development, psychological stress, etc. are the growing concerns regarding the consequences of dislocation and involuntary migration. The study explores that these variables have interrelationships and complex linkages with different factors such as social, economic, environmental, psychological and others. Additionally, displacement and migration are not an issue of only the Teesta river basin. However, it is observed in entire northern Bangladesh occupied by the mighty Brahmaputra, Dharla, Dudhkumar and other rivers. Therefore, it is essential to save the homestead, property and occupation of the affected community through different government initiatives. Likewise, the displaced community should be motivated and awaked about not resettle houses in the embankments and not damaging the river protection measures. Though flood and riverbank erosion protection initiatives have been perceived in the study area, the government should take an urgent step towards river training in the frequent erosion-affected areas along with proper maintenance of the existing embankments. Livelihood development and rehabilitation of the river displaces needs greater attention by the government as well as the non-government sector. Moreover, industrialization, modernization of agricultural and land reform accompanied by skill development programs should be executed in the study area to meet the homestead and occupational displacement challenges. Access to low-interest bank loans for the agrarian society will substantially reduce the consequences of displacement and involuntary migration. Hence, low-interest bank loans and credit facilities should be implemented in the study area to combat occupational displacement or migration of the vulnerable community. The provision of employment opportunities around the year through the establishment of the agro-based industry will be the most effective way to motivate the riverbank erosion victim to stay in their source region. Thus, the adverse impacts of dislocation and migration will be reduced.

References

Ahmed KS (1991) Distance between two populations: Displaced and non-displaced in flood and erosion hazard areas. In: Elahi KM, Ahmed KS, Mafizuddin M (eds) River bank erosion, flood and population displacement in Bangladesh. Riverbank Erosion Impact Study, Jahangirnagar University, Dhaka, pp 138–150

Arobi S, Naher J, Soron TR (2019) Impact of river bank erosion on mental health and coping capacity in Bangladesh. Global Psychiatry 2(2):195-200

Baqee, A., and Azim, S. A. (2019).Mora-Kartik to monga the acute food shortage syndrome of North Bengal. In H. Khatun, A. Baqee and H. Kabir (Eds.), *People at risk: disaster and despair*. pp. (275–293). Dhaka: Disaster Research Training and Management Center (DRTMC), University of Dhaka.

Bernzen, A., Jenkins, J. C., and Braun, B. (2019). Climate change-induced migration in coastal Bangladesh? A critical assessment of migration drivers in rural households under economic and environmental stress.*Geosciences*, *9*(1), 51.

Chindarkar, N. (2012). Gender and climate change-induced migration: proposing a framework for analysis. *Environmental Research Letters*, *7*(2), 025601.

Das TK, Haldar SK, Gupta ID, Sen S (2014) River bank erosion induced human displacement and its consequences. Living Review of Landscape Research 8(3):1–35

Das TK, Haldar SK, Sarkar D, Borderon M, Kienberger S, Gupta ID, Kundu S, Guha-Sapir D (2017) Impact of riverbank erosion: A case study. Australas J Disaster Trauma Stud 21(2):73-81

De Jong GF, Chamratrithirong A, Tran QG (2002) For Better, For Worse: Life Satisfaction Consequences of Migration 1. Int Migr Rev 36(3):838–863

Dun O (2011) Migration and displacement triggered by floods in the Mekong Delta. Int Migr 49:200–223

Elahi KM (1991a) Impacts of river bank erosion and flood in Bangladesh: An introduction. In: Elahi KM, Ahmed KS, Mafizuddin M (eds) River bank erosion, flood and population displacement in Bangladesh. Riverbank Erosion Impact Study, Jahangirnagar University, Dhaka, pp 1–10

Elahi KM (1991b) River bank erosion, flood hazard and population displacement in Bangladesh: An overview. In: Elahi KM, Ahmed KS, Mafizuddin M (eds) River bank erosion, flood and population displacement in Bangladesh. Riverbank Erosion Impact Study, Jahangirnagar University, Dhaka, pp 95–109

Elahi KM, Ara I (2008) Understanding the Monga in northern Bangladesh. Academic Press and Publishers Library, Dhaka

Elahi, K. M. and Rogge, J. R. (1990).*River bank erosion, flood and population displacement in Bangladesh.A report on the river bank erosion impact study.* Dhaka: Riverbank Erosion Impact Study, Jahangirnagar University.

Haider SKU (2010) Factors of migration on urban Bangladesh: An empirical Study of poor migrants in Rajshahi city. Pakistan J Soc Sci (PJSS) 30(2):307–323

Haque CE (1991) Human responses to river bank erosion hazard in Bangladesh: Some lessons from indigenous adjustment strategies. In: Elahi KM, Ahmed KS, Mafizuddin M (eds) River bank erosion, flood and population displacement in Bangladesh. Riverbank Erosion Impact Study, Jahangirnagar University, Dhaka, pp 191–217

Hossain MZ (1991) Displaces of river bank erosion, in urban squatter settlements in Sirajganj: the process of impoverishment. In: Elahi KM, Ahmed KS, Mafizuddin M (eds) River bank erosion, flood and population displacement in Bangladesh. Riverbank Erosion Impact Study, Jahangirnagar University, Dhaka, pp 307–322

Hutton D, Haque CE (2004) Human vulnerability, dislocation and resettlement: Adaptation processes of river bank erosion induced displaces in Bangladesh. Disasters 28(1):41–62

Ishtiaque A, Ullah MS (2013) The influence of factors of migration on the migration status of rural-urban migrants in Dhaka. Bangladesh. Human Geographies 7(2):45–52

Islam MA, Parvin S, Farukh MA (2017) Impacts of riverbank erosion hazards in the Brahmaputra floodplain areas of Mymensingh in Bangladesh. Progress Agric 28(2):73–83

Islam MF, Rashid AB (2011) Riverbank erosion displacees in Bangladesh: need for institutional response and policy intervention. Bangladesh J Bioethics 2(2):4–19

Islam MS, Khatun H (2019) River bank erosion: A prolonged vulnerability threat. In: Khatun H, Baqee A, Kabir H (eds) People at risk: disaster and despair. Disaster Research Training and Management Center (DRTMC), University of Dhaka, Dhaka, pp 231–251

Iqbal S (2010) Flood and erosion induced population displacements: A socio-economic case study in the Gangetic riverine tract at Malda district, West Bengal. India. J Human Ecol 30(3):201–211

Iva TT, Hazra P, Faisal M, Saha S, Hossain S (2017) River bank erosion and its impact on population displacement in Bauphalupazila under Patuakhali district, Bangladesh. J Sci Technol Environ Inform 5(02):371–381

Karim MR, Muhammad N, Anne DS, Narayanarao B (2017) Povertys change challenges and coping strategies of small scale farm household. Int J Agric Ext 5(1):87–96

Lein H (2009) The poorest and most vulnerable? On hazards, livelihoods and labelling of riverine communities in Bangladesh. Singap J Trop Geogr 30(1):98–113

Mahabub AQM, Islam N (1991) Urban adjustment by erosion induced migrants to Dhaka. In: Elahi KM, Ahmed KS, Mafizuddin M (eds) River bank erosion, flood and population displacement in Bangladesh. Riverbank Erosion Impact Study, Jahangirnagar University, Dhaka, pp 323–336

Mallick B (2011) Migration as an adaptation strategy and its consequences on coastal society: Experience from Bangladesh. (COMCAD Working Papers. 96). https://nbn-resolving.org/urn:nbn:de:0168-ssoar-405487

Mallick S (2016) Identification of fluvio-geomorphological changes and bank line shifting of river Bhagirathi-Hugli using remote sensing technique in and around of MayapurNabadwip area, West Bengal. Int J Sci Res (IJSR) 5(3):1130–1134

Mollah TH, Ferdaush J (2015) Riverbank erosion, population migration and rural vulnerability in Bangladesh (a case study on Kazipurupazila at Sirajgonj district). Environ Ecol Res 3(5):125–131

Mondal, M. S. H., Murayama, T., and Nishikizawa, S. (2020). Assessing the flood risk of riverine households: A case study from the right bank of the Teesta River, Bangladesh. *International Journal of Disaster Risk Reduction, 51*, 101758.

Mondal, S. H., and Islam, S. (2017). Chronological trends in maximum and minimum water flows of the Teesta River, Bangladesh, and its implications. *Jàmbá: Journal of Disaster Risk Studies, 9*(1), 1–11.

Parkins NC (2010) Push and pull factors of migration. American Rev Political Econ 8(2):6–24

Perch-Nielsen SL, Bättig MB, Imboden D (2008) Exploring the link between climate change and migration. Clim Change 91(3–4):375–393
Rabbi H, Saifullah ASM, Sheikh MS, Sarker MM, Bhowmic AC (2013) Recent study on riverbank erosion and its impacts on land displaced people in Sirajgonj riverine area of Bangladesh. World J Appl Environ Chemistry 2(2):36–43
Rahman MM, Arya DS, Goel NK, Dhamy AP (2011) Design flow and stage computations in the Teesta River, Bangladesh, using frequency analysis and MIKE 11 modeling. J Hydrol Eng 16(2):176–186
Rana MS, Nessa AM (2017) Impact of riverbank erosion on population migration and resettlement of Bangladesh. Sci J Appl Math Stat 5(2):60–69
Saleheen MU (1991) The mobility characteristics of displaces: A case study from river bank erosion hazard area. In: Elahi KM, Ahmed KS, Mafizuddin M (eds) River bank erosion, flood and population displacement in Bangladesh. Riverbank Erosion Impact Study, Jahangirnagar University, Dhaka, pp 160–169
Sarker MNI, Ali MA, Islam MS (2015) Causes and possible solutions of poverty perceived by char dwellers in Bangladesh. Int J Nat Soc Sci 2(1):37–41
Sarker, M. N. I., Wu, M., Alam, G. M., and Shouse, R. C. (2020). Life in riverine islands in Bangladesh: Local adaptation strategies of climate vulnerable riverine island dwellers for livelihood resilience. *Land Use Policy*, *94*, 104574.
Shamsuddoha, M., Khan, S. M. M. H., Raihan, S., and Hossain, T. (2012). Displacement and Migration from Climate Hotspots: Causes and Consequences. *Center for Participatory Research and Development and Action Aid Bangladesh, Dhaka: December.*
Siddique QI, Chowdhury MMH (2000) Flood'98: Losses and Damages. In: Ahmad QK, Chowdhury AKA, Imam SH, Sarker M (eds) Perspectives on Flood 1998. The University Press limited, Dhaka, pp 1–13
Sultana MR, Hassan MZ, Paul SK (2018) Mitigation of erosion-induced disaster through indigenous practices: evidence from rural community of the Teesta river bank. J Life Earth Sci 13:23–31
Sultana P, Thompson PM, Wesselink A (2020) Coping and resilience in riverine Bangladesh. Environ Hazards 19(1):70–89
Warner K, Hamza M, Oliver-Smith A, Renaud F, Julca A (2010) Climate change, environmental degradation and migration. Nat Hazards 55(3):689-715
Zachariah KC, Mathew ET, Rajan SI (2001) Social, economic and demographic consequences of migration on Kerala. Int Migr 39(2):43–71

Mst. Rebeka Sultana is an Assistant Professor with 10 years of teaching experience in the Department of Geography and Environmental Studies, University of Rajshahi. She was graduated in 2006 and completed her M.Sc. in 2008 from the same Department and University with a first-class first position in both the examination. She was awarded a gold-medal for her exemplary academic performance. Dr. Sultana was awarded 'Bangabandhu fellowship' from the Bangladesh Council of Scientific and Industrial Research (BCSIR) for Ph.D. Her research interests are in Fluvial Geomorphology, Fluvial Disaster and Climate Change. Her specialization is in river morphology change, impact and adaptation. She published articles on fluvial disaster, impact and adaptation in Northern Bangladesh. Dr. Sultana is passionate about teaching and learning and spends her time with academic research. Her teaching areas are Geomorphology, Environment, Biogeography, Oceanography, Flood and River Management.

Chapter 7
Micro-Spatial Analysis of Rural Accessibility for Imbricating Regional Development: Exemplifying an Indian District

Syfujjaman Tarafder and N. C. Jana

Abstract Good road connectivity followed by efficient transport options, which help them move and get access to themselves and the various services and opportunities, provide good access opportunity to various services and thus widen the scope of development. This access to various services and opportunities can be measured by distance, time, effort, and cost. The accessibility measured in terms of physical distance from all-weather roads is called physical accessibility. On the other hand, when accessibility is measured in terms of travel time to the highest order, centre and independent places are called relative accessibility. Both physical and relative accessibilities are important for rural people. Not only that, accessibility to basic social and administrative services is also important for balanced development. On the other hand, development has many connotations and be measured in a variety of ways. Principal Component Analysis and Z-score techniques are used to measure development. In this paper, an attempt has been made to measure the level of accessibility in various dimensions and correlated with development parameters so that micro-spatial regional planning strategies can be devised. The study finds a moderate correlation between development and accessibility parameters that need attention while considering rural development planning.

Keywords Habitation · Accessibility · Overall Accessibility Index · Weighted Facility Index · Village Priority Index · Euclidian distance

7.1 Introduction

Accessibility, which depends on mobility (ease and frequency of movements) and proximity (distance), can be improved by improving proximity to services (Starkey et al. 2002). Here the concept of basic accessibility arises. Basic accessibility is

S. Tarafder (✉)
Gour Mahavidyalaya, Malda, West Bengal, India

N. C. Jana
Department of Geography, The University of Burdwan, Burdwan, West Bengal, India

M. Mishra et al. (eds.), *Regional Development Planning and Practice*,
Advances in Geographical and Environmental Sciences,
https://doi.org/10.1007/978-981-16-5681-1_7

the minimum level of transport infrastructure and services required to sustain the socio-economic activity that can be equated with the provision of basic health and education facilities in any area (Lebo and Schelling 2001; Islam 2017). The factors that determine the accessibility are the location of facilities (distance from households), infrastructure (good roads with bridges), means of transport (walking, various IMT, e.g. cart, animal back, bicycle, small motorized vehicles, etc.), and transport service (commercial passenger and goods vehicles (Dennis 1998). The transport service increases the accessibility to the external facilities as well (Dennis 1998). The external facilities like special health service, higher education service, legal and communication services, and market need good transport service for increasing the level of accessibility and the overall agricultural development. Thus, to determine the level of accessibility to the external facilities, the transport service provider can be used as the weight for better accessibility measurements (Tarafder and Jana 2020). In this chapter, for determining the accessibility of external facilities, actual distances of the external facilities are multiplied by the Transport Service Score (TSS).

As mentioned by Lebo and Schelling (2001), accessibility may be no access, partial access or full access. In many developing countries, a substantial part of rural areas do not have motorized access, and Murshidabad is no exception. So it is urgent to measure the level of accessibility of Murshidabad district at the habitation level and correlate it with the level of various socio-economic parameters for understanding the relationship between the accessibility parameters and the socio-economic parameters. Habitation level accessibility reflects the problems faced by the villagers in accessing basic or higher needs (Shalini et al. 2015) and thus be helpful in micro spatial planning strategies to come into force (Castro et al. 2019). Tarafder and Jana (2020), in their article, investigated this relationship and found a significant relationship (Pacione 1995; Nutley 1999).

7.2 Objectives

Keeping in mind the interdependence of development and accessibility and their importance on the regional developmental planning strategies following objectives are set in the present research:

- Measuring the spatial variation of development at the village level;
- Measuring the level of accessibility in terms of physical distance, transport service provision, network distance;
- Measuring the accessibility to social and public facility services in terms of physical distance and weighted facility index;
- Finding out the relationship between accessibility and development for better regional planning strategies.

7.3 Scope and Limitation of Study

The scope of studying rural transport concerning accessibility and development is wide and there are three different components, e.g. transport, accessibility and development each of which has wider connotations and these are interconnected too. Thus the methodologies generally used to study these three components are manifold, and their results also vary widely. So there are many areas of research. From the Indian perspective, many of the methodologies developed in the urban industrial world have limited applicability. However, the methodologies developed based on third-world countries have a broader scope to test in the Indian context.

The present study provides a wider scope for regional transport study in an integrated way with development at the micro-spatial level. It is often necessary to find the pockets of development and underdevelopment at the smallest spatial level to help the planners to choose the right plan for the right areas. The present study meets that scope.

The study has limitations too. In the present study, only the distances from different road categories are considered to determine the accessibility to the basic public and social services. The quality of service provided by the service centres and the number of people served by them are kept aside due to methodological constraints. Various modes of transport used by people, the time taken to reach a facility, which depends on the type or nature of the transport service, are also not considered. For physical accessibility, Euclidian distance is used, actual transport distance and the transport impedances (like rivers and creeks without bridge) are ignored, although for determining the accessibility to various services, actual road distance is considered.

7.4 Rationale Behind the Selection of Study Area

For the present study Murshidabad district of West Bengal has been selected as the study area. The rationale behind selecting the study area is its rural background because about 80.3% of the population lives in rural areas (Census of India, 2011). The district has a third highest decadal growth rate (from 2001 to 2011) and much higher than the state's average, which is 13.8%. The literacy rate is 66.6% and ranks 16th position within the state. The work participation rate is only 36.5% in the census year 2011 and ranks 15th within the state. The literacy rate of Murshidabad is always lower than the State's average and ranks 16th since 2001. Female literacy was 47.6% in 2001, and in 2011, it is raised to 63.1%. In terms of development, the district is lagging. It ranked 15th out of 19 districts of West Bengal in human development (Human Development Report, 2004). The situation is not improved much in due course, and in terms of rural development, the situation remains the same (Tarafder and Jana 2016).

The rural connectivity by all-weather roads was not good (about 35% of rural habitation was unconnected by all-weather roads as of 2001), and slow growth of road development has been observed in public sector road construction (PWD and PW (Roads), 2018). However, considerable progress is observed since the initiation of PMGSY in the year 2000. Thus, a district with a high rural base and a low level of development struck us to research it. Besides, during our informal visits in the villages of different Blocks of the district, we found traditional inefficient transportation in both goods and passenger transport which also inspired me to study this issue further.

7.5 Database

Information in the form of maps, images, documents, reports, data sheets, etc. is used for this study and collected from various sources. For the preparation of base map at Mouza level (Revenue village) and Gram Panchayat level (Village level Local Government) Community Development Block maps of the Census of India, Block level Land use map of Land and Land Reform Department, Government of West Bengal has been used. The boundaries of the Mouzas are verified by checking and comparing all the maps collected from the above-mentioned three available sources. For the Gram Panchayat boundary, the GP boundaries demarcated by the Public Health Engineering Department, Government of West Bengal and the boundaries demarcated in the Bhuvan Panchayat Portal are considered.

Land use map of Land and Land Reform Department of Government of West Bengal, CD Block map of Census of India, 2011; and Core Network Map prepared by West Bengal State Rural Development Agency, Murshidabad have been used for infrastructure mapping taking Google Earth as base map. All block-level map-related entries are thoroughly verified in Google Earth. Besides, the list of schools of the School Education Department and the list of the hospitals of the Department of Health and Family Welfare, Government of West Bengal has been utilized in this process. The list of Panchayat of Murshidabad district has been collected from the Integrated Management Information System, Ministry of Drinking Water and Sanitation, Government of West Bengal. Field verification has been conducted as and when required.

For demographic and amenities data, census handbooks have been used. Besides, the database of the *Prime Minister's Gramin Sadak Yojana* has also been used to compute the connected habitations, the population covered by each road link etc. Village level database has been prepared and integrated with the Arc GIS database, and necessary analyses are made.

7.6 Methodology

The present study follows empirical research methods in all respects because in every case, the research outcomes are generated based on real field observation and data. Thus in the methodology, it is inductive too. Various methods and tools are used from

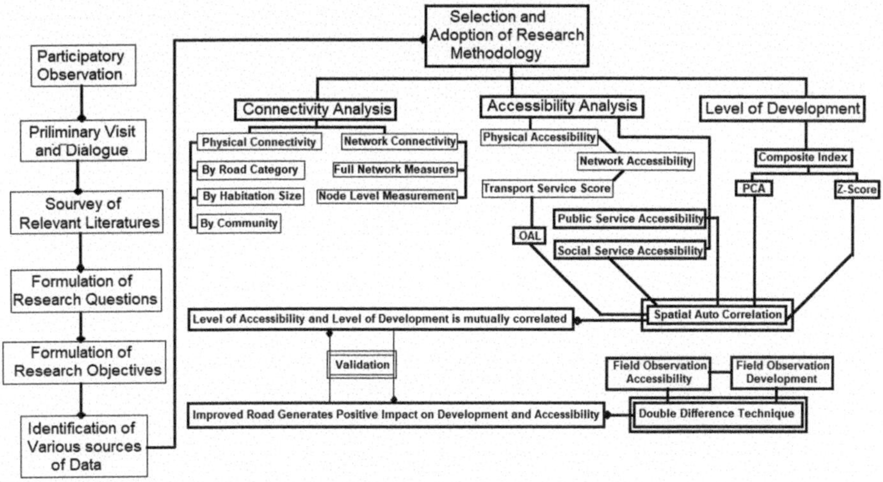

Fig. 7.1 Methodology flow chart (*Source* Prepared by the authors)

Cartography, Remote Sensing and GIS, Statistics for measuring different parameters, and developing composite indices for fulfilling the research objectives. The detailed methodology is described as a flow diagram in Fig. 7.1.

7.6.1 Measuring Level of Development

The development has many dimensions, and diverse methods are used to determine. In the present paper, the level of transport development, socio-economic development and its spatial variation are measured. Spatial variation of development or deprivation can be computed in different ways, and the findings also vary along with them (Tarafder and Jana 2020). The selection of variables is also a crucial factor in the success of any method (Tarafder and Jana 2020). In this present work, the socio-economic development of the villages is measured by two simple composite indices and correlated with the accessibility indices.

For measuring Socio-Economic Development, a composite index has been developed using three indicators with 13 variables (Table 7.1). First, all the 13 variables are converted into a standard score, and then two composite indices have been computed for assessing the level of development. The composite index has a high (0.804) mutual correlation coefficient value, which justifies its use. To compute the composite index I, the Z scores (Sarkar 2013) of all the twelve variables are summed up using the following equation:

$$CD_i = \frac{1}{k}\sum_{i=1}^{k}\frac{-Xi - \mu}{\sigma} \tag{7.1}$$

Table 7.1 Indicators and variables for socio-economic development

Indicators	Variables
1. Demographic	i. Adult literacy rate
2. Employment and occupational	ii. work participation index—total worker to total effective population (excluding population below 06 years) ratio in %
	iii. % of other workers to total workers
	iv. % of main workers to total workers
	v. % of non-agricultural workers
	vi. % household industry worker to total worker
3. Household Infrastructure & Amenities	vii. % of HH availing LPG connection
	viii. % of HH have a full concrete house
	ix. % of HH have two-wheeler
	x. % of HH have electricity
	xi. % of HH avail banking service
	xii. % of HH have latrine facility
	xiii. % of HH have liveable house

Source Compiled by the authors

where CD_i is the composite development index of the village *i*;

X_i is the actual score;

μ is the mean score of variable X;

σ is standard deviation;

k is the number of variables under consideration

Principal Component Analysis has been used to get component scores which are summed up to get the composite index-II (Mahmood 2013) using the following equation:

$$CD_i = \frac{1}{n}\sum_{i=1}^{n} PC_i(1\ldots \mathrm{n}) \tag{7.2}$$

where CD_i is the composite development index of the village *i*;

PCi is the Principal Component Score

n is the no. of Principal Components used.

Principal Component Analysis can mathematically be defined as an "orthogonal linear transformation" (Mahmood 2013), which transforms the dataset into a new set of uncorrelated components in such a way that the highest variance is experienced in the first component, the second most significant variance on the second component and so on. It is instrumental in grouping a large set of variables into few components, called principal components, which, in turn, helps in classifying the original variables.

Table 7.2 Weights of roads for calculating weighted road density

Sl. no	Road categories	Assigned weight
1	Major roads (NH, SH, MDR)	4
2	Other roads (ODR, VR)	3
3	Brick roads	2
4	Earthen road	1

Source Prepared by the authors

In the present work, it is used as a tool for measuring overall development as well as the correlation of the accessibility parameters with the components.

As already mentioned, the transport sector is one of the important aspects to be considered while measuring overall development in general and accessibility in particular. The number of factors determines the level of transport development, but to be simple, a standard score of metaled road-unmetalled road ratio, weighted road density (Table 7.2), and transport service are considered as variables to compute transport development. These are:

(i) The ratio of metalled road length to total road length,
(ii) Weighted road density, which is the sum of weighted road length per sq. km of area,
(iii) Transport service score.

7.6.2 Measuring Level of Accessibility

Accessibility in the present study has been measured in different ways keeping in mind the various dimensions of it and its spatial dependence. As Tobler (1970) rightly remarked, "everything is related to everything else, but near things are more related than distant things". First, the accessibility is measured at the habitation level and later aggregated to the revenue village level (Eq. 7.5). In all the cases, accessibility indices are converted into positive deviation following Eq. 7.4. To determine the overall accessibility level, Physical Accessibility, Network Accessibility, Transport Service Score has been computed and combined following Eq. 7.8 (modified from Sarkar et al. 2001; Sarkar 2011 and Kanuganti et al. 2014). Physical accessibility (Eq. 7.3) is the sum of weighted distances from different categories of road, and it is determined by combining the weighted distance rasters generated from different road categories (Table 7.3) using the 'raster calculator' of Arc Map v10.2 using Eq. 7.3 and extracted to habitation points. Network Accessibility Index (NAIi) is determined by nodal accessibility (Eq. 7.6) using the origin–destination matrix (L Matrix) of the nodes (Taaffe et al. 1996). Transport Service score (TSSi) measures the level of transport service accessibility. It has been determined by measuring the sum of weighted distances (Table 7.4) of the habitations from transport service location by Eq. 7.7. The reason for choosing these three indices is their role in the transport

Table 7.3 Distribution of weight of different road categories

Stage	Road categories	Weight
1	Major roads (NH/SH, MDR)	0.50
2	Blacktop roads (NH/SH, MDR, ODR, VR)	0.30
3	All-weather roads (NH/SH, MDR, ODR, VR and Brick Roads)	0.20

Source Prepared by the authors

Table 7.4 Distribution of weight of transport service class

Stage	Transport service level	Weight
1	Bus stop	0.5
2	Bus stop or major rail station	0.33
3	Bus stop or Minor rail station	0.166

Source Prepared by the authors

infrastructure. These three, together, determine the accessibility of villages in terms of transport infrastructure.

To assess the accessibility of villages towards various public utilities, Social Service Accessibility (EASS) and Administrative (Public) Service Accessibility (EAPS) has also been computed (Eqs. 7.9 and 7.10 and Tables 7.5 and 7.6) and mapped accordingly, keeping in mind the fact that accessibility of habitation and villages is not only a function of the transport distance or the transport service provision present on the road but also the location and quality of various public and social service facilities, where people get their required services for their life and livelihood, which in turn, determines the level of their development. In the latter two indices, transport service provides additional benefits. Thus transport service score is used as the weight for determining the facility service score. Distance to the nearest facility is calculated for each facility level by the 'closest facility' tool of ArcMap v10.2. To include the categories and number of facilities in the public and social service accessibilities, villages are categorized based on the weighted facility index (Table 7.7), which is calculated by the sum of weighted facilities (VFI) present in the villages (Eqs. 7.11 and 7.12). Any village that falls within a 2 km service radius of a facility will be treated as facilitated by the service. The weights of facilities are calculated using the reciprocals of the fraction of villages having the facility (Patel and Madhavan 1984).

$$AI_i = \frac{1}{j}\sum_{i=1}^{j} WjDij \tag{7.3}$$

where Al_i is the accessibility of habitation;

j is Road class;

i is the habitation;
Wj is the respective weight of road class;
Dj is the distance from a road class.

$$AI_i = \frac{Dm - Doi}{Dm} \tag{7.4}$$

where D_m is the maximum observed distance;
D_o is observed distance for habitation *i*.

$$AI_m \frac{1}{n} \sum_{i=1}^{n} \frac{AIi}{N} \tag{7.5}$$

where Al_m is the accessibility of village,
N is the number of habitation i within the village.

$$NAIi = \sum_{i=1}^{j} dij \tag{7.6}$$

$$TSSi = \frac{1}{N} \sum_{i=1}^{N} Wjdij \tag{7.7}$$

where dij is the distance of habitation i to transport service point j,
Wj is the weight of transport service class.

$$OAI_i = \frac{1}{N} \sum_{i=1}^{N} PAI, NAI, TSS \tag{7.8}$$

where OAI_i is Overall Accessibility Index of habitation *i*;
PAI is Physical accessibility;
NAI is Network accessibility;
TSS is Transport service score

$$EAP_i = \frac{1}{N} \sum\nolimits_{i=1}^{j} WjDij \tag{7.9}$$

where, EAP*i* = public service accessibility;
Wj = weight of each administrative level j;
Di = distance of habitation i from administrative service point j

Table 7.5 Distribution of weight (Sum of weights is 1) of administrative services

Sl No	Administrative Level	Weight
1	District Magistrate/ *Zilla Panchayat* Office	0.125
2	Sub-divisional Office	0.125
3	Block Development / Block Panchayat Office	0.25
4	Gram Panchayat /RI Office	0.5

Source Prepared by the authors

$$EAS_i = \frac{1}{N}\sum N\frac{1}{j}\sum_{i=1}^{j} WjDij \tag{7.10}$$

where, EAPi = social service accessibility;
Wj = weight of each facility j;
Di = distance of habitation i from facility location j;
N= total no. of facility category

$$VFI = \sum_{i=1}^{n} WFi \tag{7.11}$$

where W = the weight of facility;
n = no of facilities.

$$FW_i = \frac{1}{\left(\frac{ni}{Ni}\right)} \tag{7.12}$$

where, n = no. of the village (in all village category) in facility i;
N = Total number of the village.

Villages are arranged based on the summation of weights of selected existing facilities and mapped, taking standard deviation as the class interval. It is more fruitful to use spatial statistical tools and related analyses instead of simple non-spatial statistical analyses to assess geographical data and their spatial pattern. Besides, the subjectivity of Choropleth mapping can also be removed by using the spatial auto-correlation tool-based mapping. Thus in the present study, specific spatial statistical tools are used to analyse the spatial pattern of phenomena related to development and accessibility. This is sensible because geographical attributes have spatial inter-dependence (Tobler's first law of geography, 1970). Spatial autocorrelation (Moran's Index)[1] has been utilized for better representation of the indices as well as validation

[1] Spatial autocorrelation is a technique of analysis of spatial patterns of geographic events employing their geographic proximity as well as the attributes of the feature and assesses the spatial ordering of geographic data (Yeung et al. 2007). Not only that, it also discovers spatial covariance structure within the attribute data under consideration (Yeung et al. 2007). It helps to identify whether

Table 7.6 Distribution of weight of social services (Each service category has max weight 1)

Sl No	Type of Service	Total Weight	Level of Service	Weigh
1	Education	1	Primary School	0.33#
			High School	0.33#
			Degree College	0.33#
2	Health	1	S-1: Hospital/ RH/ BPHC	0.5
			S-2: Hospital/ RH/ BPHC / PHC/ MBBS	0.33
			S-3: Hospital/ RH/ BPHC / PHC/ MBBS / PHSC/ FWC/ MC/ Dispensary	0.166
3	Banking, Finance and Market	1	Any Bank	0.4#
			Any Market	0.4#
			Any Bank/ CSP/ ATM	0.2#
4	Communication	1	Any Post Office	1.0
5	Public Distribution	1	PDS	1.0

Source Prepared by the authors

of the local regression model. The Getis-Ord-Gi* statistic helps in identifying the clustering of high-high and low-low values over the geographical space.

Ordinary Least Square (OLS) and, Geographically Weighted Regression (GWR)[2] have been used to understand the overall relationship as well as the spatially weighted relationships between development parameters and accessibility parameters.

---

neighbouring values vary together or not and thus measure the spatial interdependence of attribute value. It is based on Tobler's first law of geography. The principle is the same as the product-moment correlation coefficient. Here the attribute x is correlated with itself across space. Moran's Index is one of the best methods of determining spatial autocorrelation, where a positive value indicates similar values nearby, a negative value indicates dissimilar values nearby and zero indicates uncorrelated and random spatial distribution (Yeung et al. 2007). Thus one can assess the distribution as a dispersed, random or clustered pattern. Getis-Ord-Gi* produces a Gi* statistic for each location is based on the interdependence with the attributes of location j in a given n number of neighbours. In Getis-Ord-Gi* statistics, the entire region is divided into n number of sub-regions and the size of the sub-region depends on the z score.

[2] Geographically weighted regression is a kind of spatial regression where several spatially weighted linear regressions are employed in different locations between the neighbours around focal locations. The logic behind this regression is that there may be a greater correlation between the variables in certain locations than others. Fixed distance band or Inverse distance weightage can be used as a spatial weight to determine the local correlations. The local regression can be set up for a fixed distance band or a fixed number of neighbours. Adaptive bandwidth, which is used to determine the geographic weighting, is set so that the software can compute best k nearest neighbours where maximum r squared is achieved or lowest aggregate prediction error occurs. This method is used more for prediction rather than fit because the regression coefficients vary geographically.

Table 7.7 Facility weight calculation

Facility Category	Facility (i)	Weight	No. of village (Census village category)#							All Total
			1	2	3	4	5	6	7	1951
Total No. of Villages (N)			265	298	456	565	232	110	25	
Educational	Pre-Primary and Primary School	1.09	161	269	435	561	230	110	25	1791
	High School	5.77	2	12	32	101	98	71	22	338
	Degree/Nursing/Technical/ Training/ Professional College	15.48	5	8	18	32	28	27	8	126
Health	PHSC	2.85	30	78	155	335	174	88	20	880
	Maternity/ FWC	11.54	14	13	26	58	35	21	2	169
	PHC	27.48	1	2	12	18	20	13	5	71
	BPHC/ RH	78.04	0	0	0	4	5	13	3	25
	Specialized hospital	97.55	0	0	0	2	3	13	2	20
Commerce	Market	6.46	11	19	44	96	64	51	17	302
	Credit society/ Other banking	5.95	9	10	31	98	88	70	22	328
	Bank	6.14	9	16	27	89	86	74	17	318
Communication	Post office	4.1	10	21	73	175	112	74	11	476
Administrative	GP/ RI office	7.8	2	1	31	63	69	62	22	250
	Block HQ	92.9	1	0	0	3	3	10	4	21
Other	PDS	2.8	28	50	149	285	109	54	21	696

#1 = Less Than 500; 2 = 500–1999; 3 = 1000–1999; 4 = 2000–4999; 5 = 5000–1999; 6 = 10,000–19,999; 7 = 20,000 or more prepared by the authors from different sources following Patel and Madhavan (1984)

7.7 Results and Discussions

7.7.1 *Level of Development and Derivation*

Levels of socio-economic development and transport development are two important indices calculated to measure the overall socio-economic development and transport infrastructure development.

7.7.1.1 Level of Socio-economic Development

First, to be clear, both z-score and principal component score produce almost identical results while determining the level of socio-economic development (Figs. 7.2 and 7.3). It is noticeable that there is wide spatial variation in the level of overall socio-economic development among the villages. It is quite obvious because there is also wide variation in accessibility and transport development. In many areas, high levels of overall accessibility and social service accessibility correspond to high levels of socio-economic development and vice versa.

Let's consider the spatial distribution of development and underdevelopment. The pockets of development and underdevelopment are visible from the Getis-Ord-Gi* statistics (Figs. 7.4 and 7.5) which is determined based on a 5.17 km fixed distance band with 30 numbers of neighbours within each cluster.

The composite score-1 shows that about 43% of villages and 54% of the total population lie above the Z-score value of 0. It looks good, but still, a large number of villages (about 48%) lie below the 0 value of the Z-score. It needs to be addressed because many of these villages are good in terms of accessibility.

The development indicators are varied widely within the villages. Some villages are advanced in some indicators, but for some other indicators, another set of villages perform better. To identify these variations, Principal Component Analysis (PCA) is an excellent measure (Tarafder and Jana 2020). Thus here, PCA has been used to categorize these indicators and four components (Table 7.8) are derived with Varimax rotation and Kaiser Normalisation. The four principal components describe the four different combinations of indicators, and based on the combination following, four groups of indicators of development have been defined.

- Principal Component-I (Non-agricultural employment with dominant household infrastructure and amenities):

The first principal component contributes about 26% variation in the level of development and constitutes the non-agricultural employment with dominant household infrastructure-amenities component. It has the highest positive loading with the percentage of households having separate latrine followed by other workers to total workers, percentage of households have two-wheelers, percentage of households

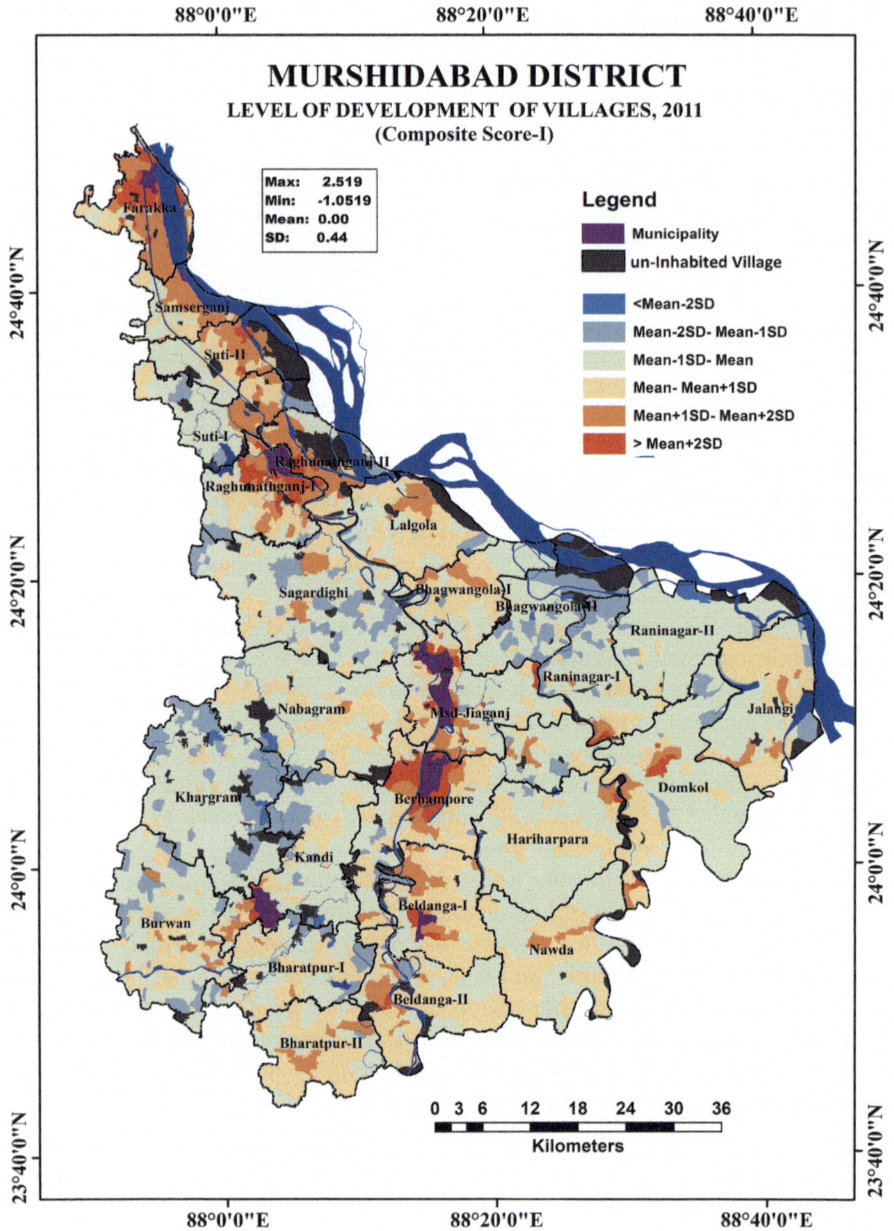

Fig. 7.2 Level of development by Z-score (*Source* Mapping sby the authors based on their own calculations)

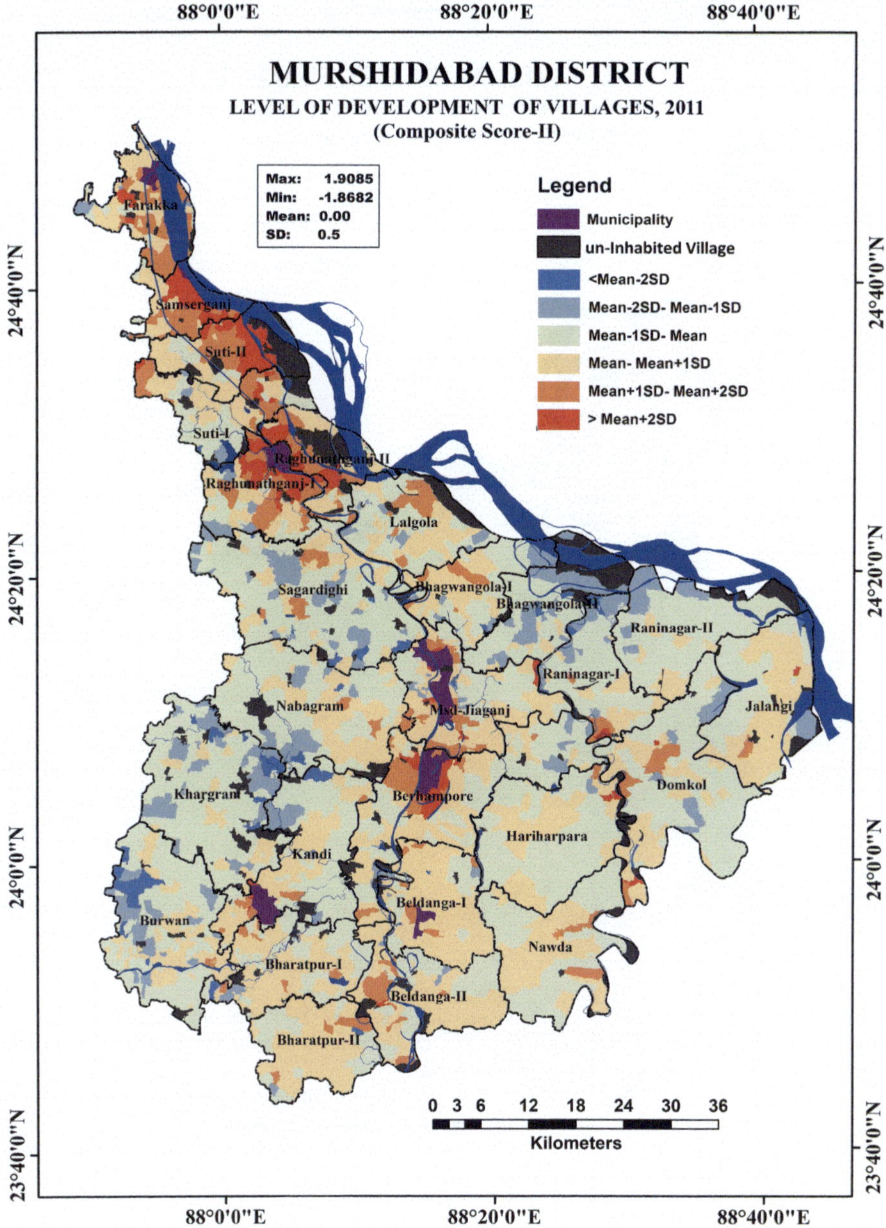

Fig. 7.3 Level of development by PC scores (*Source* Mapping sby the authors based on their own calculations)

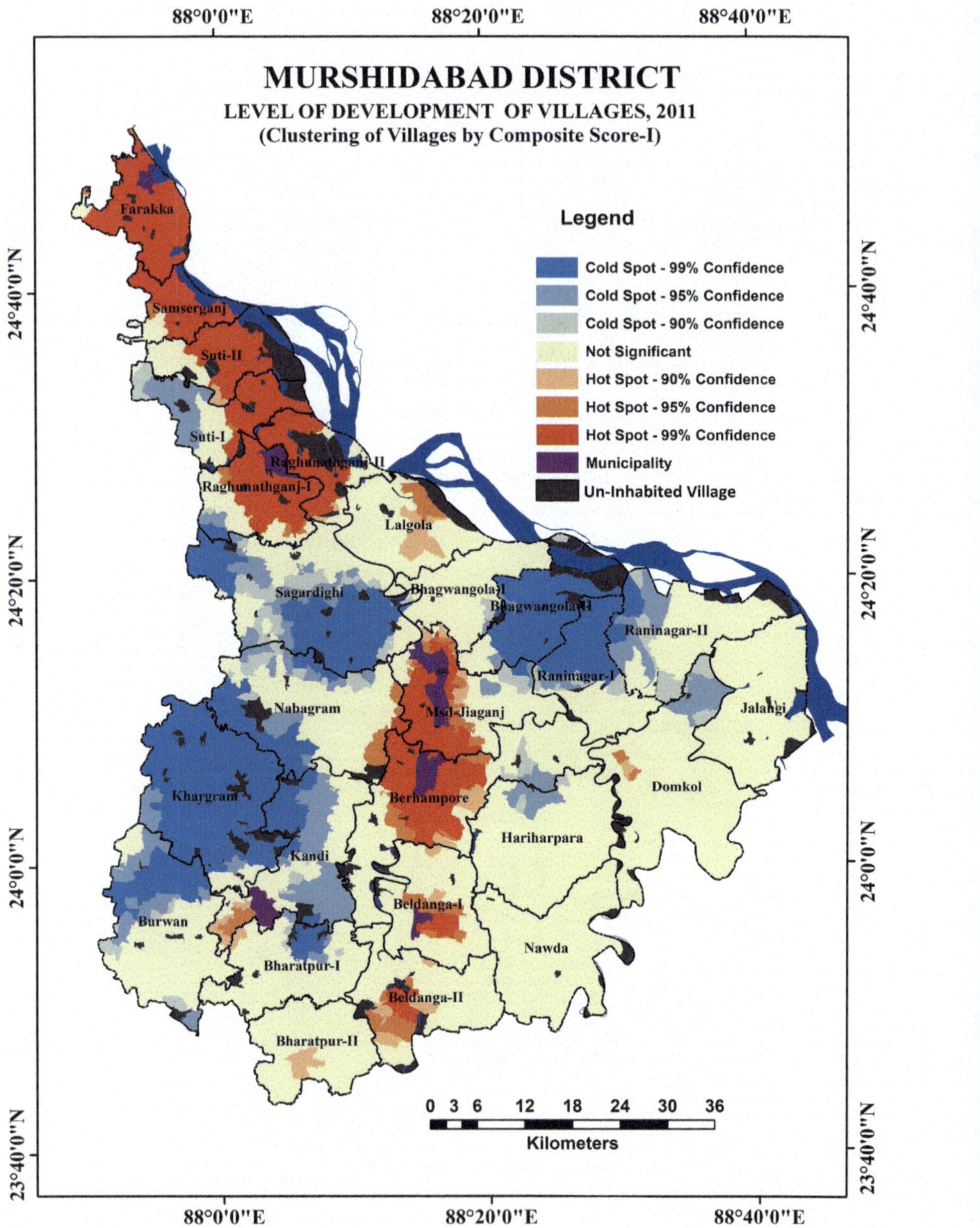

Fig. 7.4 Clustering of development (Z-Score) (*Source* Mapping sby the authors based on their own calculations)

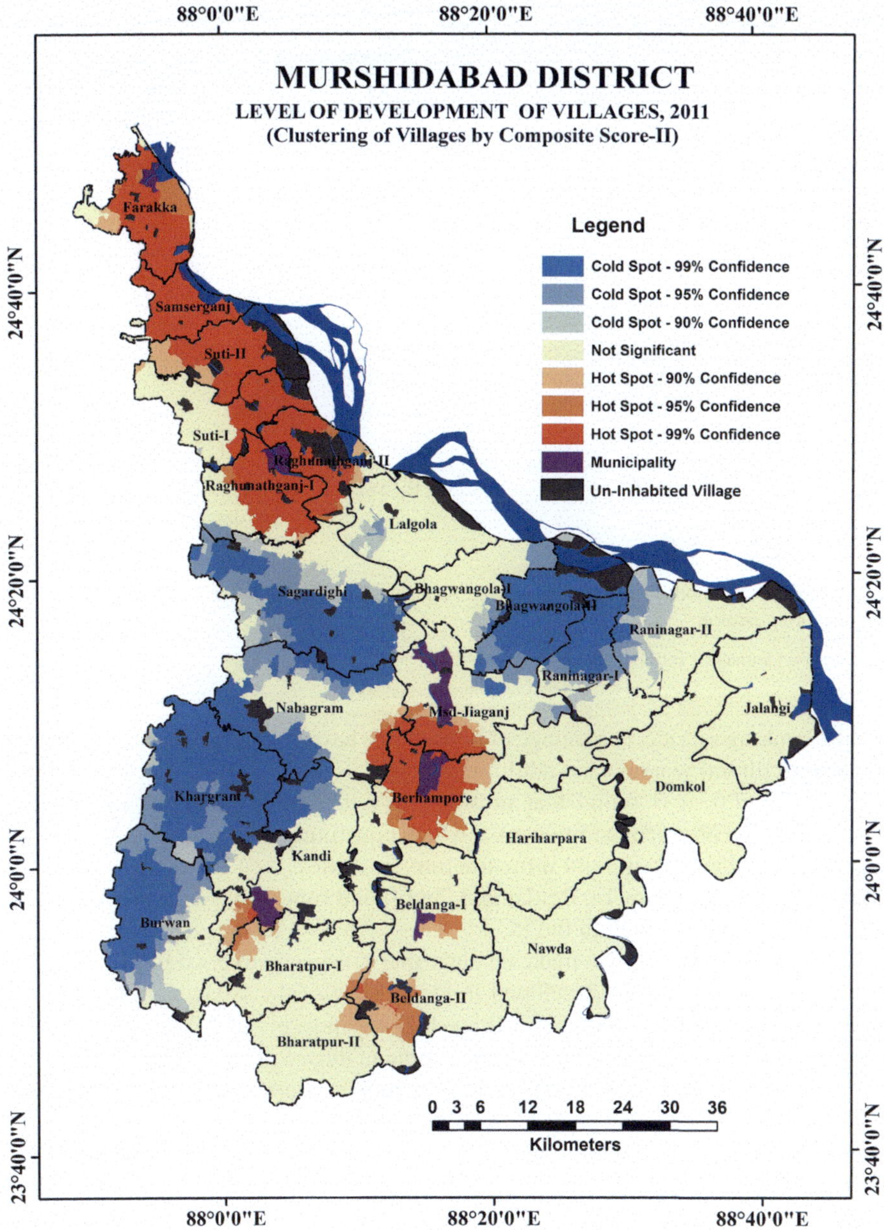

Fig. 7.5 Clustering of development (PCA) (*Source* Mapping sby the authors based on their own calculations)

Table 7.8 Principal component matrix

	Component			
	PCA1	PCA2	PCA3	PCA4
Adult_Literacy	0.306	0.518	−0.485	
Work_Participation_Index			0.760	−0.263
O_W_Tworker	0.678	0.326		
M_W_TWorker				0.962
Per_NonA_w	0.391		0.511	−0.673
LPG		0.684		
Prcnt_HHConcrete_house	0.529	0.505		
Two_WHEELER	0.595	0.453		
Electricity	0.360	0.686		
Prcnt_HH_Bank	0.569	−0.431		
HH_Latrine	0.789			
Prcnt_HH_Livable		−0.388		
P_HH_Tot_Work			0.879	

N.B. Extraction Method: Principal Component Analysis. Rotation Method: Varimax with Kaiser Normalization

Data Source Census of India

availing banking service, percentage of households have concrete houses, percentage of non-agricultural workers and adult literacy.

From Fig. 7.6, it is found that most of the villages in the Block of Farakka score high in the principal component-I, which constitutes non-agricultural employment with dominant household infrastructure amenities. Besides, many villages in Bhagwangola-I, Berhampore, Beldanga-I, and II and some villages in Nawda Block also score moderate to high in the PC-I.

A close look from Fig. 7.6 makes it clear that villages with good physical accessibility in those Blocks are developed in PC-I. Most of the villages in the western Rarh region except Bharatpur-I and II are lagging in terms of PC-I, indicating the strong agriculture-based livelihood of the people in these villages. These villages are also vulnerable in terms of household infrastructure and amenities. So life and livelihood should be improved in these villages by improving their income by developing agriculture along with other non-agricultural activities and active participation of the villagers.

- Principal Component-II: (Literacy and household infrastructure amenities)

The second principal component contributes 17% of the total variation in the level of development and constitutes the literacy and household infrastructure amenities. It has the highest positive loading with the percentage of households having electricity connection followed by the percentage of households having LPG connection, adult

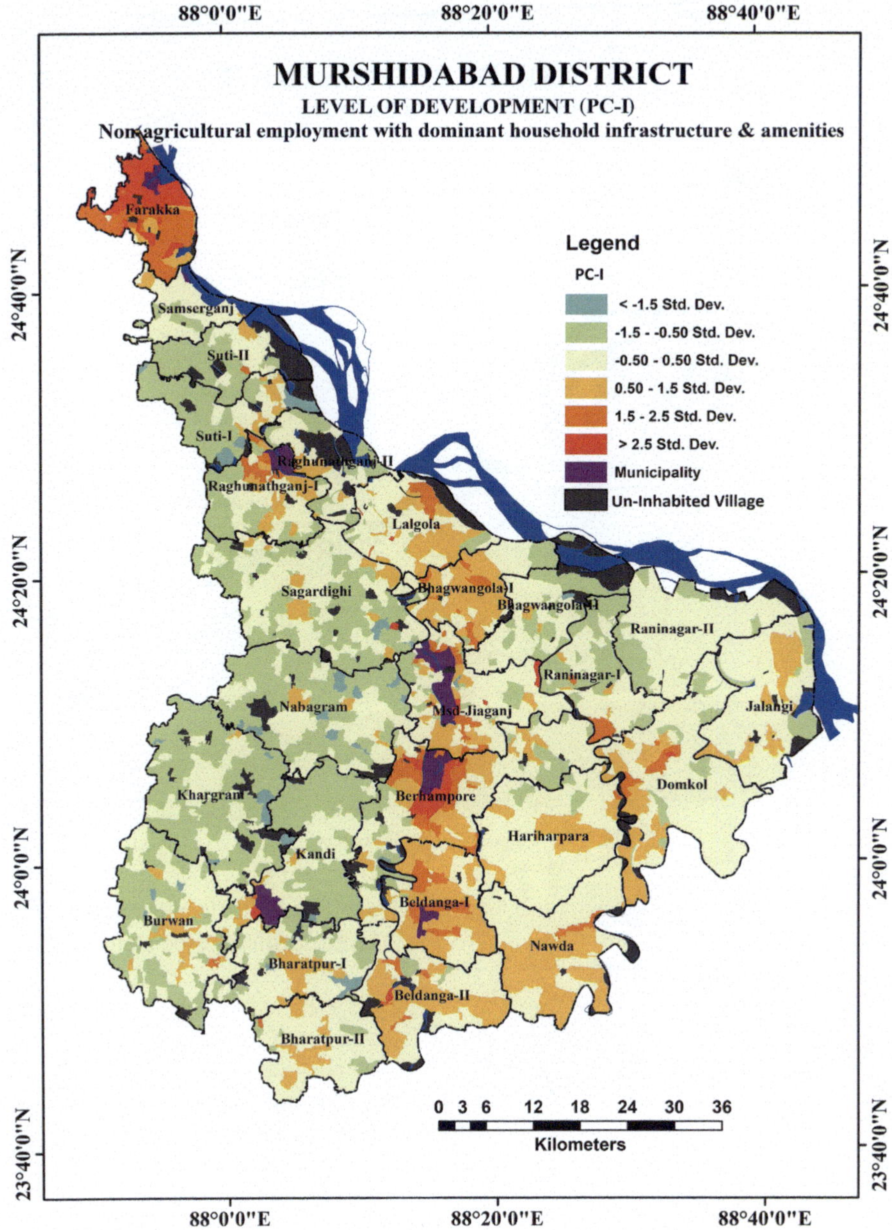

Fig. 7.6 Level of Development PC-I (*Source* Mapping by the authors based on their own calculations)

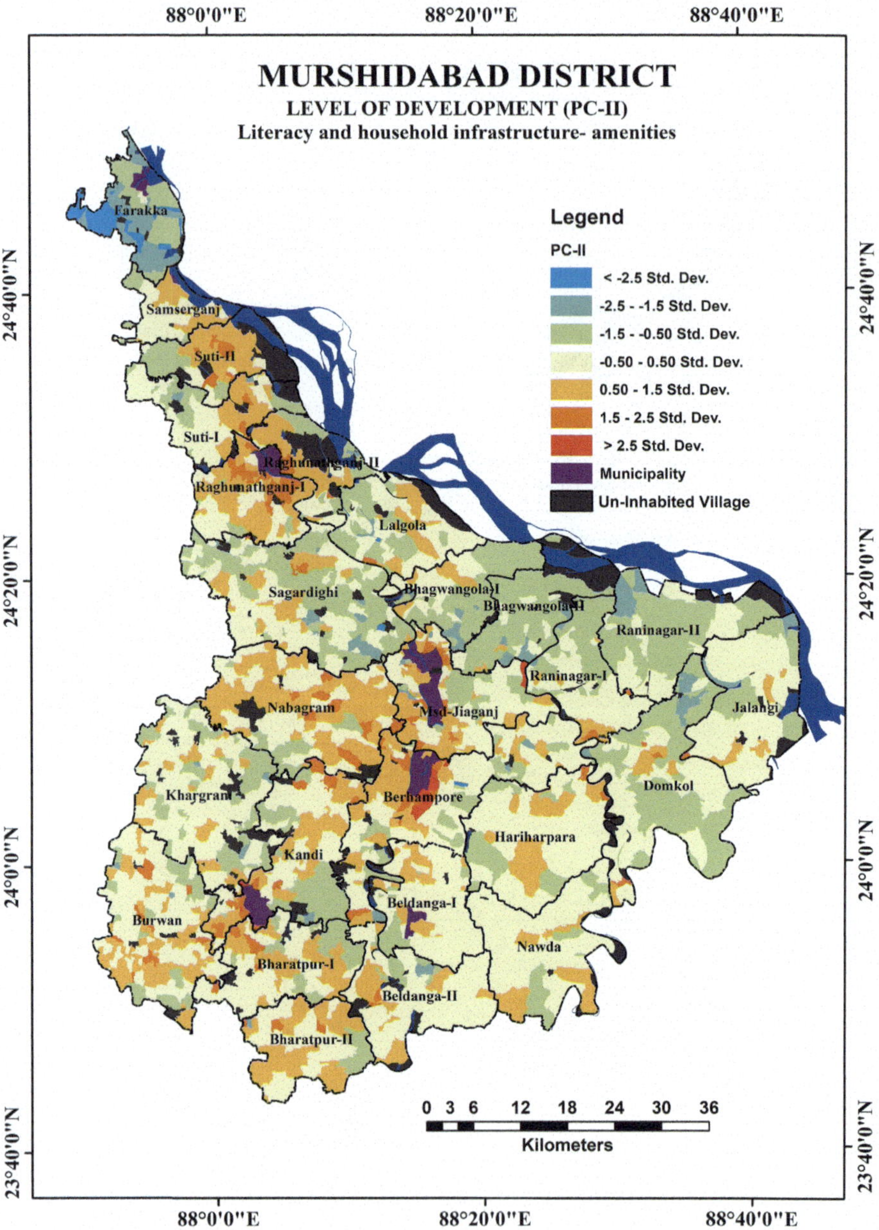

Fig. 7.7 Level of Development PC-II (*Source* Mapping by the authors based on their own calculations)

literacy, percentage of households having concrete houses, a percentage of households having two-wheelers and a percentage of other workers to total workers. Negative loading is with the percentage of households availing bank and the percentage of households having liveable houses.

Most of the villages in the Block of Raghunathganj-I, Nabagram, and some villages in Raghunathganj-II, Suti-I and II, Samserganj, Burwan, Bharatpur-I and II and Berhampore are relatively better position in terms of PC-II (Fig. 7.7). Most of these villages are in good physical accessibility, transport service provision and public service accessibility. Relatively better in terms of literacy rate, LPG and electricity connection and moderately good in terms of concrete house and two-wheeler ownership make these villages better in PC-II. The negative loading of households availing banking services and liveable houses indicates banking service is not adequate and the houses are in a vulnerable condition. So these things should be kept in mind while planning for development.

- Principal Component-III: (Work participation with dominant household Industry Employment)

The third principals component contributes 9.5% of the total variation in the level of development and constitutes the work participation with the dominant household Industry Employment component. It has the highest positive loading with the percentage of household workers to total workers followed by the work participation index and the percentage of non-agricultural workers to total workers. Negative loading is with adult literacy.

This component is dominated by participation in the household industry of the total workers, and the northern six blocks overshadowed the other 20 blocks in the district (Fig. 7.8). Ground observation shows that most of these villages are dominated by bidi (country cigar) workers and mostly females. Due to the high rate of involvement of both male and female working populations in household-based activities work participation rate is also high. But negative loading with literacy suggests low levels of literacy and needs to be improved. Basically, female literacy should be addressed.

- Principal Component-IV: (Dominantly Main worker)

The fourth principal component contributes 8.8% of the total variation in the level of development and constitutes Dominantly the main worker to total worker. It has the highest loading with the percentage of the main worker to total worker (positive loading) followed by the percentage of non-agricultural workers (negative loading) and work participation index (negative loading).

From Fig. 7.9, it is noticed that developed villages in PC-IV distributed almost all the Blocks of the District with a slightly larger concentration in the Blocks of the Bagri region. These villages have a high percentage of main workers, but these workers are involved in the agricultural sector because strong negative loading is observed with non-agricultural workers. Despite the large percentage of main workers to total workers, the work participation rate is low because it is also negatively correlated. The reason is that in agricultural fields, mainly the male working population is

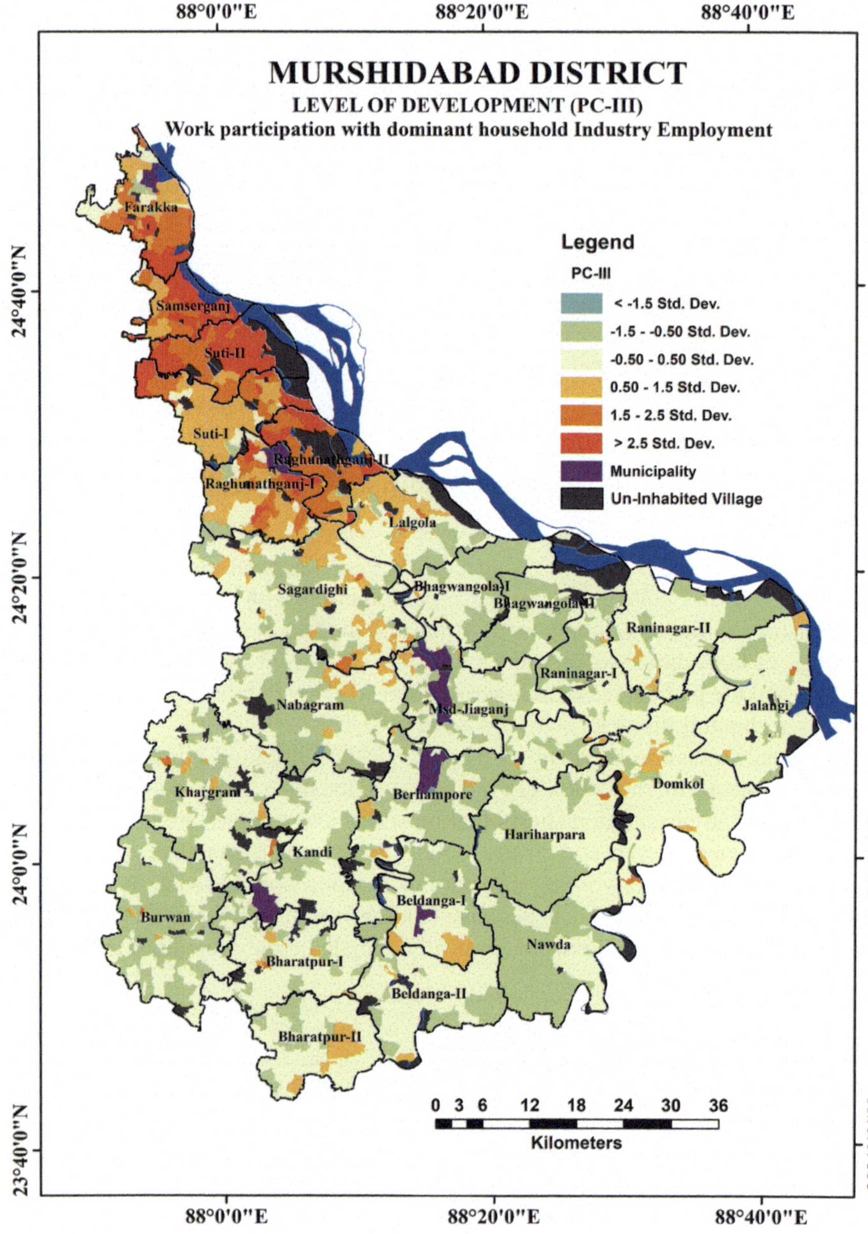

Fig. 7.8 Level of Development PC-III (*Source* Mapping by the authors based on their own calculations)

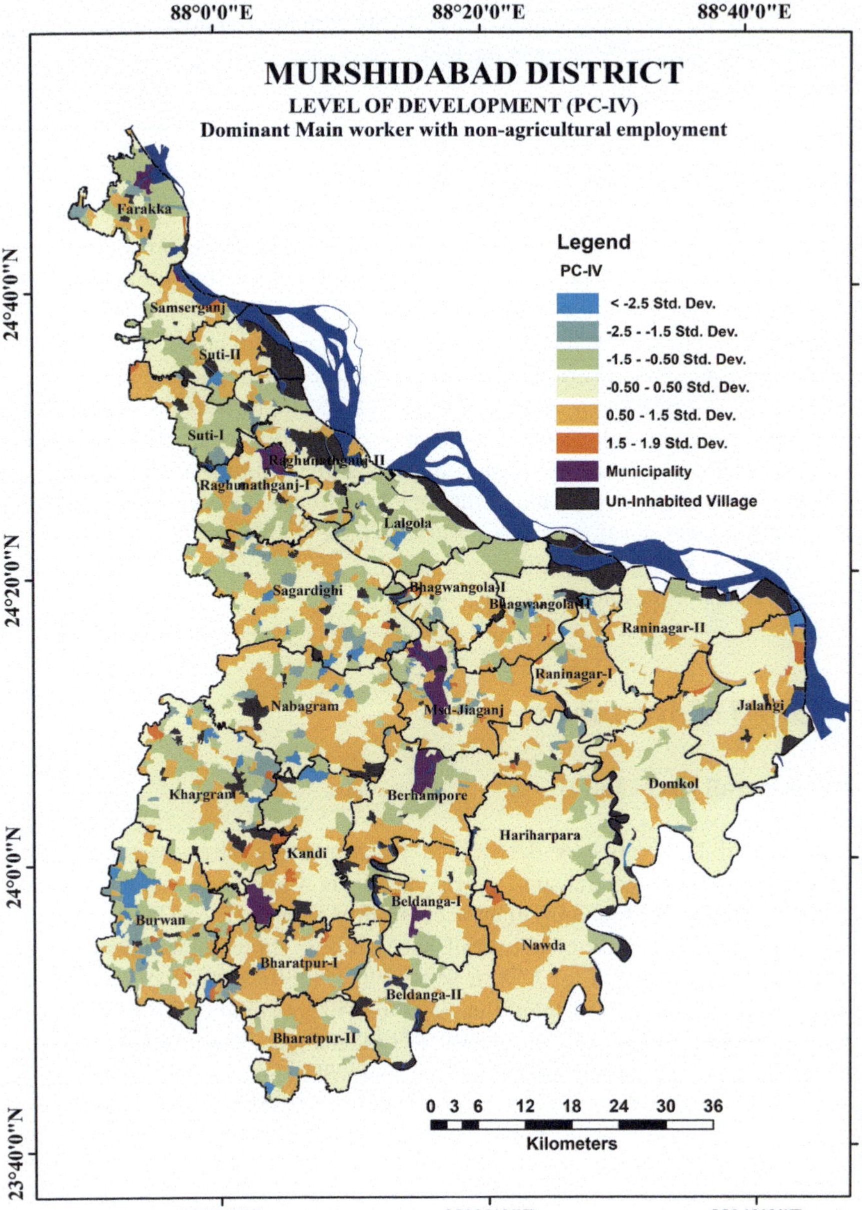

Fig. 7.9 Level of Development PC-IV (*Source* Mapping by the authors based on their own calculations)

involved. The female working population is involved in household activities as well as post-processing works of crops at the house. However, they are not considered as 'workers'.

7.7.1.2 Level of Transport Development

As already mentioned in the previous section, the level of transport development is important for the overall development of any area because transport is such an important sector that determines many components of socio-economic development and accessibility. Thus, the level of transport development will help assess the socio-economic development concerning transport infrastructure and accessibility.

In the district, significant variation of transport development is observed among villages. This variation is due to the variation of transport infrastructure and public transport provision. There are few clusters of villages in the northern part, central part around district headquarter and along the Domkol-Raninagar border of the district where good transport development is observed (Figs. 7.10 and 7.11). On the other hand, Sagardighi, Khargram, Hariharpara, etc. have few clusters of villages with a low level of transport development.

7.7.2 Level of Accessibility

7.7.2.1 Overall Accessibility

Overall accessibility, which is determined by combining physical accessibility, network accessibility, and transport service score, is a good representation of the level of accessibility as three important accessibility parameters are included.

The frequency distribution of overall accessibility shows a weak, negatively skewed distribution. From Fig. 7.12, it is clear that the overall accessibility level is quite good in the district as more than 90% of villages (and 88% of the population) have scored 0.6 or more. About 51% of villages (and 58% of the population) have scored above 0.7 out of unity. The low score of the distant villages is due to network accessibility which decreases away from the district headquarter. This is more prominent in Getis-Ord-Gi* statistic (distance band of 5.8 km is used based on the high Z-score) based mapping (Fig. 7.13).

7.7.2.2 Level of Public Service Accessibility

Ease of access to administrative services is quite good in the villages close to district headquarter, Raghunathganj-Jangipur region, Domkol-Raninagar-II border and around Kandi. The clear dominance of the Sub-Divisional headquarters is

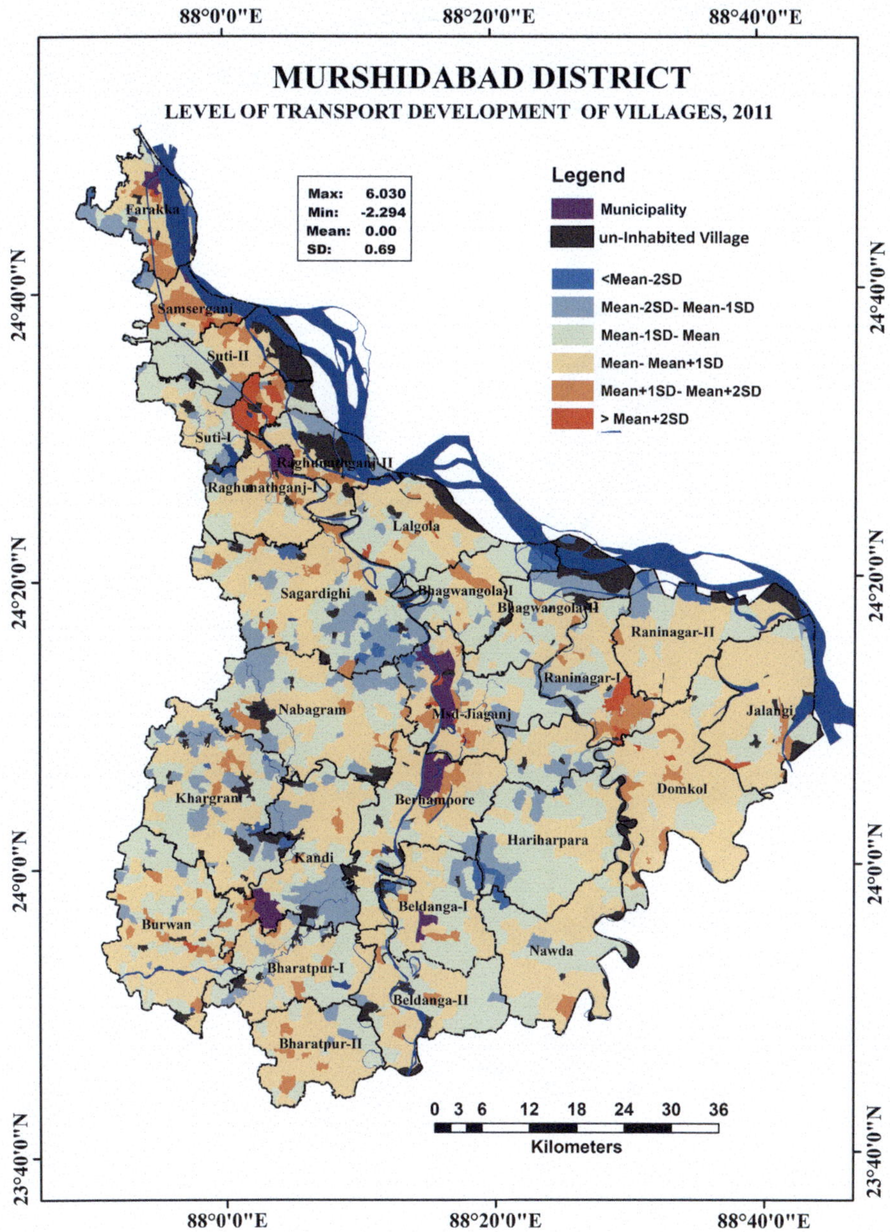

Fig. 7.10 Level of transport development (*Source* Mapping by the authors based on their own calculations)

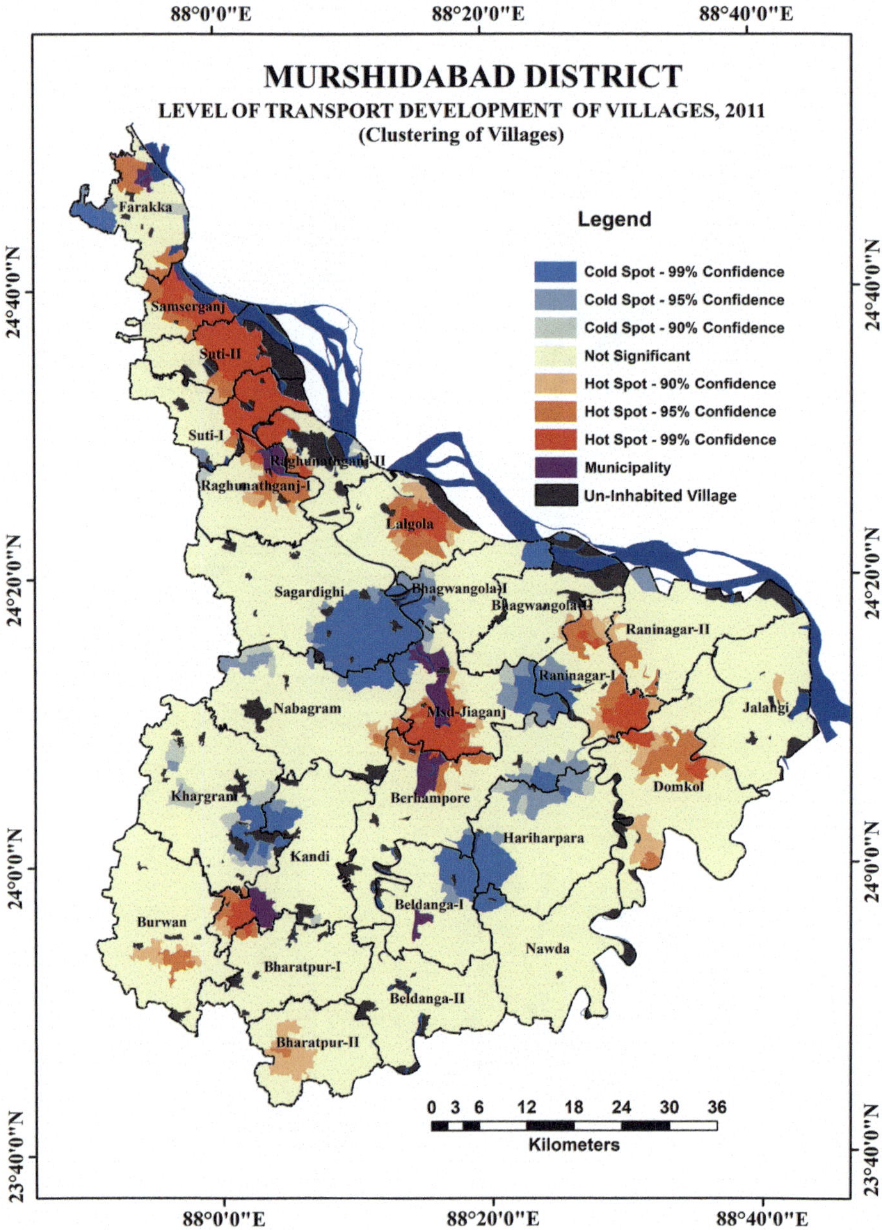

Fig. 7.11 Clustering of Transport development (*Source* Mapping by the authors based on their own calculations)

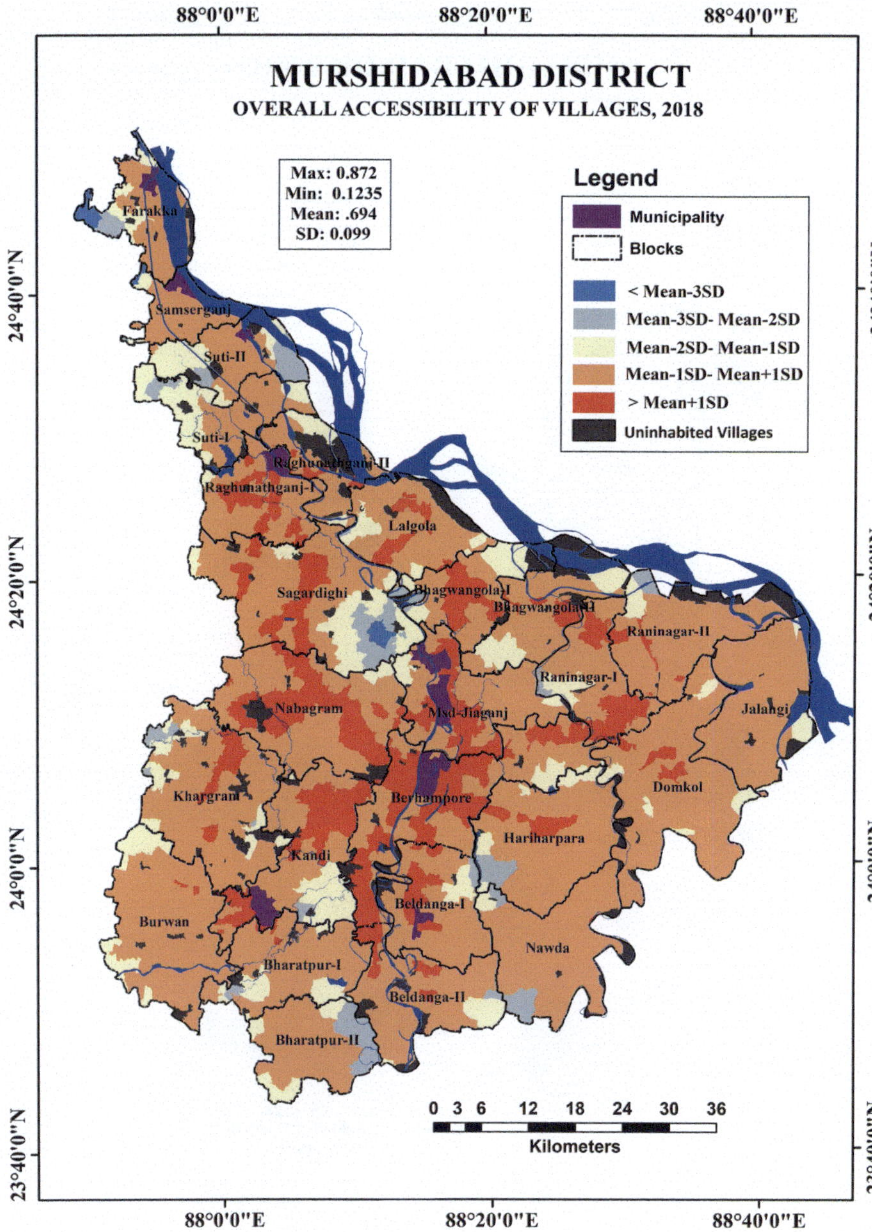

Fig. 7.12 Overall accessibility (*Source* Mapping by the authors based on their own calculations)

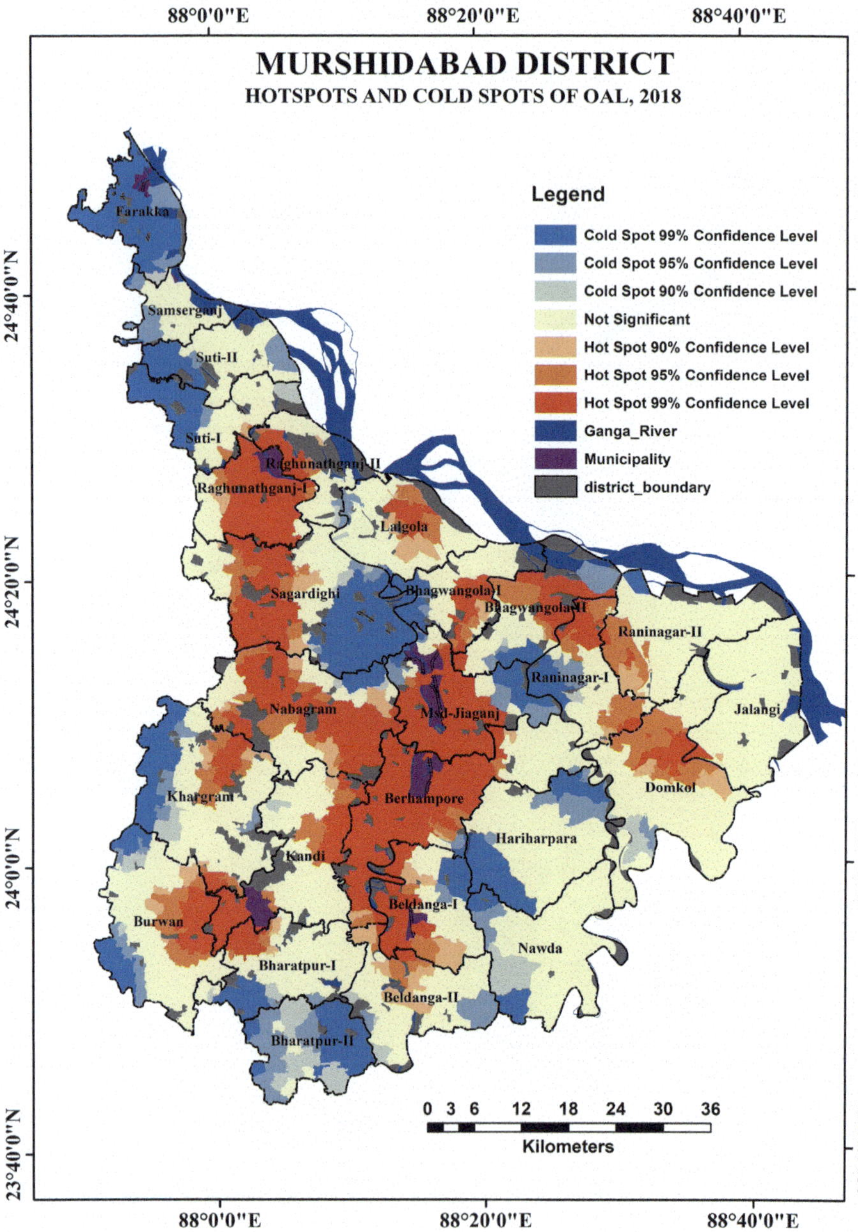

Fig. 7.13 Clustering of overall accessibility (*Source* Mapping by the authors based on their own calculations)

noticed. Accessibility is limited to those villages which are away from the Sub-Divisional headquarters in the district. Despite the low weightage given to DM and SDO office, these two types of services have largely influenced the score and is clear in Fig. 7.14. The frequency distribution shows the near-normal distribution (Fig. 7.16).

There are few clusters of villages with low accessibility scores and mostly distributed in the northern Blocks like Farakka, Samserganj, Suti, Sagardighi and Lalgola and along Raninagar-1 and Hariharpara border, Hariharpara-Beldanga-1 Block and Bharatpur-1 and Bharatpur-II border. This is more prominent in Getis-Ord-Gi* statistic (distance band of 6.6 km is used based on the high Z-score) based mapping (Fig. 7.15). Low administrative service accessibility to these pockets is due to their remote location from the DM/ SDO/ BDO office. The low score of villages on Sagardighi-Lalgola border is due to their remote location from DM/ SDO/ BDO office as well as backwardness in terms of transport service score. Slight improvement in transport service in these areas may improve their ease of accessibility to administrative services.

7.7.2.3 Level of Social Service Accessibility

It is already mentioned in the methodology section that the level of social service accessibility is measured by combining the level of accessibility to various social services. In the first section, accessibility to educational services, health services and commercial services are discussed, and at the end of this section, combined social service accessibility has been discussed.

Education service accessibility seems good in the district as the frequency distribution shows significantly negative skewness, which indicates most of the villages (56%) fall above the mean accessibility. In terms of population, it is about 67%. About 76% of villages (94% of the population) fall within the range mean ± 1*SD. Thus, overall educational accessibility is satisfactory because most of the villages have scored good accessibility scores, although there are some pockets in the district where the accessibility is very low, and about 5% of total villages (4% of the population) are under Mean-2*SD, i.e. 0.49.

As compared to the distribution of the total population of villages, spatial inequalities are noticed in educational accessibilities (Fig. 7.17), and no valid correlation is observed between the total population and educational accessibility score. Some large villages in the eastern Bagri region and the northern part of the district have moderate to low accessibility scores, which should be improved by providing new facilities in these areas. On the other hand, in the western Rarh region, villages are relatively small and scattered over the region, but the accessibility score is high to moderate irrespective of their size.

The frequency distribution of health service accessibility scores of villages shows near-normal distribution with slightly skewed distribution in the negative direction and flat in terms of kurtosis where about 90% of villages (and 92% of the total population) are above the mean score, which can be said as good accessibility. There are

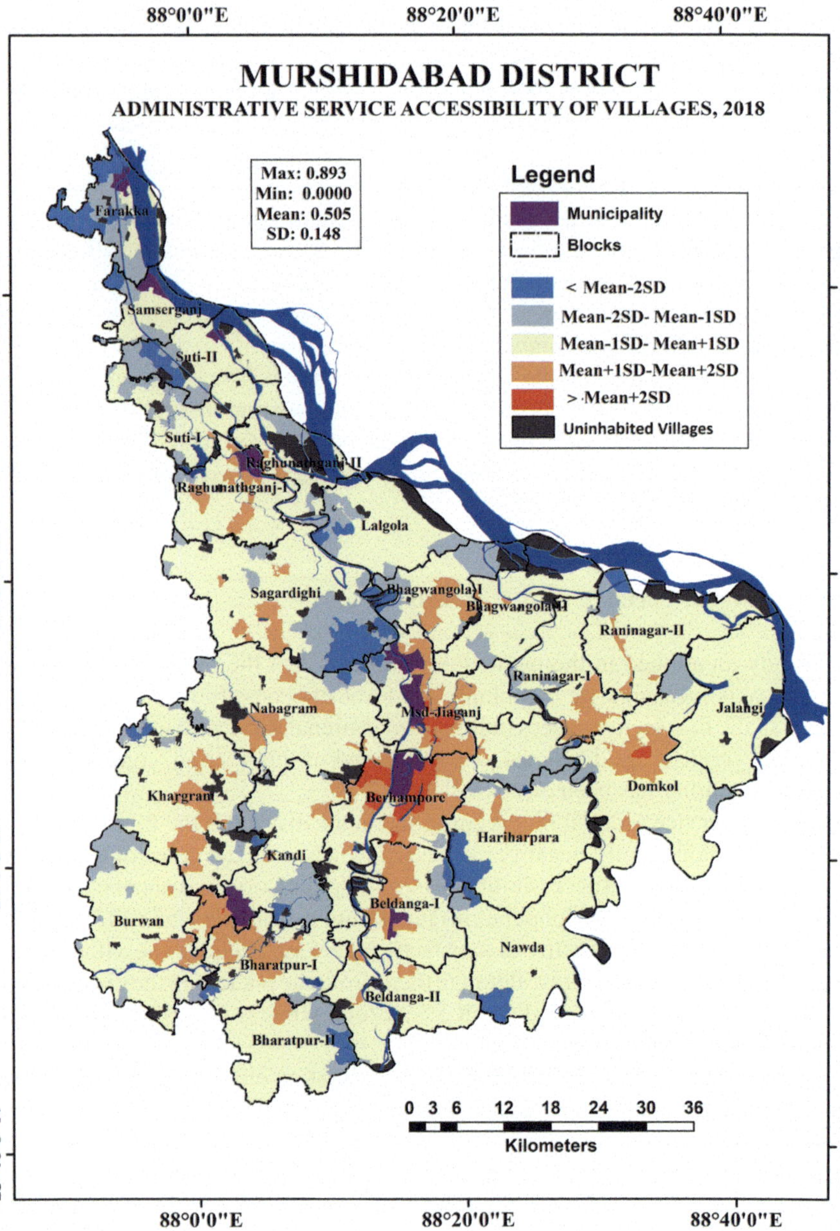

Fig. 7.14 Administrative service accessibility (*Source* Mapping by the authors based on their own calculations)

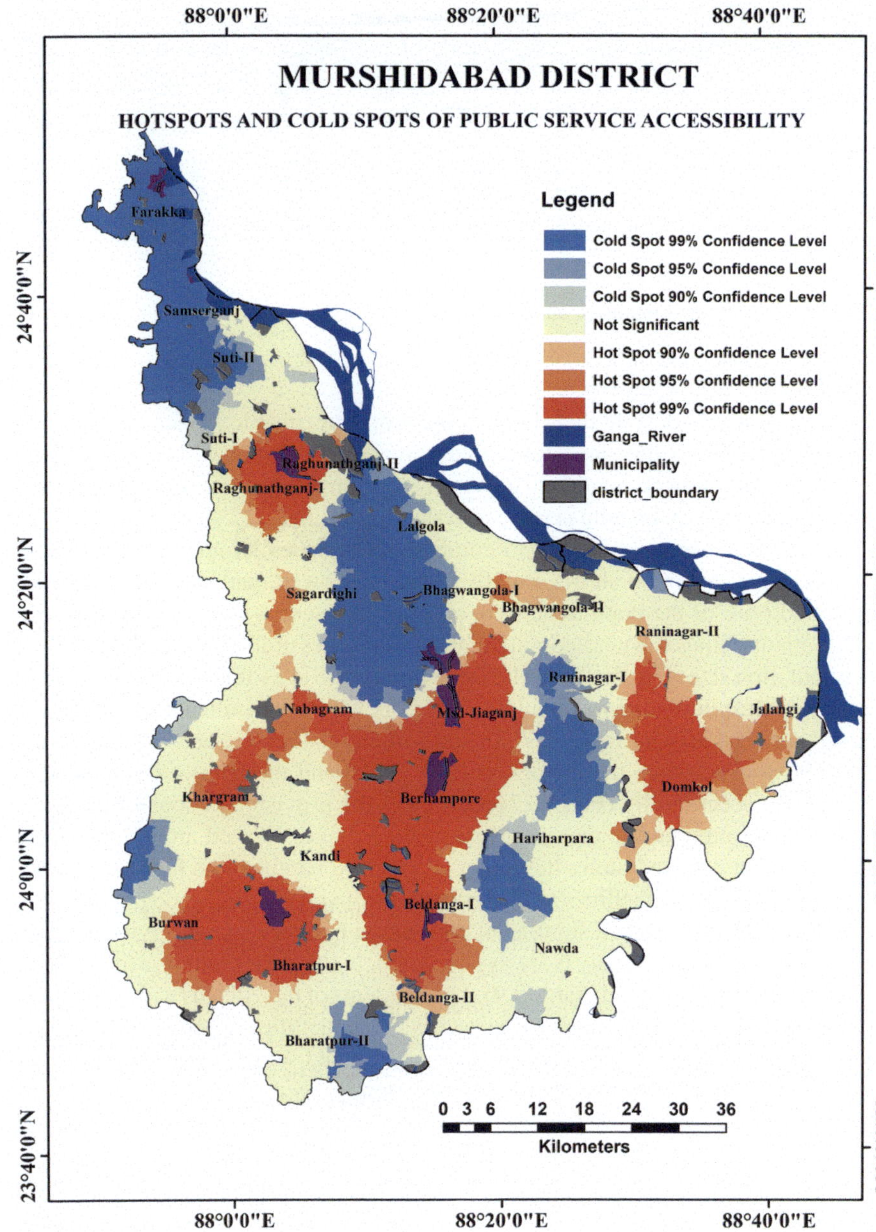

Fig. 7.15 Clustering of administrative service (*Source* Mapping by the authors based on their own calculations)

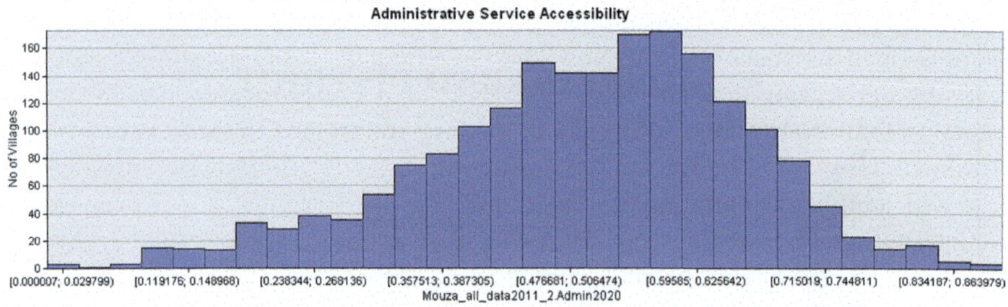

Fig. 7.16 Frequency distribution of administrative service accessibility (*Source* Prepared by the authors)

many villages where despite their large population size, health service accessibility is low (Fig. 7.18). At the same time, there are some villages with good accessibility despite their low population. This is why no significant correlation is observed between population and health service accessibility scores.

In the district, the accessibility to commercial services is quite satisfactory as the frequency distribution shows slightly negative skewed, and more than 54% of villages and 65% of the population have accessibility scores above mean. But still, there are many villages where commercial service accessibility is low despite their large population size (Fig. 7.19). At the same time, there are some villages with good accessibility despite their low population. This is why no significant correlation is observed between population (Fig. 7.20) and commercial service accessibility score.

The above discussion, the histogram (Fig. 7.21), and the map of the combined social service accessibility (Fig. 7.22) reveal that the provision of social services is not uniform throughout the district. Most of the villages located around the block headquarter and oriented along the major roads have received the highest accessibility score. Most of the facilities tend to concentrate in places with good physical and network accessibility. Thus a significant correlation between social service accessibility and physical accessibility is always expected. In the present study, a moderately strong correlation is obtained (Table 7.9). The residents of the villages with poor physical accessibility have to travel more to access those facilities. Improvement should be made in the areas with low physical accessibility and social service accessibility so that people can access those facilities conveniently.

The hotspot map of social service accessibility and population distribution shows few pockets with low clusters of social service accessibility where the population is also low. However, there is low social service accessibility in many pockets despite their high population concentration. These pockets should be appropriately addressed.

We have discussed above the level of accessibility to various social services and found it satisfactory though a very weak correlation is found between accessibility scores and population size (Table 7.9). It is known to all that the goal of public utilities is to provide social services to its people in accordance with the needs of people which varies from village to village and the location of these utilities also

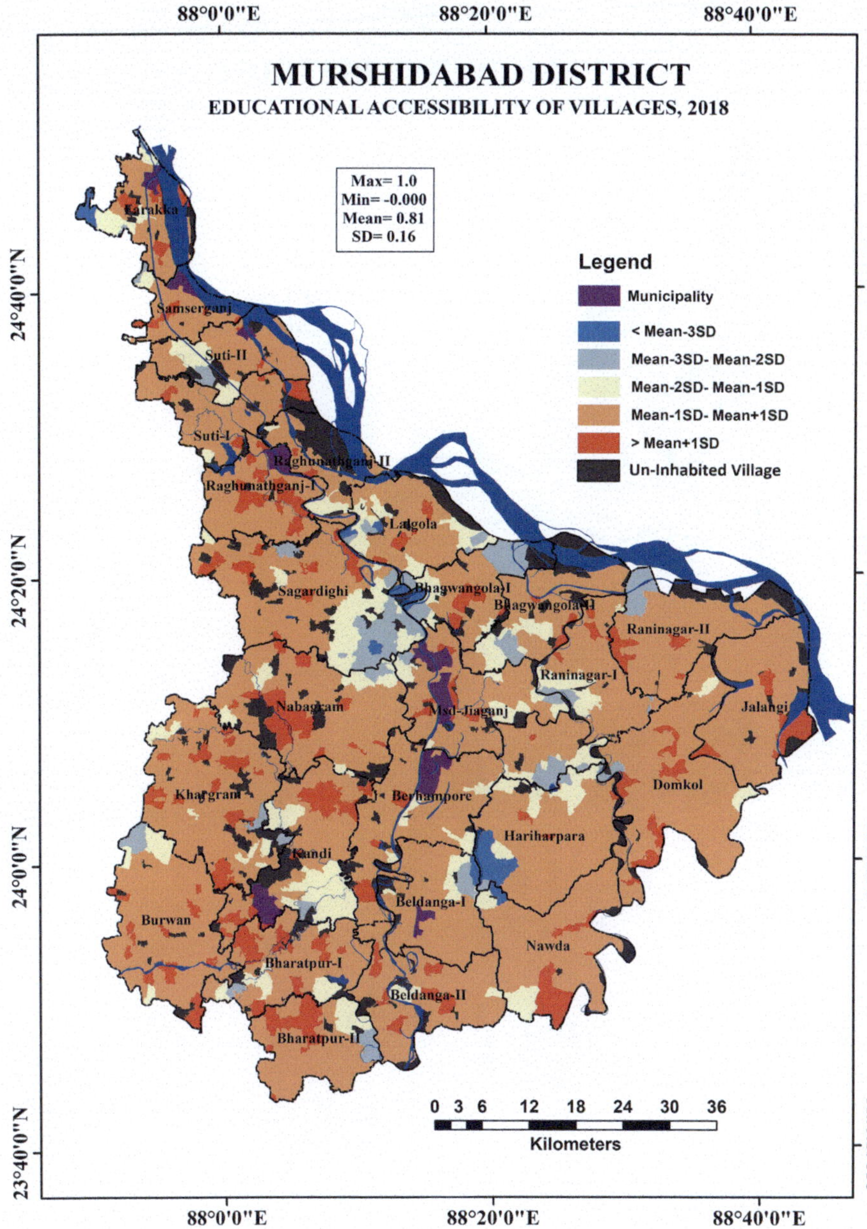

Fig. 7.17 Accessibility to education services (*Source* Mapping by the authors based on their own calculations)

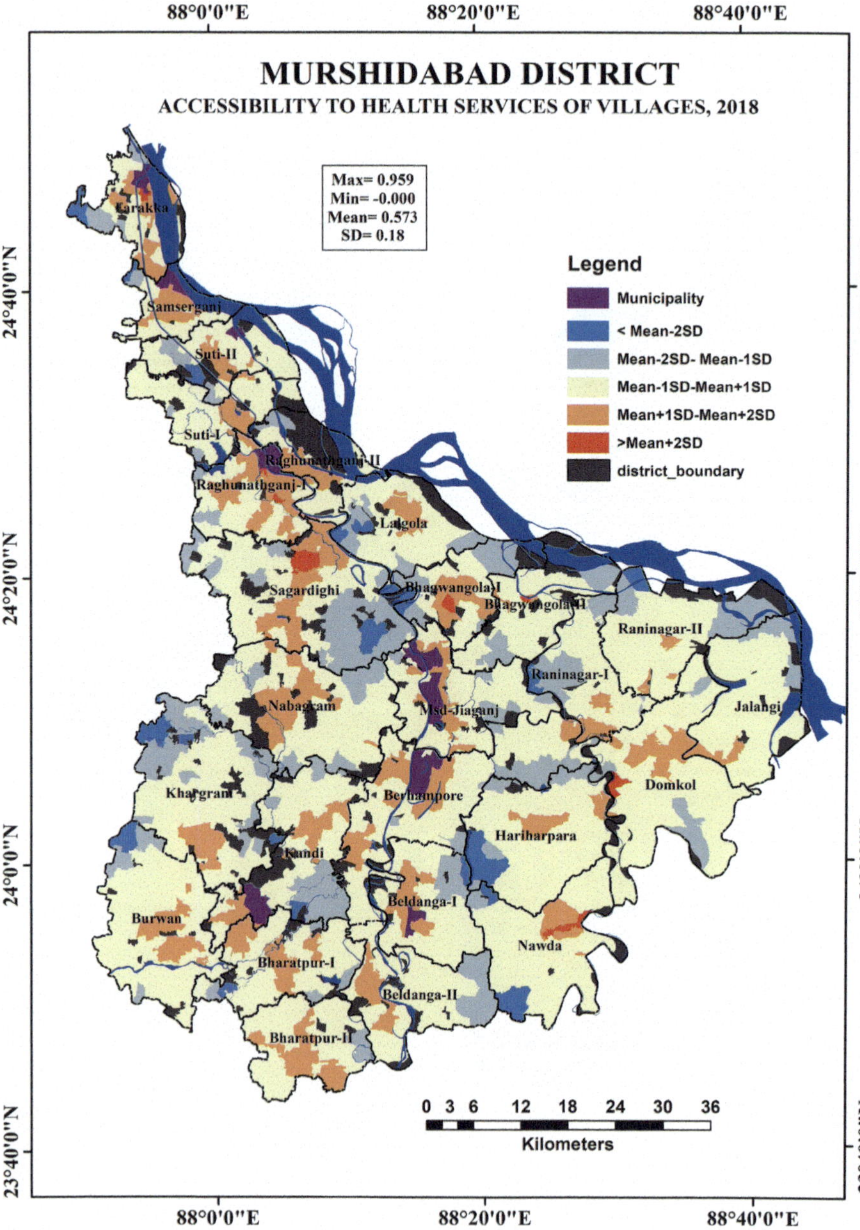

Fig. 7.18 Health service accessibility (*Source* Mapping by the authors based on their own calculations)

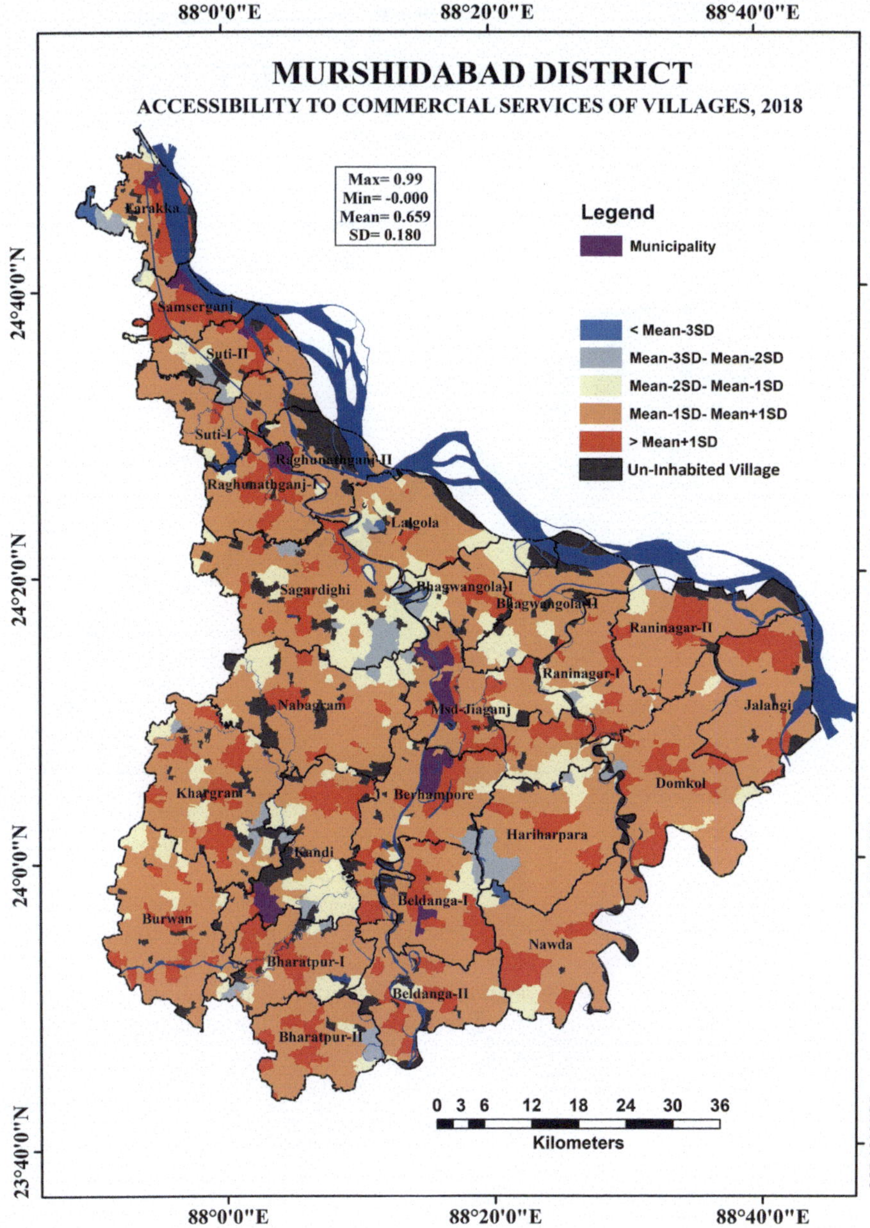

Fig. 7.19 Commercial service accessibility (*Source* Mapping by the authors based on their own calculations)

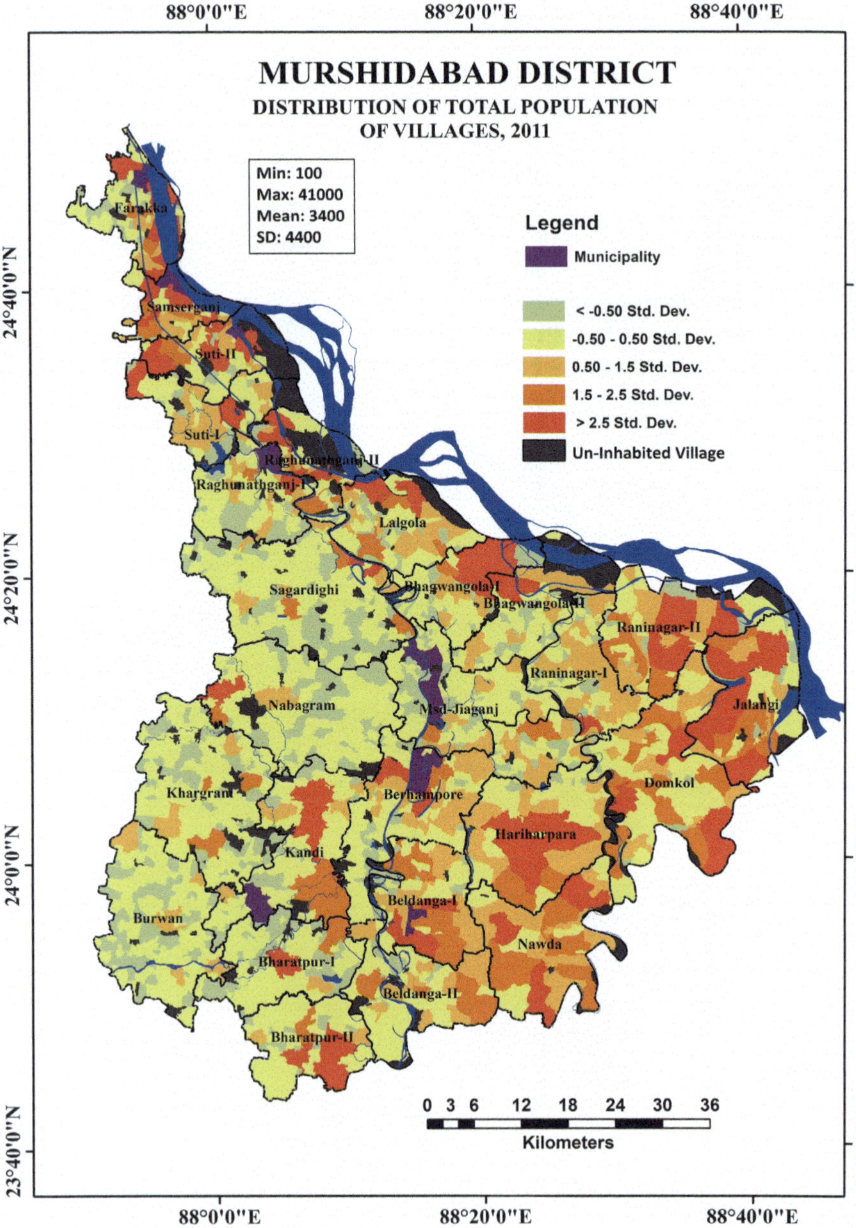

Fig. 7.20 Distribution of total population (*Source* Mapping by the authors based on their own calculations)

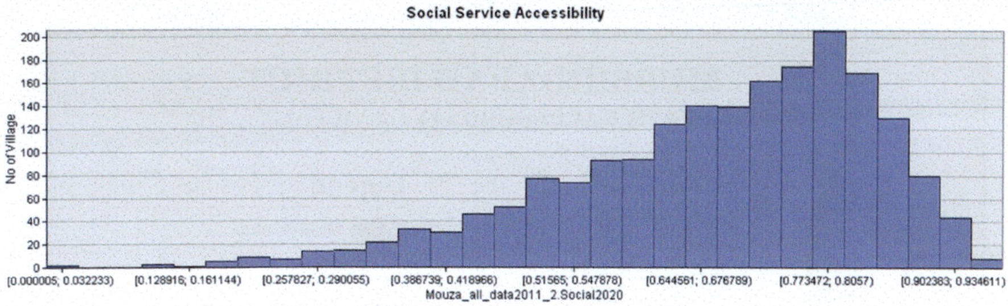

Fig. 7.21 Frequency distribution of social service accessibility (*Source* Prepared by the authors)

varies. Thus in many cases, as observed in the maps (Figs. 7.22 and 7.23), both small and large villages are benefitted similarly and in some cases, low accessibility score is obtained despite large population size and vice versa. Thus the population size, although a big determining factor of controlling the location of facilities, is weakly correlated with the accessibility parameters in the present study area (Table 7.9).

It is most common that public utilities tend to establish in areas with relatively good physical accessibility. A high correlation is always expected between social service accessibility and physical accessibility. In the present study, the correlation coefficient of social service accessibility with physical accessibility is 0.636, which is significantly good in the present context.

As already mentioned, the weighted facility index has been calculated for supplementing the public and social service accessibility index and found fruitful (Figs. 7.24 and 7.25). The frequency distribution shows an extremely skewed distribution towards the positive direction. It indicates that there are large numbers of villages with very low facility scores. There is a wide variation in the distribution of various facilities in the district, and the weighted facility index varies from as much as 365 to as low as 0. High value concentrates in and around the Block headquarters due to higher weightage assigned to block headquarter.

7.8 Findings

7.8.1 Accessibility Indices and Population Parameters

Very weak but significant correlations are found between the population and accessibility parameters (Table 7.9) which indicates good accessibility must be in the areas where population concentration is high. In addition to this, a strong positive correlation with the weighted facility index signifies our objective of using the weighted facility indices as a supplementary method to social, and administrative service accessibility.

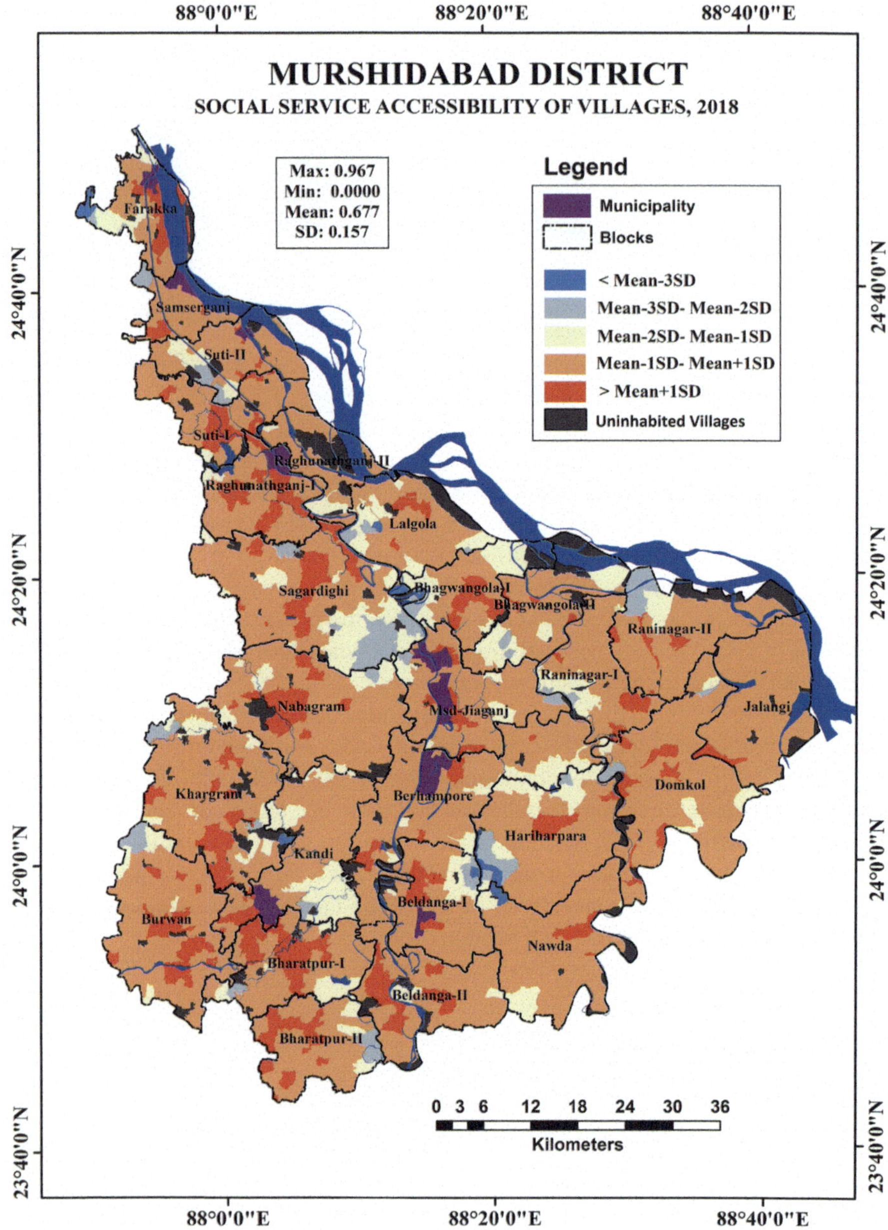

Fig. 7.22 Social service accessibility

Table 7.9 Correlation between population and accessibility parameters

Pearson's correlations

	1	2	3	4	5	6	7	8	9	10	11	12	13	14
1. Popopulation	1													
2. Phisical Accessibility	0.153**	1												
3. TSS	0.164**	0.621**	1											
4. OAL	0.120**	0.875**	0.832**	1										
5. EAS	0.166**	0.636**	0.947**	0.815**	1									
6. EAP	0.116**	0.630**	0.829**	0.856**	0.828**	1								
7. Educational Accessibility	0.222**	0.382**	0.630**	0.507**	0.609**	0.522**	1							
8. Health Accessibility	0.241**	0.531**	.615**	0.578**	0.677**	0.606**	0.889**	1						
9. Commercial Accessibility	0.257**	0.414**	0.665**	0.542**	0.681**	0.578**	0.957**	0.892**	1					
10. PDS	0.142**	0.321**	0.563**	0.445**	0.631**	0.476**	0.871**	0.801**	0.844**	1				
11. W Road Length	0.727**	0.148**	0.132**	0.144**	0.115**	0.123**	0.205**	0.192**	0.214**	0.113**	1			
12. Road Length	0.682**	0.086**	0.088**	0.098**	0.061**	0.079**	0.168**	0.136**	0.164**	0.078**	0.975**	1		
13. WF Index	0.607**	0.147**	0.174**	0.144**	0.220**	0.187**	0.199**	0.264**	0.250**	0.174**	0.567**	0.501**	1	
14. WR Density	0.095**	0.268**	0.152**	0.228**	0.185**	0.164**	0.148**	0.203**	0.185**	0.136**	0.181**	0.133**	0.095**	1
15. Road Density	0.061**	0.211**	0.101**	0.186**	0.122**	0.118**	0.107**	0.143**	0.130**	0.094**	0.167**	0.156**	0.052*	0.954**

Source Prepared by the authors
* Correlation is significant at the 0.05 level (2-tailed)
** Correlation is significant at the 0.01 level (2-tailed)

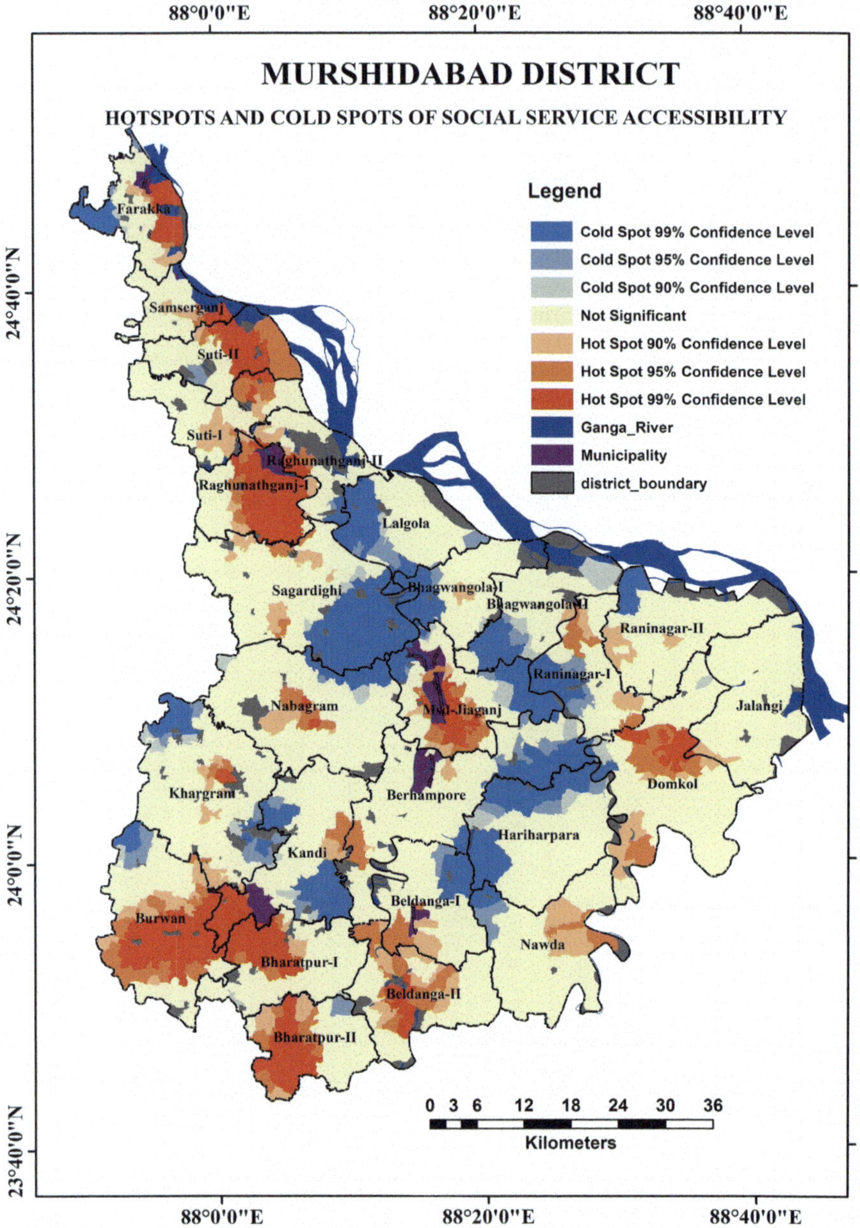

Fig. 7.23 Clustering of social services (*Source* Mapping by the authors based on their own calculations)

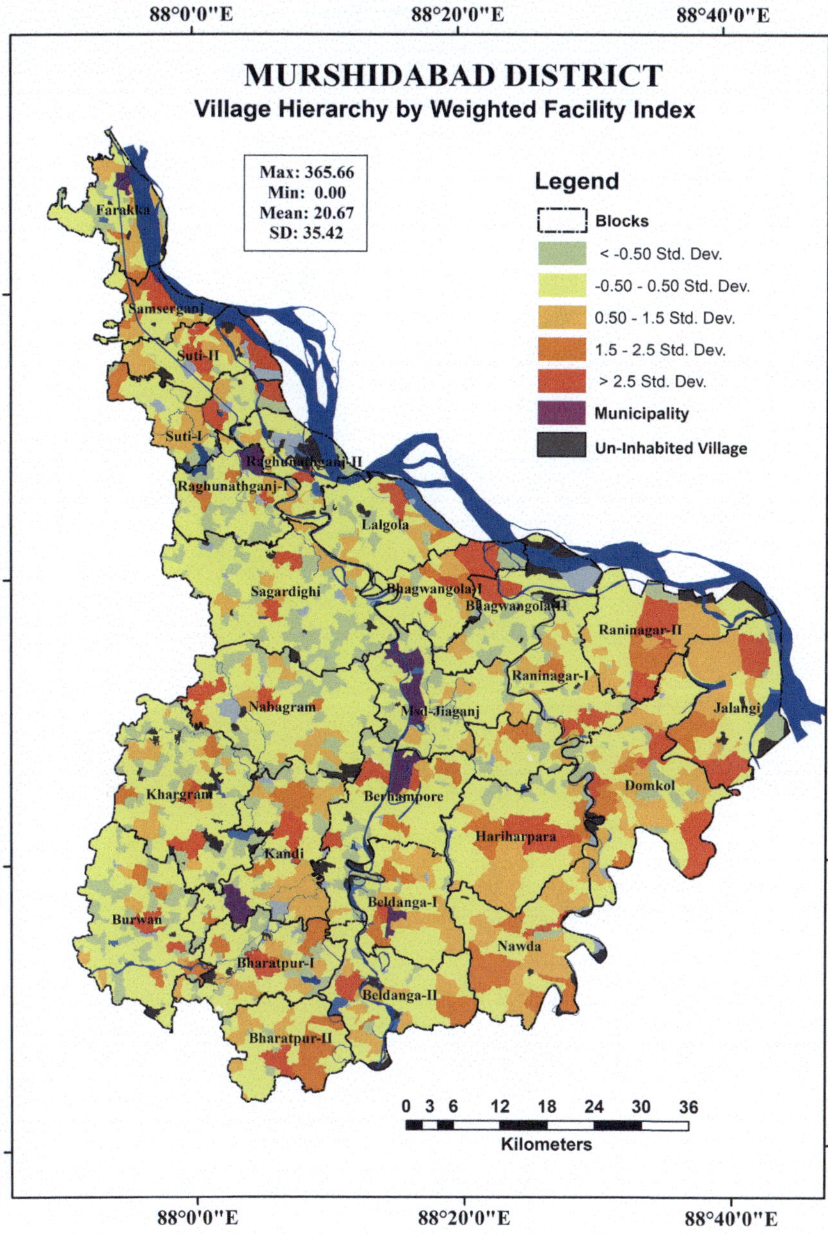

Fig. 7.24 Weighted facility index of villages (*Source* Mapping by the authors based on their own calculations)

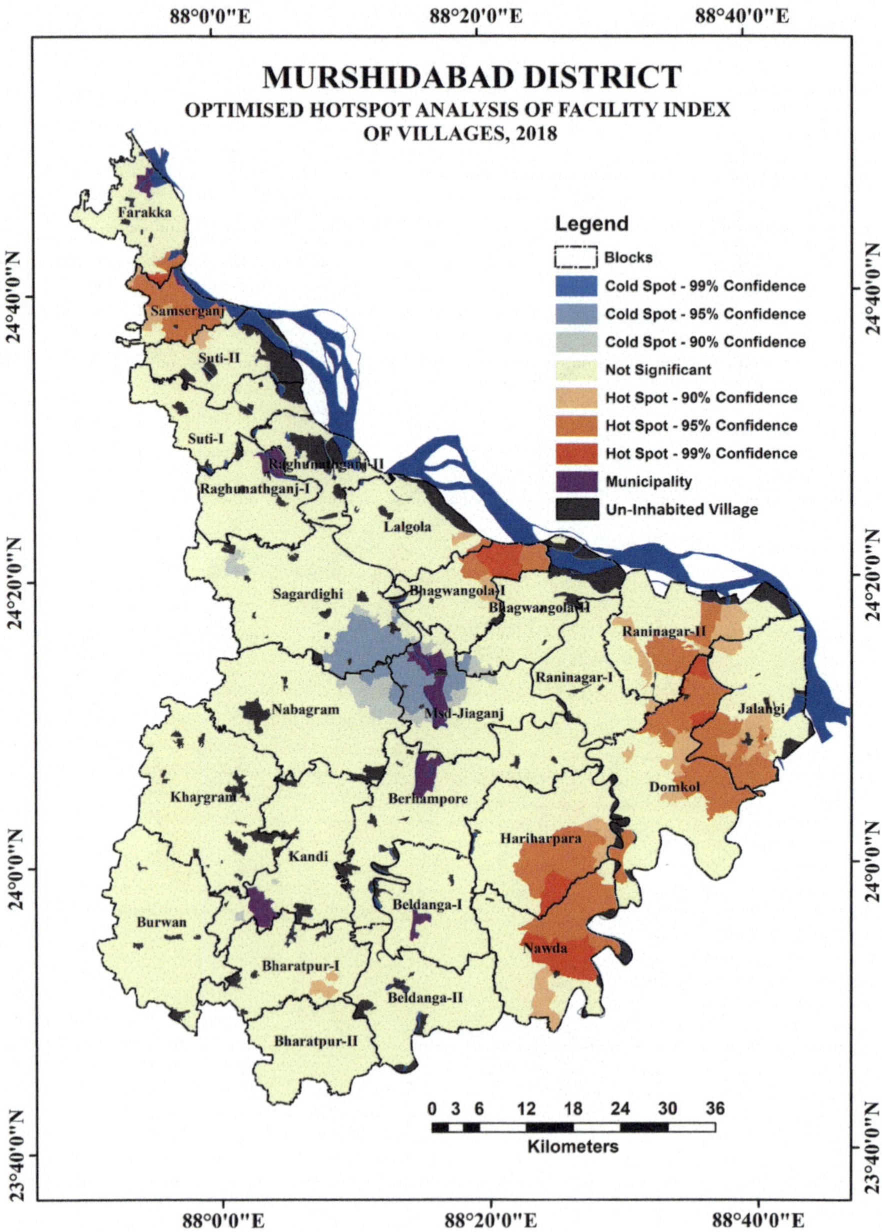

Fig. 7.25 Clustering of weighted facility index (*Source* Mapping by the authors based on their own calculations)

To understand the degree of correlation between population size and location of public utilities, both OLS (Ordinary Least Square) and GWR (Geographically Weighted Regression) are run taking Weighted Facility Index as the dependent variable and total population as the independent variable, assuming that location of the facility is a function of the population (Fig. 7.26). A positive and high correlation is always expected though in this case, a moderately positive correlation is observed from Ordinary Least Square with only 39% variation of the facility score is explained by the total population. The regression sum of squared is statistically significant at 99% confidence level, and the model is validated by running Moran's I which shows a random distribution of the standard residuals (Moran's I = 0.0105, and p-value = 0.159; Z-score = 1.406). The distribution of the residuals (Fig. 7.27) suggests that more public utilities should be established in the villages with high negative residuals in both the regression because, in these villages, the facility is not developed accordingly with the population size (Fig. 7.28).

7.8.2 Level of Accessibility Parameters and Level of Development

It is noticed that developing good road access does not ensure the accessibility of villages to basic amenities and also does not help in reducing poverty (Cervero 1990; Shalini et al. 2010). Because accessibility has many dimensions and the total improvement in all the dimensions shall produce the best impact on rural life. Besides, the impacts of good road connectivity on the overall development of people depend on their socio-economic background.

In this section, we have tried to cover all the dimensions of connectivity and accessibility. Based on this, an attempt has been made to assess the overall socio-economic development of the villagers.

In general, the developed villages lie (See Figs. 7.4 and 7.5) in the northern Blocks (many of which have converted into census towns in the Census 2011) and close to the urban centres of the district (Fig. 7.29), which clearly shows the urban influence in the villages. A significant negative correlation (–0.43) is obtained between development and distance from urban centres, which clearly states that with increasing distance from urban centre development decreases. Geographically Weighted Regression (with 51 numbers of neighbours in each kernel) is also run, and about 50% of the variation in the development is explained by distance from urban centres.

Most of these villages are also good in terms of physical accessibility and social service accessibility, although noticeable differences are observed in terms of overall accessibility. Therefore, a weak positive correlation between development with physical accessibility and social service accessibility has been obtained, but with the OAL the correlation is very weak.

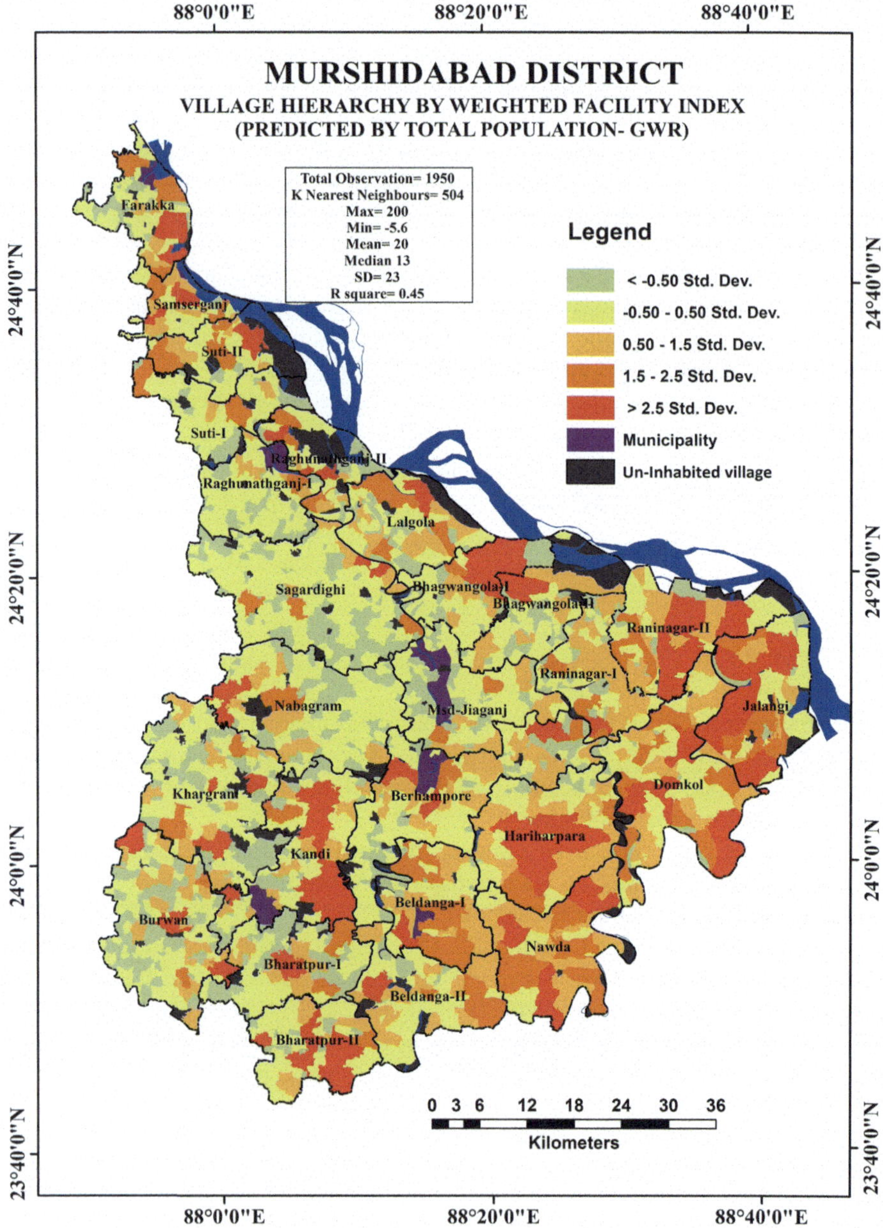

Fig. 7.26 Predicted Weighted Facility Index (Total Population) (*Source* Mapping by the authors based on their own calculations)

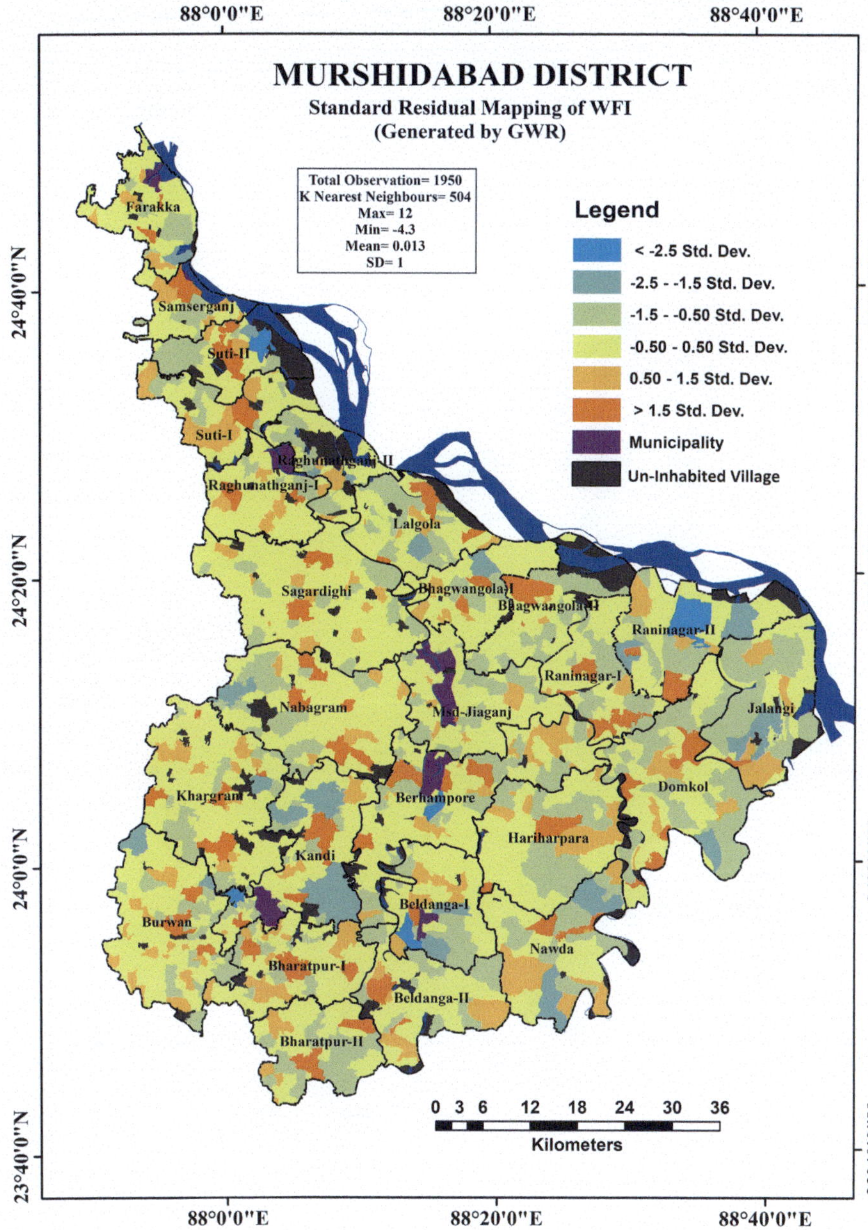

Fig. 7.27 Standard residuals (*Source* Mapping by the authors based on their own calculations)

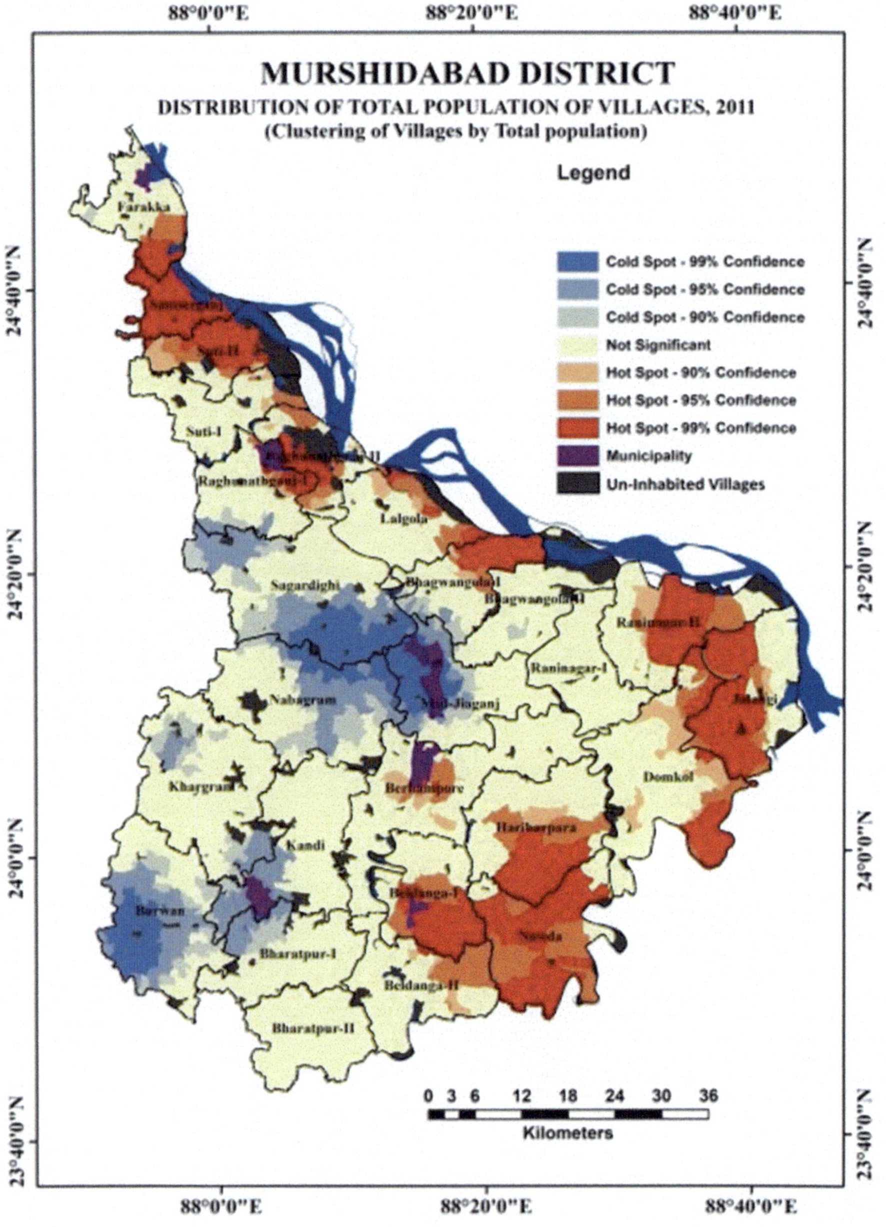

Fig. 7.28 Clustering of population (*Source* Mapping by the authors based on their own calculations)

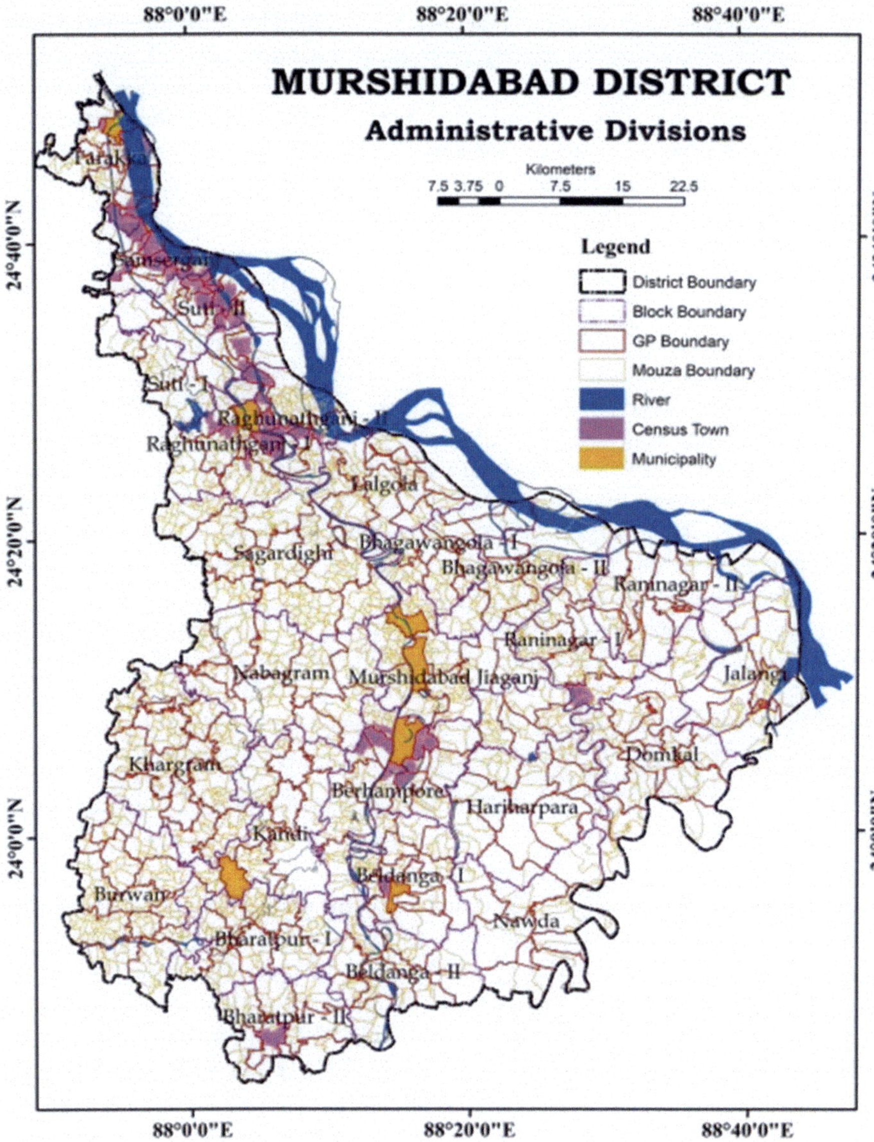

Fig. 7.29 Location of urban centres (*Source* Mapping by the authors based on their own calculations)

This may be explained by the fact that OAL is computed by combining three indicators of transport accessibility, namely, physical accessibility, network accessibility and transport service score. Network accessibility greatly influenced it by giving extra weightage to the villages around the district node, i.e. Berhampur. Besides, network accessibility is negatively correlated with development, and the axes of development lie towards the northern blocks. Therefore, a very weak correlation has resulted between development and overall accessibility.

The GWR with 444 numbers of neighbours in each kernel shows that 38% of the variability in development is explained by overall accessibility, social service accessibility, and ease of access to public service, where social service accessibility solely responsible for 34% (444 neighbours in each kernel) of the variability in the level of development and OAL solely contribute 32% (444 neighbours in each kernel) of the variability in development.

The residual map (Fig. 7.30) of GWR between the level of development and accessibility parameters which contribute about 38% of the variation of development, shows that about 50% of the total habited villages are low in standard residuals and can be said as developed per the level of accessibility. But there are some villages in the northern, western, and north-eastern part of the district where both high negative and high positive residuals are obtained.

A moderately strong and positive correlation is observed between the level of socio-economic development and level of transport development (Table 7.10) which is also clear from the figures and of clustering of villages with relative development and underdevelopment in terms of socio-economic and transport development.

7.9 Conclusion

There is wide spatial variation in the level of overall socio-economic development among the villages and there is also wide variation in the level of accessibility and transport development. In many areas, high levels of overall accessibility and social service accessibility correspond to a high level of socio-economic development and vice versa.

Here measuring the accessibility helps the decision-makers to prioritize the regions or habitations where accessibility should be improved for further developmental incentives. Thus, in this papser, the Village Prioritization Index (VPI) has been devised to prioritize the villages for further development purposes.

For this, a simple quantitative technique has been devised using the three indices-Overall accessibility, Social Service accessibility and population. The reciprocal of ranks are used as the weight of the population of the villages. The average value of the two accessibility indices is multiplied by the population weightage (reciprocals of ranks given in ascending order of census categories) to get the village prioritization index (VPI), which indicates high priority to low values and vice versa.

From the VPI, the folslowing categories of villages can be recognized as per the priority:

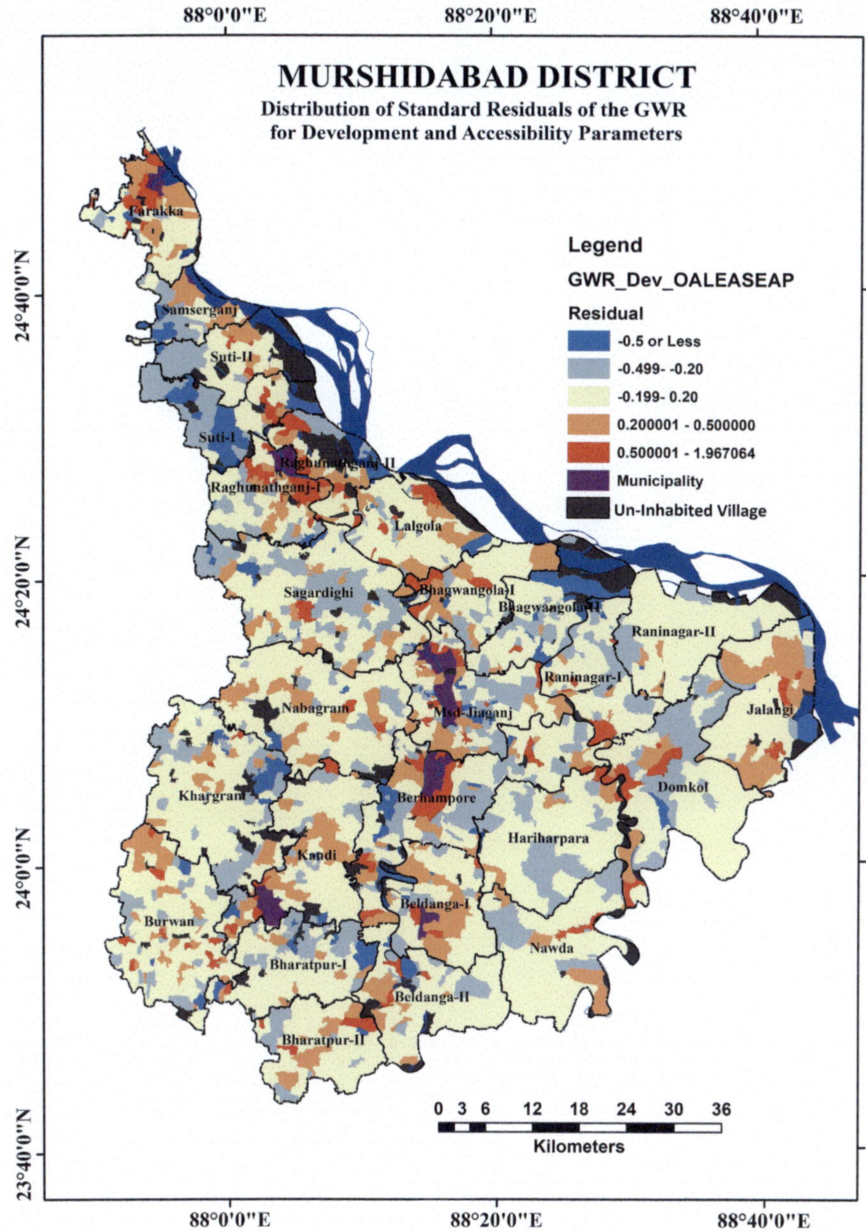

Fig. 7.30 Residual map of development (*Source* Mapping by the authors based on their own calculations)

Table 7.10 Correlation between development and accessibility parameters

Pearson's Correlations

	1	2	3	4	5	6	7	8	9	10	11	12	13
1. Physical Accessibility	1												
2. Network Accessibility	0.312**	1											
3. TSS	0.621**	0.112**	1										
4. EAP	0.630**	0.437**	0.829**	1									
5. Weighted Facility Index	0.147**	−0.035	0.174**	0.187**	1								
6. Composite Index Z	0.283**	−0.144**	0.218**	0.222**	0.221**	1							
7. PCA1	0.198**	−0.172**	0.089**	0.116**	0.212**	0.770**							
8. PCA2	0.238**	0.178**	0.286**	0.376**	0.115**	0.479**							
9. PCA3	0.022	−0.360**	0.030	−0.151**	0.045*	0.320**							
10. PCA4	0.028	0.086**	0.006	0.061**	0.017	0.039							
11. Composite Score PCA	0.243**	−0.134**	0.205**	0.202**	0.195**	0.804**							
12. Transport Development	0.565**	0.077**	0.659**	0.567**	0.219**	0.335**	0.229**	0.267**	0.102**	0.010	0.304**	1	
13. OAL	0.875**	0.531**	0.832**	0.856**	0.144**	0.191**	0.078**	0.314**	−0.094**	0.044	0.171**	0.619**	1
14. EAS	0.636**	0.122**	0.947**	0.828**	0.220**	0.295**	0.162**	0.312**	0.046*	0.004	0.262**	0.673**	0.815**

Source Computed by the authors

** Correlation is significant at the 0.01 level (2-tailed)

* Correlation is significant at the 0.05 level (2-tailed)

- Villages with low accessibility, low development but a high population (Lowest VPI)—need improvement in overall accessibility and provision of economic opportunities, and provision of social services should be the topmost priority.
- Villages with high accessibility, low development but a high population (Low VPI)—need a revision of the economic forces running in these villages. Economic forces should be modified accordingly so that new and viable economic opportunities might arise.
- Villages with low accessibility, high development, and a high population (Lowest VPI)—need special attention to improve accessibility. These villages can give the idea of developing the backward villages.
- Villages with high accessibility, high development but the low population (High VPI)—can be the potential area for development, and new economic forces might come up in these areas, which may attract people from surrounding villages.

Development and Village Priority Index are not at all correlated in the district of Murshidabad, which means development is not per the population and accessibility criteria of the villages, and in many cases, high priority villages are least developed, which needs to be addressed. Accessibility should be improved in such a manner that it should reach a standard level where the people of every village irrespective of their size can take part in their development activities.

The regression analysis (Fig. 7.31) helps to reach the conclusion that both high negative and high positive residuals exist which means development is not following the level of accessibility in certain villages. High negative residuals are produced due to the low observed composite development score despite their higher observed accessibility score. These villages are to be addressed for finding out the causes of their relative underdevelopment and necessary actions should be taken accordingly.

High positive residuals are the result of high observed composite development scores despite their moderate to low level of accessibility. These villages can be observed to find out the factors, other than the accessibility, which are responsible for their development and similar actions can be taken in those villages with high accessibility but low levels of development.

Thus, the suggestion in this regard is that accessibility could be improved in such a manner that it reaches a standard level where people of every village irrespective of their size can take part in their development activities. New facilities should be introduced to raise their accessibility level for those villages which are lagging behind in terms of social service accessibilities. Provision of urban amenities, along with good transport linkages and services, in those villages with population size 10,000 or more may help themselves to develop as semi-urban centers with rural marketing hubs.

The present spatial analysis thus provides us a solid framework for identifying the spatial pattern of accessibility and development as well as the distribution of population based on which proper planning strategies can be designed.

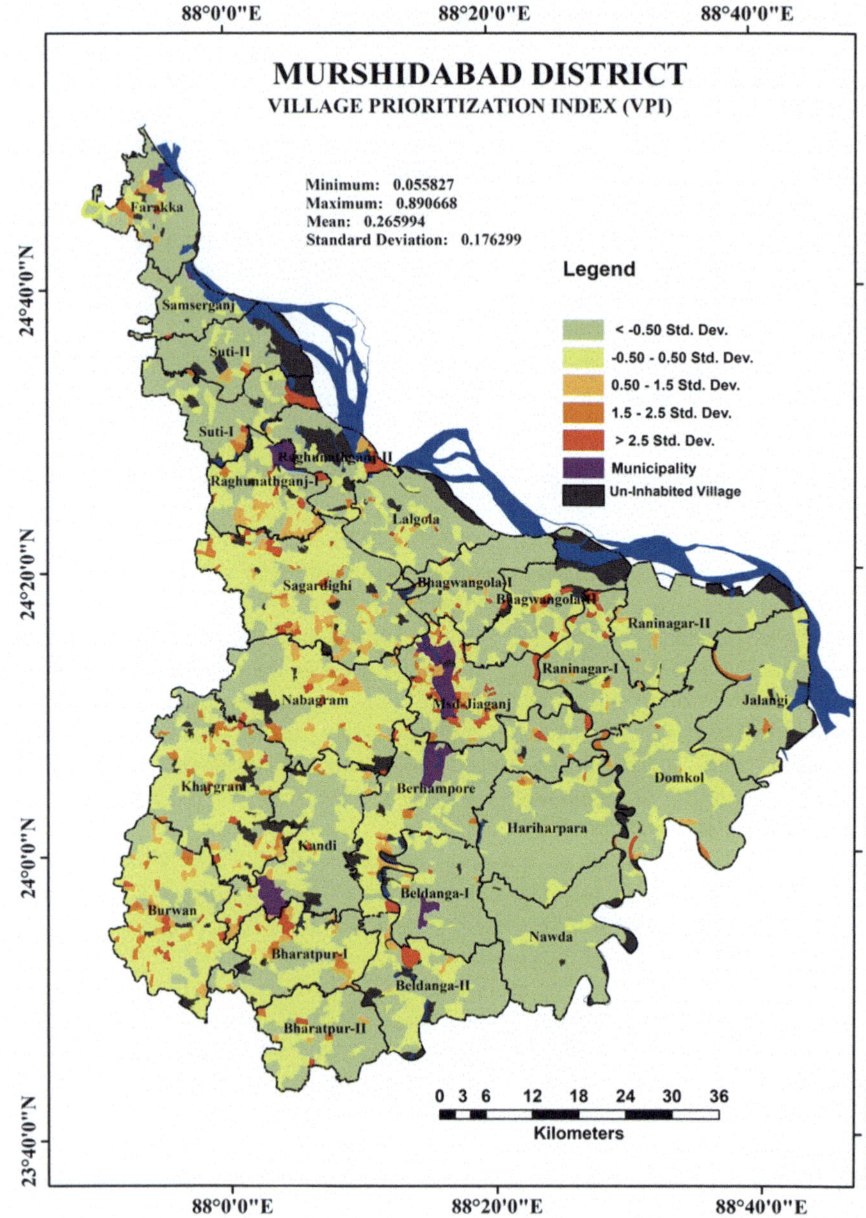

Fig. 7.31 Village prioritization index (*Source* Mapping by the authors based on their own calculations)

References

Castro J, Vistan EL (2019) A Geographic Information System for Rural Accessibility: Database Development and the Application of Multi-criteria Evaluation for Road Network Planning in Rural Areas. In: Castro J, Vistan EF (ed) Urban and Transit Planning (pp. 277–288). Springer.

Cervero R (1990) Accessibility and third world rural development: a case study of sumatra. Rev Urban Region Develop Stud 2(1):125–138

Dennis R (1998). Rural transport and accessibility—A synthesis paper (RATP No. 1). International Labour Office, Geneva

Islam A (2017) Assessment of Rural Transport Accessibility in Bangladesh. Unpublished Master Dissertation of the Department of Civil Engineering, BUET

Jhonston DC (1989) Transport in rural development: toward a specification of the relationship. In: Rao DP (ed) Dimension of rural transportation (pp I.161–I.174). Inter India Publishers, New Delhi

Kanuganti S, Sarkar AK, Singh AP, Arkatkar SS (2014) Quantification of accessibility to health facilities in rural areas. Case Stud Transp Policy:311–320.

Lebo J, Schelling D (2001) Design and appaisal of rural transport infrastructure: ensuring basic access for rural communities. World Bank, Washigton DC

Mahmood A (2013) Statistical Methods in Geographical Studies. Rajesh Publications, New Delhi

Nutley S (1999) Rural Accessibility and Transport. In: Pacione M (ed) Applied Geography: Principles and Practices. Rutledge, New York, pp 474–485

Pacione M (1995) The Geography of Multiple Deprivation in the Clydeside Conurbation. Tijdschr Econ Soc Geogr 86(5):407–425

Patel NR, Madhavan T (1984) Planning for rural roads in India. IIM, Ahmedabad

Sarkar AK, Mashiri M (2001) Quantification of accessibility level of rural areas: a case study in the Northern Province-South Africa. Trans Res Board Week 347–376

Sarkar AK, Dash M (2011) Quantification of accessibility and priritization of villages for local level planning. Trans Commun Bulletin Asia Pacific 1–22

Sarkar A (2013) Quantitative geography: techniques and presentation. Kolkata, Orient Blackswan

Shalini P, Vajjhala, Walker M (2010) Roads to Participatory Planning: Integrating Cognitive Mapping and GIS for Transport Prioritisation in the Rural Lesotho. J Maps:488–504

Shalini K (2015) Quantification of rural accessibility and development of a need based approachfor rural road network planning. Pilani, Unpublished PhD Thesis of BIST

Starkey P, Ellis S, Hine J, Ternell A (2002) Improving rural mobility: options for developing motorized and nonmotorized transport in rural areas. World Bank, Washington DC

Taaffe EJ, Gauthier HL, O'Kelly ME (1996) Geography of transportation (second ed), London, Prentice Hall Inc

Tarafder S, Jana NC (2016) Level of rural development in burdwan and murshidabad districts, west bengal: a comparative study. Space Culture, India 4(1): 65–80

Tarafder S, Jana NC (2020) Explaining spatial variation of rural development concerning accessibility parameters: a case study in Beldanga-I block of Murshidabad District in West Bengal. Ensemble 2(1):36–51

Tobler W (1970) A computer movie simulating urban growth in the detroit region. Eco Geograph 46(2):234–240

Part III
Regional Imbalance in India Revisited

Chapter 8
Managing the Regional Inequalities in India with Particular Reference to the Transformation of Aspirational Districts Programme

Tanmoy Sarkar, Mukunda Mishra, and R. B. Singh

Abstract Empowering local governments in the planning process is a long-waited goal of the decentralized planning mechanism in India as well as it remains as one of the most-debated issues in the planning process under the federal structure of the governance. The regional social, economic and cultural diversities in the country have remained challenging to the policymakers in framing the policies for development as the planning initiatives without percolating up to the grass-root level have been actively promoting to widen the inequalities between the advanced and backward sectors and regions. At this backdrop, the Transformation of Aspirational District Programme (TADP), launched by the Government of India in 2018, looks path-breaking as it links the government's development scheme with the Sustainable Development Goals (SDGs) to upthrust the backward districts through utilizing its potential resources for development through regular monitoring, recording of progress, outcome assessment and re-organizing the planning directives as per the requirement within a specific timeframe. This chapter reviews the TADP to analyse its prospect in transforming the districts through utilizing the 'power' of federal linkages and making districts resilient for balanced development.

Keywords Local governments · Planning · SDG · Inequalities · Federalism · Development

8.1 Introduction

India's growth trajectory since independence has shown a remarkable improvement through time, leaving an enormous scope to focus on development. Economic growth

T. Sarkar
Department of Geography, Gazole Mahavidyalaya, Malda, West Bengal, India

M. Mishra (✉)
Dr. Meghnad Saha College, Uttar Dinajpur, Itahar, West Bengal, India

R. B. Singh
Department of Geography, Delhi School of Economics, University of Delhi, Delhi, India

M. Mishra et al. (eds.), *Regional Development Planning and Practice*,
Advances in Geographical and Environmental Sciences,
https://doi.org/10.1007/978-981-16-5681-1_8

is a means for achieving human development, but higher economic growth does not always represent the overall up-gradation of life. India geographically covers 32,87,263 km^2 with an extensive divergence in physiography, culture, language, and traditions. India stands as the seventh largest country and second-largest population with 1,210,193,422 (Census of India 2011) total population distributed in 28 states and 8 Union territories. In such an extensive geographical area, spatial inequality in natural resources and economic prosperity is inevitable. However, the country has a role in accelerating the actual development goal by gearing up the human well-being of the entire population and providing the positive outcome of the higher economic growth trajectory to each section of the society. The impact of aspiring GDP growth through the decades must be shown in the lives and living standards of the people across the country. Consistent economic growth through the decades must be a means to lift millions of people from the gloomy state of poverty. However, the real story tells something else! The UNDP human development report 2016 shows the rank of India as 131 out of 188 countries according to the Human Development Index (HDI) when India is dreaming of being a five trillion-dollar economy in the world. Moreover, the introspection reveals that the inter-district and inter-state disparities in living standards and human well-being must be taken care of to improve the overall ranking in HDI.

8.2 India's Development Scenario

After the British colonial period for more than a century, India emerged as an independent country in 1947, and the journey of the newly independent India primarily struggled with various social, economic, political and demographic issues. Fight against hunger and the development of a self-sustaining economic structure was the primary concern (Shaban 2016). Extensive diversity in culture, geography, society, along with different political beliefs and ideologies, growing population pressure, guided the government to take multifaceted strategies for the country's development in this pre-reform period (1947–1990). India's unique social, political, and demographic character has had been guided the economy to balance between public and private sectors with varying facets through time. Adopting economic policies to develop a self-sustaining economy and economic growth was the primary target since independence. On the eve of independence, the Indian economy had to face the legacy of an exploitative colonial economy where India was considered the major source of raw material for British industrialization and was the consumer of Britain's finished products. Hence, the states and cities with the port facility, e.g. Calcutta, Chennai, Mumbai, etc., have had some locational benefit and the historical development legacy. The developmental strategies and planning were executed through the Five Years Plan taken by the Planning Commission of India. The linear thinking of 'economic growth led development' dominated all the planning decisions and programs after independence to develop a stable economic base, loaded with a large population below the poverty level (Shaban 2016).

The Indian economic policy was followed by a balanced economy between the socialist and capitalist modes of the economic model, though a slight leaning towards socialist principles. In the first two decades after independence, experimentation with various theoretical models and principles for planning and development decision-making by Indian planners was not potent enough to address the regional imbalances. However, the Second Five-Year Plan followed the Mahalanobis model for rapid capital-intensive industrialization with particular emphasis on heavy industries (Mukherji 2009) and also targeted to reduce the regional disparity (Government of India [GoI] 1956). The Third Five-Year Plan also emphasized the importance of balanced regional development (Government of India [GoI] 1961), and the next five-year plan emphasized the issue of resource allocation and decentralization of economic activities to reduce the regional disparity (Government of India [GoI] 1969). Planning Commission's role and importance in planning and development decision-making were very fluctuating in different periods after independence with the change of Prime Ministers and allied government policy principles (Mukherji 2009). And the changing balance between the role and importance of state over private sectors had contributed to the issue of regional imbalance and inequality. Sardar Patel's demise in 1950 and Nehru's command on the Indian National Congress and the Central Government led to more robust control of the State over private sectors; and, the Planning Commission's power was increased. The Nehru period's legacy was changed when Lal Bahadur Shastri became the Prime Minister after Jawaharlal Nehru's death in 1964. Furthermore, these changes in policy and principles made some noises in long-term stable economic policy for development. The absence of proper addressing of regional imbalances and inequality since the Second Five Years Plan resulted in a sharp increase in inequality—state-wise and region-wise. The crisis experienced due to the severe droughts in 1964–65 and 1965–66; Indo-Pak war in 1965, and the sudden demise of Prime Minister, Lal Bahadur Shastri in 1966, had a severe blow to this grim economic situation. The slow growth rate was contributed by the political instability, economic hardship, social unrest against the personalization of politics by Mrs. Indira Gandhi and the last straw was the declaration of National emergency in 1975. India's growth rate was moderately accelerated with an average of greater than 5% from 1975 to 1990 in the context of the steadily strengthened private sector, and trade orientation started in the 1970s and got an acceleration since the 1980s. But, as the Indian economy struggled with different economic principles, approaches in searching for a stable economic growth, and getting a stable economic base, there were inadequate attempts and limited scope to address the regional inequality and spatial development imbalances.

There was a twist and turn in the Indian economy in the 1990s with a focus on liberalization. The Indian economy has been experienced a paradigm shift in the early 1990s with economic reforms. India's remarkable shift in economic structure and acceptance of economic liberalization and consequently economic growth trajectory has had been appreciated globally since the 1990s. However, the growth outcome was not percolated equally in all socio-economic dimensions and geographical spaces. The development programs undertaken so far, targeting employment generation, income enhancement, augmentation of physical quality of life, betterment of public health, etc. with the financial sponsorship of both State and Central Government, have succeeded in partial fulfillment of the target. In most cases, the lack of sufficient magnitude of public participation in the planning has been exposed as the formidable challenge in gaining holistic success, and failure of a plan in this respect would have every possibility to carry the benefits to a narrow part of the population by depriving the rest, and the inequality tends to increase away. Unchecked and uncontrolled process of growth leads to these regional disparities (Rao 1984). The growing regional imbalances, elevating social and class inequalities, raise the issue of growth versus development. The target of inclusive growth by the government of India was taken in the Eleventh Five-Year Plan, and it was continued to the next five-year plan.

8.3 District as a Planning Unit

The concept of empowering local government in planning processes was sown when the Grow More Food (GMF) Enquiry Committee in 1952 suggested the community development block as a remedy to the problems of multiple rural development agencies working in scattered, without a sense of shared objectives[1] (Table 8.1). Based on the committee's recommendations, the Government of India launched the community development program on a pilot basis in 1952 as a step toward a substantial increase in the country's agricultural program along with the improvements in systems of communication, rural health and hygiene, rural education for transforming the social and economic life of the rural communities. Independent India witnessed the rapid implementation of community development programs. By the end of the First Five-Year Plan period, there were 248 blocks, covering around about 20% of the population. However, by the end of the Second Five-Year Plan period, there were 3,000 blocks covering 70% of the rural population, and the entire country was covered by 1964.

The first Administrative Reforms Commission (ARC)[2] speculated the planning at the district level was pertinent to address the local variations in development patterns. The Planning Commission of India issued its first guidelines for district planning in 1969. Although it led to several states formulating district plans, the major drawback of this initiative was that it did not incorporate urban planning processes as part of the district planning mechanism. Ultimately, this first initiative of the district-level

---

[1] Government of India in 1952 appointed the GMF Enquiry Committee to examine the working of the Grow-More-Food. The Committee recommended to bringing about an appreciable improvement in the standards of rural life to make it fuller and richer that would awaken the rural masses to take interest in not only increasing agricultural production but also improving their own conditions and creating a will to live better (For details, see the complete Report at https://indianculture.gov.in/report-grow-more-food-enquiry-committee).

[2] The first ARC was established by the Ministry of Home Affairs, Government of India vide resolution no. 40/3/65-AR(P) on 5 January 1966. The Commission was initially chaired by Morarji Desai. Latter, K. Hanumanthaiah became its chairman as Mr. Desai became the Deputy Prime Minister of India.

Table 8.1 A chronology of attempts to empower districts the planning unit

Year	Leading role	Effort
1952	Grow More Food (GMF) Enquiry Committee	Suggested the implementation of Community Development Blocks to enable multiple rural development agencies to work with common objectives
1952	Planning Commission	The community development program was launched on a pilot basis to break up the planning exercise into national, state, district and local community levels
Second plan (1956–61)	Planning Commission	3,000 blocks were constituted by the end of the Second Five-Year Plan period that covered 70% of the rural population
1957	Balwant Rai Mehta Committee	Recommended for the establishment of the Village, block, District Panchayat institutions
1964	Planning Commission	The entire country was covered under Community Development Blocks
1967	Administrative Reforms Commission	Recommended for the resources to be given/local variations accommodated, purposeful plan for the area
1969	Planning Commission	Formulated guidelines; detailed the concept of the district plan and methodology of drawing up such a plan in the framework of annual plans, medium-term plans, and perspective plans
1978	Prof. M. L. Dantwala Committee	Recommended for the block-level planning to form the link between the village and district-level planning
1983–84	CSS/Reserve Bank of India	Strengthen district plan/District Credit Plan
1984	Hanumantha Rao Committee	Decentralization of functions, powers and finances; Setting up of district planning bodies and district planning cells

(continued)

Table 8.1 (continued)

Year	Leading role	Effort
1985	G. V. K. Rao Committee	Administrative arrangements for rural development; District Panchayat to manage all development programs

Source Adopted from the Planning Commission of India (2008)

planning ends with a dissatisfactory outcome as the local planning exercise was not linked to the annual planning process in most of the states (Planning Commission of India 2008).

The Fourth Five Year Plan (1969–74) of India is marked for the first effective effort towards decentralization of national planning by introducing the district-level plan. Further, the district-level plan was decentralized at the block level during the Sixth Five Year Plan (1980–85). The seventh plan (1985–90) emphasized local area planning by stressing demographic decentralization (Siddiqi and Hussain 2010). However, till then, the overall control on planning processes, decision-making function, and strategy formulation remained centralized around the two political levels, namely the Union and the states, whose spheres of responsibility had been defined in the Constitution. However, India's planning processes between the late sixties and the mid-eighties were highly administered by a strongly centralized mechanism. The growth and multiplication of sectoral departments and parastatal bodies, in contrast with the weak self-governance institutions, got the space to drive the planning as the centralized process. Though several states constituted the Panchayats after the Balwant Rai Committee's recommendation, the Panchayats were still not considered a permanent feature of a multi-tiered government system. The development schemes, mostly being formulated as Centrally Sponsored Schemes (CSS) and implemented through line departments, resulted in the district planning process to technically collapse.

It was the enactment of the Constitution (73rd[3] and 74th[4] Amendment) Acts in 1992 which ushered in the new era of the decentralized process where district became a key unit in the multi-level planning mechanism, bestowing a constitutional status on rural (Panchayat) and urban (Municipality) local bodies to enable them

---

[3] A new part 'IX', which consists 16 Articles and the Eleventh Schedule, has been added to the Constitution vide the Constitution (73rd Amendment) Act, 1992. It envisages the Gram Sabha as the foundation of the Panchayat Raj System for performing the functions and powers as conferred by the State Legislatures. A three-tier Panchayat Raj System at the village, intermediate and district levels have also been enacted by this amendment. The Gram Sabha has been authorized to exercise such power and perform such functions at the village level as the Legislature of a state (may be law provide) by the Article 243A of the Indian Constitution.

[4] Part IXA has been added to the Constitution vide the 74th Constitutional Amendment Act (CAA), 1992. It was enacted in 1993 that constitutionally recognizes the municipal governments by making it a formal part of the three-tier governing system. This Act mandates the setting up of municipal governments in urban areas, however, the details of their constitution has been left to the state governments to follow their respective Municipal Acts.

function as effective democratic self-government institutions (Qaiyum 2004). These amendments mandated the establishment of Panchayats at the district, intermediate and village levels clearly envisaged a reversal of the hitherto centralized approach to district planning. The 74th Constitutional Amendment Act mandated the establishment of the District Planning Committee (DPC), which would be entrusted to prepare the Draft District Plan by consolidating the plans prepared by Panchayats and Municipalities in the district.

Consequently, the eighth plan (1992–97) emphasized building and strengthening people's institutions and making people actively participate within the liberalization and privatization framework. The main focus was on population control, environmental protection, and infrastructural development. The ninth plan (1997–2002) aimed at growth with social justice and equity. In its tenth plan (2002–2007), the Planning Commission advocates the area approach and aims to strengthen decentralization of planning (Mishra and Puri 2002). Thus, the decentralized planning policy procedure was adopted to prepare village plans by collecting village requirements at block levels, and finally, they were put together at the district level for district plans. However, such initiatives were mostly confined only to documents. Removal of regional imbalances in development has remained the avowed goal of planning in India (Mohan 2005). By the beginning of the new century, India's decentralized planning efforts started functioning through the allotment of aids and resources to the micro-level administrative departments and their expenditure through the local governments, keeping the rationale of the decentralized plans mostly unfulfilled (Reddy 1988).

Amidst this confusing scenario of the role of local governments, the Government of India (GOI) constituted the Second Administrative Reform Commission (Second ARC)[5] with the mandate to 'suggest measures of achieving a proactive, responsive, accountable, sustainable and efficient administration for the country at all levels of the government' (Government of India 2004). A Group of Ministers (GoM) was constituted on 30 March 2007 under the Chairmanship of the then External Affairs Minister to review the recommendations of the Second ARC and issue guidelines to the concerned Departments for their implementation. Core Group on Administrative Reforms under the Chairmanship of the Cabinet Secretary has considered fifteen reports where the Sixth Report on 'Local Governance: An Inspiring Journey into the Future', which was submitted in October 2007[6] contains the Commissions speculations on the empowerment of the local governance. The Government of India accepted the commission's recommendation on 'Devolution of Powers and Responsibilities (Para 3.3.1.7)' in its Sixth Report. It speaks:

> a. There should be clear delineation of functions for each level of local government in the case of each subject matter law. […] (20)

[5] The Second ARC was set up with a resolution no. K-11022/9/2004-RC of the Government of India on 5 August 2005 with the Chairmanship of Veerappa Moily. Mr. Moily resigned from the Commission on 1 April 2009 and V. Ramachandran was appointed as the Chairman.

[6] See the Complete Report here: https://darpg.gov.in/sites/default/files/local_governance6.pdf.

b. Each subject-matter law, which has functional elements that are best attended to at local levels, should have provision for appropriate devolution to such levels – either in the law or in subordinate legislation. […] (21)

c. In the case of new laws, it will be advisable to add a 'local government memorandum' […] (22)

d. In case of urban local bodies, in addition to the functions listed in the Twelfth Schedule, the following should be devolved to urban local bodies:

– School education;

– Public health, including community health centres/area hospitals;

– Traffic management and civic policing activities;

– Urban environment management and heritage; and

– Land management, including registration. […] (23)[7]

Meanwhile, the effort of the Ministry of Panchayati Raj to constitute an Expert Group chaired by Shri V. Ramachandran in 2005 is worthy to be mentioned here. The expert group was mandated to study and make recommendations, inter-alia, on the 'formulation of District and Sub-District Plans at all levels of Panchayats aimed at delivery of basic minimum needs to citizens at the grassroots levels' (Planning Commission of India 2008, p. 5). The report of the Expert Group was accepted and the Planning Commission issued its detailed guidelines for district plans in the Eleventh Five Year Plan through its circular dated 25 August 2006.

With this backdrop, the Planning Commission of India launched Integrated District Planning in 2008 to 'transforming a vertical planning process into a horizontal process' (Planning Commission of India 2008, p. 7). The district administration had been involving in the preparation of the sector-wise plans (e.g. District Health Plan, the District Watershed Plan, and the Education Plan, and many more) to address the sector-wise thrusts in development and to making planning mechanism fine-tuned with the increased specialization across the sectors. 'While some level of planning on these lines cannot be avoided, it is important, particularly from the view of increasing accountability to local governments and people, and optimally using resources, that the vertical planning process is transformed into a horizontal planning process, where local governments and other planning entities work together and explore the possibilities of planning together' (Planning Commission of India 2008, p. 7).

The Integrated District Planning 2008 aimed to delineate specific roles of different local bodies and the line departments. Erstwhile 'vertical' planning mechanism sought for the district heads to have complete knowledge about a plan. The 2008s integrated planning architecture aimed to harmonize the vertical planning flow, so essential to ensure that sectoral expertise is well used, with the horizontal overall planning process, equally essential to ensure efficiency and accountability (Fig. 8.1). In the context of urban planning, a multiplicity of agencies has led to segmented, fragmented, and sub-optimal planning. The local governments are empowered to

[7] Details of the Government's decisions on the recommendations and 'Action taken' thereon Administrative Reforms Commission's 6th Report titled 'Local Governance—An Inspiring Journey into the Future'. Available here: https://darpg.gov.in/sites/default/files/decision6.pdf.

play decision-making in the district planning as they are the most appropriate level in creating linkages and building synergies.

8.4 Transformation of Aspirational Districts Programme (TADP)

Hon'ble Prime Minister of India launched the Transformation of Aspirational Districts Programme in January 2018 with the target of the rapid, time-bound transformation of the districts that are seriously lagging on certain development parameters.

8.4.1 The Working Principles

The broad contours of the TADP are '3C'—'Convergence', 'Collaboration' and 'Competition'. TADP is one of the most extensive experiments on real-time outcome focussed governance to achieve a radical improvement in the sectors of health, education, infrastructure, and economy for the most socio-economically challenged districts in India. The government's commitment regarding good governance and raising every citizen's living standards with the slogan, '*Sabka Saath, Sabka Vikas*' (*i.e. the development of each citizen with the participation of everyone*), tunes the objectives of the TADP. It aims to enable every individual to participate in the vibrant economy as per the maximum utilization of their potential. The program focuses on the areas of Health and Nutrition, Education, Agriculture and Water Resources, Basic Infrastructure, Financial Inclusion and Skill Development (Table 8.2) with 49 key indicators to assess the progress of a district (Fig. 8.2). TADP primarily includes 115 aspirational districts from 28 states 8603 g-panchayet to improve the living standard and quality of life of the people of these lagging regions.

Technically, the project recalls the importance of collaborative and competitive federalism. 'Convergence' among the central government and state government schemes, 'Collaboration' of central-state-level '*Prabhari*' officers[8] and district magistrates/collectors, 'Competition' among districts based on real-time progress on socio-economic parameters of lives and livelihoods.

[8] Senior IAS bureaucrats have been roped in as '*Prabhari* officers' (i.e. Officer-in-Charge). They are additional and joint secretary ranked officers who will coordinate the efforts of the Centre and the states by addressing specific developmental needs of the districts.

Table 8.2 Indicators and data-points for TADP

Sl	Monitorable data points	Weight in composite index (%)
Health and nutrition		
1.1	Percentage of Pregnant Women receiving four or more antenatal care check-ups against total ANC registrations	0.6
1.2	Percentage of ANC registered within the first trimester against total ANC registrations	0.9
1.3	Percentage of pregnant women (PW) registered for ANC against estimated pregnancies	0.9
2	Percentage of pregnant women taking Supplementary Nutrition under the ICDS programme regularly	0.9
3.1	Percentage of Pregnant women having severe anaemia treated against PW having severe anaemia tested cases	1.5
3.2	Percentage of Pregnant Women tested for Haemoglobin 4 or more than 4 times for respective ANCs against total ANC registration	1.2
4.1	Sex Ratio at birth	0.9
4.2	Percentage of institutional deliveries out of total estimated deliveries	1.2
5	Percentage of home deliveries attended by an SBA (Skilled Birth Attendance) trained health worker out of total home deliveries	0.9
6.1	Percentage of newborns breastfed within one hour of birth	1.2
6.2	Percentage of low-birth-weight babies (Less than 2500 gms)	0.9
6.3	Proportion of live babies weighed at birth	0.9
7	Percentage of underweight children under 5 years	2.1
8.1	Percentage of stunted children under 5 years	0.6

(continued)

8.4.2 TADP: Pathways for 'A New India by 2022'

TADP is merely not a classification scheme. It bridges the government-sponsored development schemes' 'targets' and 'outcomes' to streamline the planning strategies for upgrading the performance, which is the 'transformation' process intended by the TADP. In its working principle, TADP assumes the States as the main drivers. It records and analyses each district's strength; identifies the low-hanging fruits for immediate improvement; measures progress; ranks the districts (Fig. 8.3).[9] The

[9] See the NITI Aayog website http://niti.gov.in/about-aspirational-districts-programme.

Table 8.2 (continued)

Sl	Monitorable data points	Weight in composite index (%)
8.2	Percentage of children with Diarrhoea treated with ORS	0.6
8.3	Percentage of children with Diarrhoea treated with Zinc	0.6
8.4	Percentage of children with ARI in the last 2 weeks taken to a health facility	0.6
9.1	Percentage of Severe Acute Malnutrition (SAM)	1
9.2	Percentage of Moderate Acute Malnutrition (MAM)	0.5
10.1	Breastfeeding children receiving adequate diet (6–23 months)	1
10.2	Non-Breastfeeding children receiving adequate diet (6–23 months)	0.5
11	Percentage of children fully immunized (9–11 months) (BCG + DPT3 + OPV3 + Measles1)	3
12.1	Tuberculosis (TB) case notification rate (Public and Private Institutions) against estimated cases	0.75
12.2	TB Treatment success rate among notified TB patients (public and private)	0.75
13.1	Proportion of Sub centres/PHCs converted into Health and Wellness Centres (HWCs)	1.8
13.2	Proportion of Primary Health Centres compliant to Indian Public Health Standards	1.5
13.3	Proportion of functional FRUs (First referral units) against the norm of 1 per 5,00,000 population (1 per 3,00,000 for hilly terrain)	0.45
13.4	Proportion of specialist services available in District hospitals against 10* core specialist services	0.6

(continued)

initiative is unique of its kind, as, after the independence, TADP is the first effort to assesses the development schemes, analysis its outcome, and setting the driving direction afresh if required (i.e. the transformation stimuli) to fetch the targeted goal of the decentralized planning process (Fig. 8.4).

The NITI Aayog speculates that despite the Indian economy running with a high growth trajectory, the country ranked at the rear panel in UNDP's Human Development Index. While the country has lifted 271 million people out of poverty between

Table 8.2 (continued)

Sl	Monitorable data points	Weight in composite index (%)
13.5	Percentage of Anganwadis centres/Urban PHCs reported to have conducted at least one Village Health Sanitation and Nutrition day/Urban Health Sanitation and Nutrition day/respectively in the last one month	0.6
13.6	Proportion of Anganwadis with own buildings	0.6
13.7	Percentage of First referral units (FRU) having labour room and obstetrics OT NQAS certified (i.e. meet LaQShya guidelines)	0.45
Dimension weight (%)		**30**
Education		
1.1	Transition Rate—Primary to Upper Primary level	4.2
1.2	Transition Rate—Upper Primary to Secondary level	1.8
2	Toilet access: Percentage schools with functional girls' toilets	1.5
3	Learning outcomes 3.1 Mathematics performance in Class 3 3.2 Language performance in Class 3 3.3 Mathematics performance in Class 5 3.4 Language performance in Class 5 3.5 Mathematics performance in Class 8 3.6 Language performance in Class 8	15
4	Female literacy rate (15+ age group)	2.4
5	Percentage of schools with functional drinking water facility	1.2
6	Percentage of schools with functional electricity facility at secondary level	0.3
7	Percentage of elementary schools complying with RTE specified Pupil-Teacher Ratio	2.4

(continued)

2005 and 2015, it still accounts for a large number of the world's 'multidimensional' poor and, currently, home to 28% of the world's poor (UNDP 2019).[10]

It is a clear indication that the performance of the country in terms of improving the quality of life for its citizen is not consistent with its growth story. In its vision statement of TADP, campaigned by the NITI Aayog, is clearly stated:

> While different States have their unique strengths, they face challenges in improving health, education, basic infrastructure etc. for their citizens. Within states too, there are major variations. Some districts have done well whereas some have faced challenges. It is important

[10] See the report of the Down To Earth dated 09 December 2019, available here: https://www.downtoearth.org.in/news/governance/india-home-to-28-of-world-s-poor-human-development-index-2019-68336.

Table 8.2 (continued)

Sl	Monitorable data points	Weight in composite index (%)
8	Percentage of schools providing textbooks to children within 1 month of start of the academic session	1.2
Dimension weight (%)		**30**
Agriculture and water resources		
1.1	Percentage of net sown area under micro-irrigation	3.5
1.2	Number of water bodies rejuvenated under MGNREGA	2.5
2	Crop Insurance—Percentage of the net sown area under Pradhan Mantri Fasal Bima Yojana	3
3.1	Percentage increase in agricultural credit	2
3.2	Certified quality seed distribution	1.5
4	Number of district mandis linked to electronic market	2
5	Percentage change in Price Realization—the difference between Farm Harvest Price (FHP) and Minimum Support Price (MSP)	1
6	Percentage share of high value crops to total sown area in district	0.5
7	Agricultural productivity of 2 major crops	1
8	Percentage of animals vaccinated	1.5
9	Artificial Insemination coverage	1
10	Number of Soil Health Cards distributed	0.5
Dimension weight (%)		**20**
Financial inclusion		
1	Total Disbursement of Mudra loan (in rupees) per 1 Lakh population	0.83
2	Pradhan Mantri Jeevan Jyoti Bima Yojana (PMJJBY): Number of Enrolments per 1 Lakh population	0.83

(continued)

that a concerted effort is made to improve performance of districts which are pockets of under-development. This will result in a dramatic rise in ranking of the country in terms of HDI and would also help the country in achieving SDGs. This will be a step towards creation of a New India by 2022.[11]

[11] See Nity Aayog's primer on 'Transformation of Aspirational Districts—A New India 2022', available in the URL: http://niti.gov.in/sites/default/files/2018-12/Transformation-of-Aspirational Districts-Primer-ANew-India2022.pdf.

Table 8.2 (continued)

Sl	Monitorable data points	Weight in composite index (%)
3	Pradhan Mantri Suraksha Bima Yojana (PMSBY): Number of Enrolments per 1 Lakh population	0.83
4	Atal Pension Yojana (APY): Number of Beneficiaries per 1 Lakh population	0.83
5	Percentage of accounts seeded with Aadhaar as a percentage of total banking accounts	0.83
6	Number of accounts opened under Pradhan Mantri Jan Dhan Yojana	0.83
Dimension weight (%)		**5**
Skill development		
1	Number of youth certified in short term and long-term training schemes/no. of youth in the district in the age group 15–29	1.25
2	Number of certified youth employed/No of youth trained under short-term and long-term training	0.75
3	Number of Apprentices completing/Total number of trainees registered on the portal	1.25
4	Number of people certified under Recognition of Prior Learning/Non-formally skilled workforce	1.25
5	Number of Vulnerable or Marginalized youth certified trained under short term and long-term training 5.1 Women—certified trained 5.2 SC—certified trained 5.3 ST—certified trained 5.4 OBC—certified trained 5.5 Minorities—certified trained 5.6 Differently abled—certified trained)/Total number of youth certified trained	0.5
Dimension weight (%)		**5**

(continued)

In partnership with the Government of Andhra Pradesh, NITI Aayog has created a dashboard for monitoring the districts' real-time progress. District Information Officers have been trained to be tuned with the system regarding how to enter data to the dashboard and generate the Management Information System (MIS) reports. Districts are being ranked based on progress made (i.e. 'delta ranking') on a real-time basis. The dashboard is open to the public to observe the aspirational districts' progress and use the data for research and development purposes.

Table 8.2 (continued)

Sl	Monitorable data points	Weight in composite index (%)
Basic infrastructure		
1	Percentage of households with electricity connection	2
2	Percentage of Gram panchayat with internet connection	0.5
3.1	Percentage of habitations with access to all-weather roads under PMGSY	0.75
3.2	Cumulative number of kilometres of all-weather road work completed as a percentage of total sanctioned kilometres in the district under PMGSY	0.75
4	Percentage of households with Individual Household Latrines	1.5
5	Percentage of Rural habitations with access to adequate quantity of potable water—40 lpcd drinking water in rural	2
6	Coverage/establishment of Common Service Centres at Gram Panchayat level	0.5
7	Number of pucca houses constructed for households who are shelter-less or having one room with kuchha wall and kuchha roof or having two rooms with kuchha wall and roof	2
Dimension weight (%)		**10**

Source Adopted from NITIAayog (2018)

8.5 A Critical Review

The UNDP Human Development Report (2018) has raised the very pertinent question against development scenarios and illuminates the inconsistent well-being of a large section of people with a high growth trajectory in India. The government has realized that without the overall up-gradation of living standards of the deprived section of peoples and escaping the development of spatially marginalized sections, the dream of becoming a $5 trillion economies would be a fallacy. The upliftment of the standard of living in the largely deprived districts and villages may be a potential solution to upgrade India's overall rank in UNDP Human Development Index. And a new kind of outcomes-focused governance has come in the form of the Transformation of Aspirational Districts' Programme (TADP) in January 2018 to address the districts with poor socio-economic indicators. The 115 aspirational districts from 28 states have been selected to monitor the real-time progress that will contribute to the overall up-gradation in India's human development status. TADP is based on 49 indicators to address the five large dimensions of Health and Nutrition, Education, Agriculture

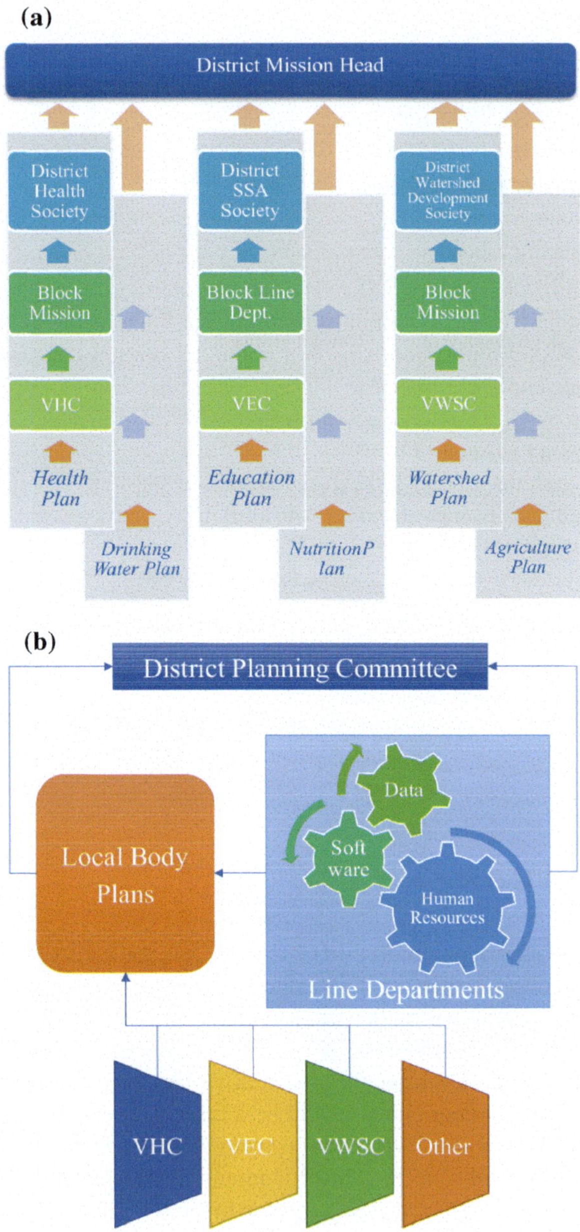

Fig. 8.1 Transforming (**a**) the vertical planning process into (**b**) the horizontal process (*Sources* Prepared by the authors after Planning Commission of India 2008)

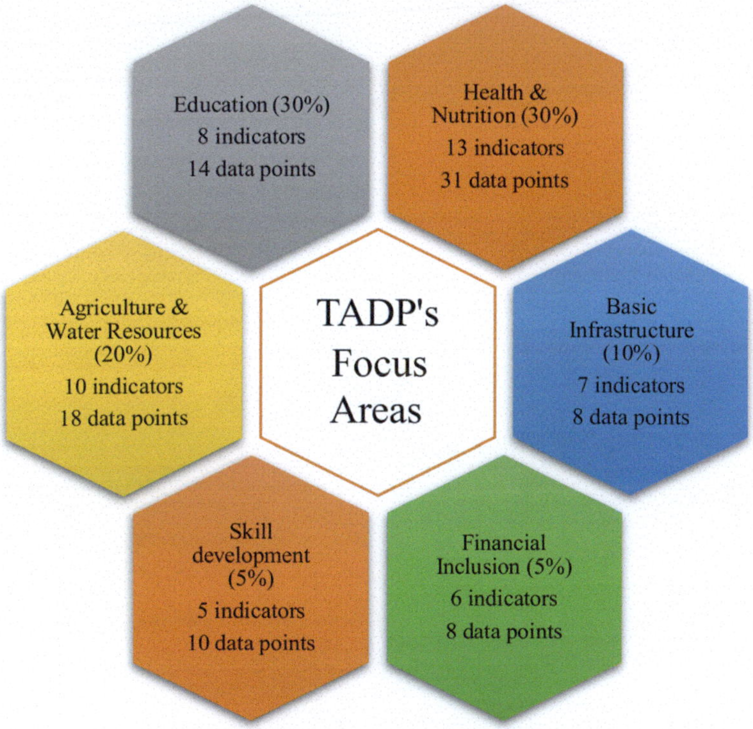

Fig. 8.2 TADP's focus area and monitoring indicators (*Source* Prepared by the authors after NITI Aayog)

and Water Resources, Financial Inclusion and Skill Development, and Basic Infrastructure. The scheme is based on three principles: convergence, collaboration, and competition. The TADP's vision comprises two principles: regional equality in development and real-time monitoring. Moreover, the broad spectrum of this initiative is the convergence of various Central and State government schemes; Collaboration of National-level, State-level Officers, and the District Collectors; Competition among the participating districts. Considering states as the key drivers, the TADP focuses on every district's strength and analyse the lagging outcomes to prioritize the immediate improvement by a real-time monitoring system.

8.5.1 Selection of the Districts

The 115 districts were selected from 28 states on the basis of composite index considering various sectors of health and nutrition, poverty, education, and basic infrastructure with variable weight in the composite index. Poverty has been given the highest

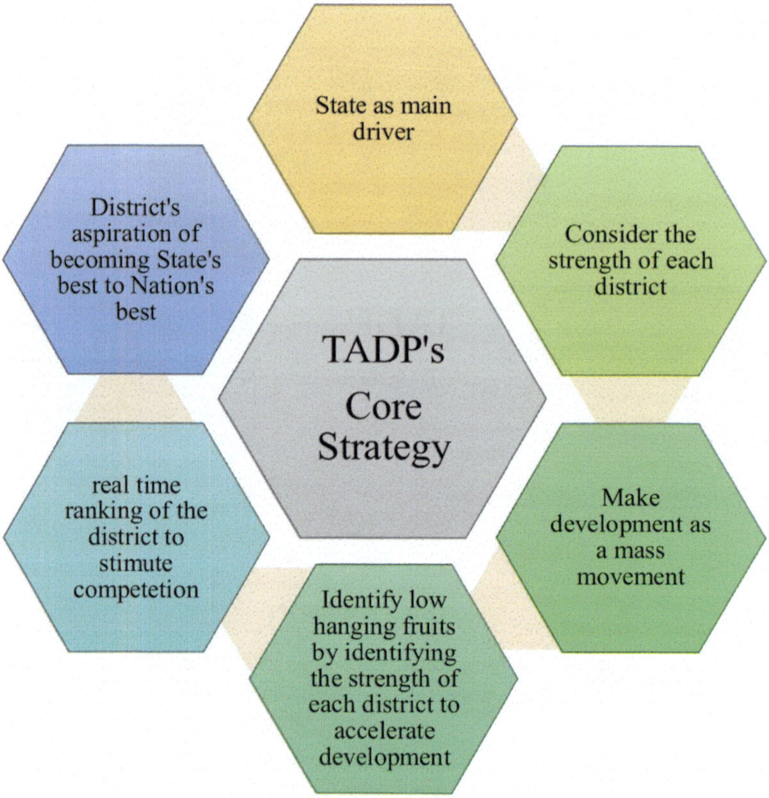

Fig. 8.3 Core strategy of transformation of aspiration districts (*Source* Prepared by the authors after NITI Aayog)

weightage with 25%, as it has had a multidimensional impact on the overall status of life and livelihood. Out of 115 districts, 35 were selected by the Ministry of Home Affairs based on the extremist left-wing violence. 55 districts were selected by the Central Ministries of Govt. of India on the basis of extreme deficiency in the districts, and the remaining districts (25) were selected by the NITI Aayog.

However, it is to be mentioned that when there is a composite index in exercise to select the districts, the baseline selection of the districts by different agencies with particular emphasis on different standpoints of deficiency or sensitivity must raise some questions.

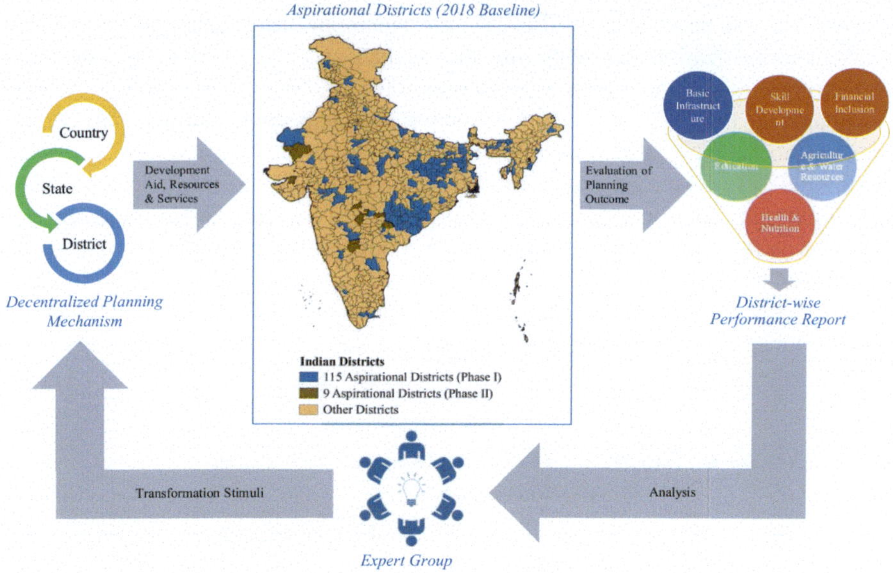

Fig. 8.4 Possible pathways of transformation of Indian districts (*Source* Authors' illustration)

8.5.2 'Sab ka Sath, Sab ka Vikash': Addressing the Diversity in All Sphere

India is a country with immense diversity in all spheres of nature, natural resources, society, culture, language, race, and religion. Equal, balanced, and sustainable development is really an extremely hard task to achieve. Each spatial unit with its embedded population have retains some unique potential and distinct causes of its lagging outcome or of its success story. Now, the Government of India desires to address every district's aspiration and to resolve their own issues to upgrade the overall quality of life. There are a lot of issues of lagging regions, including Left Wing Extremism (LWE), dense forest areas, extensive rugged surfaces with infertile soil, areas with a high concentration of tribal population, hilly regions, issues from extensive international border lines, etc. Moreover, the very positive aspect of the TADP is to bridging the regional imbalances and disparity through the district level planning to emphasize the 'local'.

Amidst the Indian federal structure and planning procedure, the centralized planning and decision structure is the real barrier to address the local district level development as per local needs. Aiyar (2021) has mentioned that a district is not permitted to use a government fund allocated for toilet construction under Swachh Bharat Mission to promote waste management that must be considered a primary step towards healthy sanitation. However, TADP aims to bypass the issue of institutional intervention by appointing *Prabhari* officers at the central and district levels by giving the authority to converge and combine the central and state funds and to execute as per district-level priorities that can be assessed by the real-time data monitoring dashboard. Flexibility in scheme prioritization and funds allocation is the key of the TADP to be considered as an unprecedented step towards development for all.

Though, the flexibility in fund allocation has already been questioned! The central government has advocated a flexible budget for Sarva Shiksha Aviyan for district 'innovation'. However, the state's attempt to provide 25% of central scheme funds as 'discretionary' has been ignored by the Central government, raises some obvious questions on the 'flexibility' of the budget. A robust federal partnership could be the key to make the TADP a real flexible and dynamic execution. A *Prabhari* officer can allocate funds according to the district's aspiration and needs when there would be an environment of mutual cooperation between central and state governments.

8.5.3 Needs a 'Reform': It is Knocking at the Door

There are several schemes and programs of central government and state government running simultaneously at the district level in India. Moreover, the district administration is the implementing authority of these schemes. In the Indian democratic system, every elected government has a fixed tenure, and the political-party-driven government has to be contested in the next election. So, most of the schemes and programs have a political priority-based vision, and the priorities are mostly taken for the short outcome. There is a frequent clash between the central and state government regarding the schemes where two opposing political forces rule the Centre and the State. TADP promotes the concept of competition among the states and allocation of funds as per competing needs. Here lies the real challenge. Allocation of government funds by the appointed *Prabhari* officer as per districts' prime needs in education, health, and infrastructure is very too tough to believe. In this ecosystem, a reform in governance, planning procedures, and implementation priorities is the need of time.

Autonomy to the local government to make the funding decision through *Prabhari* officers is a very inspiring part of the TADP. Competitive and cooperative federalism demands the provision of a decentralized environment in every aspect of governance, and the '*Panchayati Raj*' system[12] sets one of the best examples of decentralization of powers in the Indian federal structure.

However, in TADP, the planning and funding decision by the *Prabhari* officer as per the needs of the districts, escaping the role of *Panchayats* for intervention in development planning and scheme execution, has started raising questions over the role of the central government in the decentralized planning exercises. Although TADP aims to focus on the local aspirations and needs for up-grading human well-being, the undermining of *Panchayets'* role somehow ignores the relevance of the *Panchayat* system as per the principles of decentralization.

---

[12] Panchayet is the village-level administrative unit elected by the democratic rights of the people.

8.5.4 Criteria and Indicators: SDG is the Bull's Eye

TADP depends on 49 key performance indicators from six dimensions: education, health and nutrition, agriculture and allied activities, financial inclusion and skill development, and basic infrastructure. NITI Aayog has finalized the key performance indicators based on the discussion with all stakeholders, and various weightage has been assigned to assess every citizen's inclusive growth and living standards.

Health and nutrition have been prioritized with 30% weightage of the overall composite score. The health and nutrition standards of the citizens have been aimed to estimate by 13 indicators covering antenatal and postnatal care for the women, gender parity at birth, the health of new-borns and continuous health assessment of the newborn, growth of children, child immunization, contagious diseases, and complete health infrastructure. Indicators have been set to achieve the target of the Sustainable Development Goals (SDG). Indicators of antenatal care have been taken to control the maternal mortality rate that addresses SDG 3.1. Indicators to consider newborn health and infant mortality reduction have been set according to SDH 3.2. Moreover, the indicators addressing the primary health services and institutional support have been set according to SDG 3.

The education sector has been given 30% weightage to the overall composite score. Eight indicators have been used to estimate the education standard of the citizens in a district. TADP has emphasized outcome-based assessment. Transition rates from primary to upper primary and upper primary to secondary standards have been taken to assess the drop-out condition. Students' performance in mathematics and languages in different standards has been taken to assess the learning outcomes. Lack of girls toilet in the functional condition is a significant parameter for girls' drop-out from school, leading the TADP ranking to consider the percentage of function girls' toilet as an important indicator. Functional drinking water facilities in schools and the percentage of schools with functional electricity facilities are considered as other infrastructural indicators. Timely delivery of textbooks and RTE-specified student–teacher ratio has been considered to assess the standard of the academic environment. Most importantly, female literacy has been considered as an important indicator to address the gender aspect in the education sector.

Agriculture is the backbone of the Indian economy, evident by its 17.76% share in Indian GDP for the financial year 2019–20 (Ministry of Statistics and Programme Implementation 2021). Agriculture and allied activities contribute 41.49% of India's total workforce in 2020 (International Labour Organization 2021). TADP aims to assess the agricultural sector's condition with 10 indicators addressing input, output and institutional support. The extent of institutional support in agriculture is measured by crop insurance, growth in agricultural input and supply, quality seed distribution, market support, change in price realization, animal vaccination and artificial insemination coverage.

Financial inclusion and skill development jointly share 10% of weightage to the overall composite score. Six indicators have been taken to measure the financial inclusiveness and ease of government institutional financing for small-scale businesses. The growth in beneficiaries for Atal Pension Yojana (APY) ensures the minimum financial support to sustain life in old age has been considered. Multiple financial support measures have been taken by considering the coverage in Pradhan Mantri Jeevan Jyoti Bima Yojana (PMJJBY), Pradhan Mantri Suraksha Bima Yojana (PMSBY), etc. Indicators like the rate of bank account opening and AADHAAR[13] linking to bank accounts represent financial institutionalization and must act as a bridge between the government and the beneficiaries. TADP uses five indicators to track youth skill development, employment and employability of informally skilled youth. Empowerment of the marginalized and vulnerable section of the youth has also been considered in the indicator prioritization.

'Basic infrastructure' shares 10% weight in the composite score with seven critical indicators. Access to adequate quantity of drinking water, access to houses and shelter, electric facility, road and transport connectivity and availability of common service centre in gram *panchayat* are the priority indicators to track the infrastructure. The 2019 HDR talks about the 'enhanced capabilities' that include access to modern facilities like access to the internet that is the demand of the twenty-first century. Internet is the inevitable communication demand of the twenty-first century, and it has been taken as an important indicator for infrastructure quality assessment.

The earlier debate on the concept of development as a purely economic phenomenon had been changed over time. During the 1970s, the concept of development came to be redefined in the context of poverty, inequality, and unemployment. The concept of human development replaced the popular concept of economic development. UNDP advocated the importance of people's participation in development (UNDP 1993). Further, the importance of human security in life (UNDP 1994), gender inequalities (UNDP 1995), and poverty (UNDP 1997) were emphasized in the discussion of development (Sagar and Najam 1998).

Now human security is a critical approach considered in Agenda 2030, and most importantly, the goals of sustainable development cannot be achieved without sustaining peace. The concept of human security has a multidimensional implication that demands inclusive and peaceful societies and a world free of poverty, hunger and have access to education, good health, food and safe drinking water. In this context, the assessment of freedom of expression and freedom of speech, and the ability to control one's own life are extremely important aspects to be considered in human development measures. So, there is ample scope to incorporate the human security and freedom aspect in TADP procedure and ranking.

---

[13] AADHAAR is a unique number which is having 12 digits, implemented by the Unique Identification Authority of India (UIDAI), authorized by the Government of India. A card containing this number is issued to all residents of India. Aadhaar is linked to the basic biometric and demographic information like photograph, fingerprints and iris of each individual. It aims to eliminate the large number of duplicate and fake identities in government and private database.

8.5.5 Data-Driven Monitoring of the Districts: The Gamechanger

Emphasis on the poorest districts and their real-time monitoring based on regular district-level data is a commendable step towards good governance. The fixation of criteria and the identification of indicators combining the output and output focus have shifted from the typical tendency to conflate output with outcomes. TADP introduces the outcomes-focused performance assessment and governance management system (Aiyar 2021).

However, the data-centric assessment does not tell the truth always. The time-bound data-driven outcome does not always substitute the administrative reform at large. For a particular district or administrative unit, for a short span of time, data may not represent the inherent change or real up-gradation of living standard, e.g. on-time delivery of textbooks has been considered as a process level indicator for socio-economic condition assessment, but this may or may not reflect the actual learning standard up-gradation for a specific administrative unit.

The percentage of children with Diarrhoea treated with ORS and zinc has been considered an important indicator for health condition assessment. Undoubtedly, it is extremely important and relevant to assess a child's health status. However, the data regarding the treatment with ORS and zinc are available for those who are admitted to the hospital or govt. Health Centres, but a large section of people are out of the government-run health system, certainly escaped from the real-time database. Moreover, the radical improvement in data for a short span of time looks promising, but the real story may or may not say the same.

Actually, the development of education and health infrastructure in rural areas on one side and to spur a sense of mass awareness to take the institutional health and education facilities can work as a catalyst for achieving TADP's goals.

Partnership with Tata Trusts and Bill & Melinda Gates Foundations for data validation and to assist the district TADP teams to improve data quality and to upgrade the socio-economic standard of living through real-time assessment by the household survey is really a very promising aspect of TADP.

Though, the district database on a real-time dashboard does empower the district and its administration to identify the real-time position compared to others. A district can plan its targets as per its distinct aspiration and needs.

The evidence-based ranking procedure of the districts, along with the *delta* ranking system, is a very promising aspect of the TADP initiative. The ranking is dynamic in character through time. This dynamic ranking of the districts is aimed to act as a catalyst to make positive competitiveness among states. The TADP report shows that the districts that had ranked poor in the baseline ranking (April 2018) have already shown remarkable improvements in the next time ranking procedure.

8.6 Needs a Second-Order Classification of the Districts

From a planner's perspective, 'transformation' requires a long-term effort as the transformation of the districts envisages a process of developing a strategic plan for the modification of policies, procedures, and processes to move the districts from an 'existing' state to a 'projected' state. All good transformation initiatives start with a solid strategic foundation, and the TADP seems robust enough as the cornerstone of the future transformation of Indian districts. A development scheme is customized, updated, and upgraded through its working experiences. However, at this phase, some operational tools may be recommended to include for making the TADP more effective.

Reaching the benefits of a development scheme to most of the population needs to nourish those economic sectors that nurture those populations. Agriculture is the primary source of livelihood for about 58% of India's population. Agriculture, forestry, and fishing have contributed an estimated Gross Value Added (GVA) of Rs. 19.48 lakh crore (i.e. US$ 276.37 billion) with growth in GVA at 4% in the Financial Year 2020 (IBEF 2021). India has the potentiality to place itself as the top five exporters of agro-commodities if it focuses on cultivation and effectively handholding farmers.[14] It is evident by the statics placed by the India Brand Equity Foundation:

> Total agricultural export from India grew at a CAGR of 2.27% during FY16-20 to reach US$ 35.09 billion in FY20. India's spices export increased by 23% to Rs. 2,690 crore (US$ 359 million) in June 2020 from Rs. 2,190 crore (US$ 292 million) in the same month last year. The Agriculture Export Policy, 2018 aimed to increase India's agricultural export to Rs. 4,19,340 crore (US$ 60 billion) by 2022. Government aims to raise fishery export from India to Rs. 1 lakh crore (US$ 14.31 billion) by 2024–25. According to Department for Promotion of Industry and Internal Trade (DPIIT), Indian food processing industry attracted Foreign Direct Investment (FDI) equity inflow of about US$ 10.15 billion between April 2000 and June 2020. (IBEF 2021)

Delineation of agro-climatic zones based on soil, water, rainfall, temperature, etc. is found instrumental for sustainable agricultural development plans in maximizing the production from the available resources and prevailing climatic conditions. However, to construct a need-based and location-specific-technology-based agro-economic development, the aspirational districts need to be undergone the second-order classification system in compliance with different agro-climatic regions (ACR)[15] or more specific agro-ecological regions (AER).[16]

---

[14] See the report of the Economic Times dated 29 July 2020, available here: https://economictimes.indiatimes.com/news/economy/agriculture/india-can-be-among-top-5-agri-goods-exporters-with-effective-policies-report/articleshow/77244405.cms?from=mdr.

[15] An 'Agro-climatic zone' is a land unit in terms of major climates, suitable for a certain range of crops and cultivars.

[16] An agro-ecological region (AER) is the land unit on earth's surface, carved out of agroclimatic region by superimposing climate on landforms and soils, which are the modifiers of climate and length of growing period.

> The AER is designed to address the issues related to agricultural production by delineating agriculturally potential areas suitable for particular genotype and is essentially needed for achieving the optimum production potential of a crop and crop variety. AER is capable in delineating agro-ecologically comparable regions for generating and transferring various agro-technologies.[17]

The second-order classification of the district, following the ACR or AER, would enable the district with deficiencies in agriculture production to undergo further investigation for suitable cropping pattern, crop choice, farming system and allocation of resources and technologies. These districts should consider *Krishi Vikas Kendras* (i.e. Agricultural Development Centres) as important resource centres to disseminate the training and information on modern and innovative agricultural mechanisms to the agro-based communities in the district.

8.7 Addressing Rural–Urban Inequalities

India is a country that, amongst many others in the South-Asia, has been experiencing rapid urbanization. In 2019, approximately a third of the total population lived in cities. Historical data shows an increase in urbanization by almost 4% in the last decade. It indicates that people have been moving away from rural areas toward the cities to find work and live in the urban areas.[18] However, while rapid urbanization is in place, the rural–urban disparity is evidently increased. The labour productivity in rural areas rises at a slower rate than in urban areas; the disparity between rural and urban areas widens even if the ratio of the population residing in rural and urban areas remains constant. While there happens an increase in the proportion of the population or labour force in urban areas, overall inequality tends to increase much faster.[19]

As the TADP traces on districts as the planning unit, there should be adequate tools to record and address the rural–urban differences because of the co-existence of the rural and urban economies within the administrative jurisdiction of a district. The TADP scheme, at its present uniform format of data capturing for the rural and urban units, may not be able to address the rural–urban inequality issue that could leave the space for rural–urban differences to widen in the absence of strict surveillance. While the TADP concentrates on a district for its transformation as a whole, different speeds of change in labour, productivity and population in rural and urban areas result

[17] See e-publication on 'ICAR-NBSSLUP Technology: Delineating Agro-Ecological Regions' in the NBSS & LUP Website, available here: https://www.nbsslup.in/assets/uploads/clinks/Delineating%20Agro-Ecological%20Regions.pdf.

[18] See the Statista data and also the report by Aaron O'Neill dated 31 March 2021 here: https://www.statista.com/statistics/271312/urbanization-in-india/.

[19] See the study on 'Asia's rural–urban disparity in the context of growing inequality', by Katsushi S. Imai of the University of Manchester and Bilal Malaeb of University of Oxford which was funded by y the Asia and the Pacific Division (APR) of the International Fund for Agricultural Development (IFAD). URL: https://www.ifad.org/documents/36783902/40280999/.

in different patterns of change in inequality at the sub-district level. In the course of transformation, the changes are obvious for both the rural and urban economies. However, these changes must be sustainable for both areas. The uncontrolled process of transformation has ample shreds of evidence to build an urban economy at the cost of its rural counterpart. Here lies the challenge of the TADP to formulate the transformation process, ensuring balanced and equitable sustainable development.

8.8 A Long Way Ahead

The idea of 'inclusive growth' is now at the centre of development thoughts, notably more pertinent for developing nations. Emphasizing the 'equality of opportunity,' inclusive growth's working principle takes care of both the outcomes and processes through participation and benefit-sharing. The pre-1970s concept in development theories, which used to speak about 'development follows a pre-determined path', was challenged by the evidence from the countries where growth produced significant inequality. The emergence of the understanding that the growth process should ensure equity started driving the sustainable development thoughts; gradually, the scholarly world agrees that reducing inequality contributes to making growth sustainable.

However, the 'growth' remains the most 'trustable' component of government-sponsored poverty reduction schemes in India (say the entire South-Asia) until the 1990s. Rising concern about 'redistributive growth' gradually replaced the 'distribution-neutral growth' idea as the newer one is more likely to be effective than the former. A methodological shift in planning mechanism was observed, which traced on safeguarding the interest of the poor who needs to be benefited from growth. These 'pro-poor growth' policies focused on improving the conditions of the poor to grow, ensuring that the poor are benefited more than the rich. Furthermore, as poverty appeared as a multidimensional phenomenon, it sought to stress the importance of adding non-income dimensions to policies.

A broader version of the 'pro-poor growth' policies set the foundation of the 'inclusive growth policies' though they are oftentimes used interchangeably in literature. Inclusive growth extends the opportunities for everyone to participate in the growth process and ensures that benefits are shared. Transformation of the Aspirational Districts Programme is a remarkable experiment and a bold step of the Government of India on outcome-focused development that talks about inclusive growth. The revolutionary vision on development has been focused on the core strategy of the TADP. Considering the role of the state as the main driver and focus on each district's strength in the planning process and implementation of schemes, TADP streamlines inclusive development in the right way. The shift in the planning process from vertical to the horizontal mode of attempt will actually contribute to the decentralization of power and governance. The principles of convergence and collaboration among various schemes and stakeholders strengthen cooperative federalism. Acknowledging the role of the district as a planning unit and sector-wise planning at the district level as per its strength and weakness and its implementation to explore the opportunities must enhance the possibility of better people-centric development.

However, the program was launched just in 2018, and there is a long way to go to make a final remark on its performance, but the inclusive approach of development, the actualization of decentralized planning, and its improvement in the implementation of schemes by inter-district competitiveness based on real-time monitoring have set a landmark in the planning procedure. Now, the government has to focus on the issues of insufficient budget allocation, lack of coordination among different ministries and resolve the issues arising out of the differences of opinion between the Central and the State governments. The district ranking index must be dynamic and flexible enough to incorporate new dimensions and indicators through time and space.

References

Aiyar J (2021) Hindustan Times, Published on 2 Jan, 2021 7:2 PM IST

Government of India (2004) Resolution no. K-11022/9/2004-RC. https://mdoner.gov.in/contentimages/files/Resolution_2ARC.pdf

Government of India (GoI) (1956) Second five year plan. Planning Commission, New Delhi

Government of India [GoI] (1961) Third five year plan. New Delhi: Planning Commission

Government of India [GoI] (1969) Fourth five year plan. New Delhi: Planning Commission

IBEF (January 2021) Indian agriculture and allied industries industry report: January 2021. India Brand Equity Foundation. https://www.ibef.org/industry/agriculture-presentation

International Labour Organization, ILOSTAT database. Data retrieved on January 29, 2021

Mishra SK, Puri VK (2002) Indian economy, 20th edn. Himalaya Publishing House, Delhi

Mohan K (2005) Addressing regional backwardness: an analysis of area development programmes in India. Manak Publication, New Delhi

Mukherji R (2009) The state, economic growth, and development in India. India Rev 8(1):81–106. https://doi.org/10.1080/14736480802665238

NITI Aayog (2018) Transformation of aspirational districts: Bae line ranking and real time monitoring dashboard

Planning Commission of India (2008) Manual for integrated district planning, vol 1. New Delhi

Qaiyum A (2004) Decentralized planning in India: issues and options. J Inst Town Plan India (ITPI J) 1(2):1–2

Rao H (1984) Regional disparities and development in India. Ashis Publishing House, New Delhi

Reddy MG (1988) Participation of rural poor in rural development and delivery system: a case study of integrated rural development programme in Andhra Pradesh. Ph.D. Thesis submitted to University of Mysore, Mysore. http://www.isec.ac.in/Participation_of_rural_poor_in_rural_development.pdf

Sagar AD, Najam A (1998) The human development index: a critical review. Ecol Econ (Elsevier) 25:249–264

Shaban A (2016) Development strategies in India. In: Ramachandran H, Vyas RN (eds) Economic geography: urbanisation, industry and development. Oxford University Press, New Delhi

Siddiqi FA, Hussain N (2010) Analysis of micro level socio-economic disparities in Malda District, West Bengal. Asia-Pacific J Soc Sci II(1):41

UNDP (1993) Human development report 1993. Oxford University Press, New York

UNDP (1994) Human development report 1994. United Nations Development Programme (UNDP) Oxford Oxford University Press, New York

UNDP (1995) Human development report 1995. Oxford University Press, New York
UNDP (1997) Human development report 1997. Oxford University Press, New York
UNDP (2019) Human development report 2019. United Nations Development Programme 1 UN Plaza, New York, NY 10017 USA

Tanmoy Sarkar is an Assistant Professor in the Department of Geography of Gazole Mahavidyalaya (College) in Malda, West Bengal, India. The college is affiliated with the University of Gour Banga. Mr. Sarkar completed his postgraduate studies in Geography and Environment Management with a specialization in Remote Sensing & GIS at Vidyasagar University. His research chiefly focuses on the application of geospatial tools and techniques and multi-criteria predictive models in addressing physical and human processes on the spatial framework. He has been teaching Geographical Science at the undergraduate level for more than ten years. Mr. Sarkar has published five research papers in reputed journals and contributed book chapters, published by Springer Nature.

Dr. Mukunda Mishra is an assistant professor in the Department of Geography and designated vice principal of Dr. Meghnad Saha College, affiliated with the University of Gour Banga in West Bengal, India. Dr. Mishra completed his postgraduate studies in geography and environmental management at Vidyasagar University and holds a Ph.D. in geography from the same university. He was selected for the National Merit Scholarship by the Ministry of Human Resource Development, Government of India. His research chiefly focuses on analyzing unequal human development and creating multi-criteria predictive models. Dr. Mishra, so far, has authored one monograph and edited 3 contributory volumes from the house of Springer Nature and published more than 25 research articles and book chapters. He has more than 10 years of hands-on experience in dealing with development issues at the ground level in various districts of eastern India.

Dr. R. B. Singh is a professor of geography at the Delhi School of Economics, University of Delhi; the secretary general and treasurer of the International Geographical Union (IGU); and the chair of the Council for Scientific and Industrial Research (CSIR)–Central Food Technological Research Institute of the Government of India. He is also a member of the International Council for Science (ICSU) and the Scientific Committee for Urban Health and Well-Being. He was awarded the Japan Society for the Promotion of Scientific Research Fellowship and has presented papers and chaired sessions in more than 45 countries. He has published 14 books, 36 edited research volumes, and more than 240 research papers. He has supervised 36 Ph.D. and 81 M.Phil. students. In 1988 the UNESCO/International Social Science Council awarded him research and study grants in social and human sciences.

Chapter 9
Understanding the Disability Divides in India: A Spatio-Temporal Analysis with District-Level Datasets

Baikunth Roy

Abstract The study quantifies and compares the prevalence of each type of disability across gender, regions, and social groups of India. The findings show that the incidence of disability is distributed across every nook and corner. The decadal growth of disability is more among vulnerable sections such as scheduled castes and women. Further, the regression results suggest that improvement in educational standards among women, creating employment opportunities, and improving average living conditions can significantly reduce the burden of disability. However, it is a matter of grave concern that there is a growing elderly population, and increasing age is positively associated with a rising level of disability. Although there is a significant improvement in data collection in the 2011 Census, concerns have been raised in the present study about the enumeration process. There is an apparent underestimation of disability estimates because of deep-seated stigma among women and inadequate coverage of mental health issues in the country. As the burden of disability falls disproportionately across geographic regions and socioeconomic groups, public health policies in India should take this variation into account. Thus, providing access to basic capabilities may be strategically important to reduce the challenges for the disabled in India.

Keywords Correlates of disability · Census of India · Social groups · Gender · Regions · Aging · GIS mapping

9.1 Introduction

Persons with disabilities (PWDs) are the largest minority groups in the world (Khanday and Akram 2012). Jha (2016) argued that.

> disability cuts across class, caste, gender, race, religion, ethnicity, and nationality, but mostly a differently-abled person's first identity among their other identities is their disability (p. 5).

B. Roy (✉)
PG Department of Economics, College of Commerce, Arts and Science, Patliputra University, Patna, India

M. Mishra et al. (eds.), *Regional Development Planning and Practice*, Advances in Geographical and Environmental Sciences,
https://doi.org/10.1007/978-981-16-5681-1_9

According to Mont (2007), the disability prevalence rates mostly depend on the kind of data sources used, and the methodological approaches followed. The author reported that global estimations from international agencies like the World Bank or the UN show that about 10–12% of the world's population suffers from at least one disability.

The WHO Global Burden of Disease Study (2017) reported that the global burden of disability increased by 52% between 1990 and 2017. Further, World Health Organization (WHO 2018), in its report on "*Disability and Health*" highlights that.

> over a billion people, about 15% of the world's population, have some form of disability. The rates of disability are increasing due to population aging and increases in chronic health conditions, among other causes.

Amartya Sen (2004) noted that persons with disabilities are not only the most deprived but also most neglected. United Nations (2018) in the "Disability and Development Report" suggests that the PWDs experience much higher hardships in life even to meet their basic needs. DFID (2018) reported that:

> the global situation and wellbeing of the majority of people with disabilities has not improved. In low and middle-income countries, people with disabilities and their families are poorer than people without disabilities in nearly all socioeconomic indicators. People with disabilities face intersecting and compounding forms of discrimination on the grounds of gender, sexuality, impairment type, age, race, ethnicity, religion or belief, and location which all contribute to disability-related exclusion (p. 5).

In the context of India, it is evident that the disabled communities are disadvantageously placed in society. Disability brings in a socioeconomic burden, and the dynamics of life undergo a significant change as disability sets in. It not only has its implications on the individual but also his household. Saikia et al. (2016) showed that there is also a wide socioeconomic and regional divide in the prevalence of disability in India. Mishra and Gupta (2006) examined that individuals born with a disability or acquiring disability after birth are bound to begin a life with limited opportunities and also starved of services and facilities available to persons who are not disabled; therefore, disability acts as a hurdle for their self-development. World Bank (2009) reveals that disabled people are among the most excluded in Indian society, facing widespread social stigma, income poverty, and limited access to education and jobs.

However, the availability of robust data in India has been a stumbling block for substantive research in this area. It is as low as 2.21% as per the 2011 Census of India. Alternative estimates using robust and sophisticated measures present a higher prevalence of disability in India. World Bank (2009) reported that.

> while estimates vary, there is growing evidence that people with disabilities comprise between 5 and 8 percent of the Indian population, around 55–90 million individuals (p. xi).

During the 11th Plan[1] (2007–12), the erstwhile Planning Commission suggested that the disability figures are possibly 5–6% of the total population, which is much higher than the official estimates of India.

[1] The Eleventh Five Year Plan, Planning Commission, Government of India. A section on Empowering Persons with Disabilities.

Jeffery and Singal (2008) argued that there is a lack of precise and reliable estimates of the number of persons with disabilities in India. It raises serious concerns that, given the deep-seated stigma surrounding disability, many persons with severe impairments, mainly rural disabled and women, are excluded from Census and surveys. In this context, Reddy and Sree (2015) put forth that.

> any attempt at the enumeration of persons with disabilities is a challenging task for the simple reason that the constructions on impairments, both physical and mental, are deeply embedded in the social fabric.

In light of these insights, the study makes an effort to examine temporal and spatial variations in the prevalence of disability and the socioeconomic-demographic factors that affect disability prevalence across the districts of India.

9.2 Data and Methods

The study uses the district level secondary datasets to cover entire India. It covers disability data available in the Census 2001 and 2011 for different types of disabilities, namely, disability in seeing, hearing, speech, movement, mental retardation, mental illness, any-other, and multiple disabilities. Five types of disability data were collected in the 2001 Census of the government of India, namely, persons with seeing, hearing, speech, movement, and mental disability. The list expanded to eight types of PWD categories in Census 2011. In 2001, data were collected only for mental disability; however, it was bifurcated into two groups, mental retardation (R) and mental illness (I), while collecting data in Census 2011. For the sake of simplicity and comparative analysis, mental (I) and mental (R) have been clubbed to make a comparative study. In addition, data on two new categories, multiple disabilities, and any other disability, were collected for the first time in the 2011 Census. Therefore, technically precise comparison between the two time periods is difficult to ascertain. Multiple disabilities cover as many as three types of disabilities. Any-other category has emerged as the fourth largest (one-fifth of total PWDs) category of disability in Census 2011. This category was included in the Census to overcome the problem of estimation or to count the numbers of persons with disabilities. In other words, for those PWDs who are not listed in the Census question, the informants were required to report in this category.

The paper comprehensively examined the temporal and spatial variations in the prevalence of the various type of disability and differentials by social groups, age-cohorts, genders, and sectors (rural and urban) in India. The study carried out a comparative analysis and examined decadal growth in the prevalence of disability between these two periods. It attempted to understand the links between social groups and types of disabilities. Further, the analysis is drilled down to the state and district level. ArcGIS (ESRI) software is used to understand the holistic scenario in the incidence of disability. In this context, the maps showing the prevalence of male and female disabilities are drawn across the districts of India. Descriptive statistics have been presented to understand the district profiles which are relevant to our study.

9.2.1 Specification of the Disability Determinant Model

Cross-sectional regression analysis was carried out to identify and estimate the association between disability prevalence and socio-economic-demographic characteristics at the district level in 2011. The methodology to build socio-economic-demographic determinants of disability is largely borrowed from Saikia et al. (2016) paper. The present study performed multiple linear regression in STATAMP 13.0 statistical software. The disability determinant model was specified as follows:

$$\begin{aligned} Dis = {} & \beta_0 + \beta_1 SC + \beta_2 ST + \beta_3 Female\ Literacy + \beta_4 Urban\ Popn \\ & + \beta_5 Workers + \beta_6 Ln\ Housing + \beta_7 Water + \beta_8 Dwelling\ Rooms \\ & + \beta_9 Clean\ Fuel + \beta_{10} Banking + \beta_{11} Toilets + \beta_{12} Female \\ & + \beta_{13} Elderly + \beta_{14} Children + Ei \end{aligned} \tag{9.1}$$

where,

Dis = Proportion of persons with disability across districts of India.
SC = Proportion of the population who are members of Scheduled Castes.
ST = Proportion of the population who are members of Scheduled Tribes.
Female Literacy = Proportion of the population who are females.
Urban Popn = Proportion of the population living in urban areas.
Workers = Proportion of the population who are main workers.
Housing = Proportion of the households who were living in a dilapidated building.
Water = Proportion of the population having access to safe drinking water.
Dwelling Rooms = Proportion of households with two or more dwelling rooms.
Clean Fuel = Proportion of households who used clean fuel for cooking.
Banking = Proportion of households accessing banking services.
Toilets = Proportion of households without access to toilet facilities within the premises.
Females = Proportion of population who are females.
Elderly = Proportion of Persons in the Age-Group of 60 Years and above.
Children = Proportion of Persons in the Age-Group of 0–14 Years.

9.2.2 Explanation and Justification of the Selected Variables

The dependent and independent variables of the regression equation and the logic behind the selection of these variables are explained below.

9.2.2.1 Outcome Variable

The dependent variable in the regression model is the total persons with disabilities. The data was collected for all the 640 districts and 35 states and union territories available in Census 2011. The prevalence of disability was defined as the proportion (percentage) of total PWDs out of a total population of that particular district.

9.2.2.2 Explanatory Variables

Several socio-economic-demographic factors affect the prevalence and incidence of disability. Justification of these variables is self-explanatory in general. However, the logic for the inclusion of these variables has been provided wherever needed. After examining the available literature on the subject related to health and disability in India, the selected district level contextual variables are listed as follows: (1) The proportion of the population who are females (as nutritional outcome among women is poor); (2) The proportion of the population whose age is over 60 years (elderly population tend to suffer from higher illness and bodily impairments); (3) The literacy rates among females (higher educational standards should lead to a reduction in the burden of disability); (4) The proportion of the population who are members of SCs; (5) The proportion of the population belonging to STs communities. The constitution of India recognizes the facts that scheduled castes and scheduled tribes are disadvantaged communities. Therefore, these groups also have poor health outcomes than the general population.

The Census of India does not have data related to the income or expenditure of these population groups used in the present study. Hence, a few alternative variables were used as proxies for these indicators. (6) The proportion of the population living in urban areas; (7) The proportion of the population who are main workers (working for 6 months or more with regard to the reference period). Economic mobility should lead to a lower burden of disability. Average living conditions of the households were examined under the following heads: (8) The proportion of the population having access to safe drinking water (i.e., Tap water from treated sources + Hand Pump + Tube Well); (9) The proportion of the households who were living in a dilapidated building; (10) The proportion of households who did not have access to latrine facilities within the premises (lack of toilet facilities may cause illness and disease); (11) The proportion of households who used clean fuel for cooking (arrived at by adding Kerosene + LPG/PNG + Electricity + Biogas). Unavailability of clean fuel for cooking is often linked to a number of disabilities; (12) The proportion of households with two or more dwelling rooms; (13) The proportion of households accessing banking services (improvement in banking facilities and financial inclusion are associated with better access to finance when in need as well as access to various government welfare schemes).

9.3 An Analysis of Disability Prevalence in India: Evidence from Census 2001 and 2011

It is reflected in Table 9.1 that out of a total 121.10 crore population of India in 2011 Census, 2.70 crore persons have been enumerated as disabled, which was 2.20 crore in the year 2001, it shows a decadal growth of 22.4% during 2001 and 2011. As per Census data of 2011, on an average, one out of each tenth family includes a disabled member. The decadal growth is the highest in urban areas than the rural areas. Persons belonging to the vulnerable section of society (SCs and STs) have registered a much higher increase in decadal growth. In urban areas, there is a drastic increase in the decadal growth among the persons belonging to ST communities (81.75%), it could be the result of urbanization, displacement, and migration. As far as gender analysis is concerned, female PWDs have recorded higher decadal growth than their male counterpart. Again, it is higher in urban areas than the rural regions. The growth is the highest among STs, followed by SCs and the overall. In a nutshell, it is important to note that the decadal growth is the highest among SCs followed by STs. However, in urban areas, STs recorded significantly higher growth rates than SCs and the total. The same pattern can be observed among female STs.

Table 9.2 shows that there were 2.12% persons with disabilities in India, which increased to 2.21% in 2011. The highest proportion of individuals with disabilities can be seen among SC communities (2.45% in 2011), and least among STs (2.05%). However, it is important to highlight that decadal growth among SCs is much higher than the total and almost equivalent to STs. The proportion of persons with disabilities is lower in urban areas (2.17%) as against (2.24%) in rural areas. However, decadal growth is much higher in urban areas. It means a higher rate of urbanization is linked to the higher prevalence of disability. The gender analysis of females with disabilities suggests that there is an increase in the proportion of female PWDs in 2011 relative to 2001. The proportion is higher in rural areas than the urban ones. It is the highest among SCs who are female PWDs and living in rural areas (2.22%). However, the proportion in the 2011 Census has declined among males and has increased among females relative to 2001 Census figure.

It has been reported by many experts of the disability studies that the intensity of disability is more among marginalized groups and vulnerable sections of the society like SCs and STs. In general, persons born with disabilities in the SC and ST categories face the double brunt of caste-based discrimination on the one hand and deprivation of opportunities and discrimination on the other hand. In this context, Pal (2010) views that these communities are more likely to be afflicted with the trauma of disability and caste identity. Therefore, such intertwined problems may cause multiple disadvantages, which are emotionally very traumatic and can lead to permanent psychological damage. Further, Mehrotra (2013) mentions that:

> The intensity of discrimination, deprivation and exclusion experienced by persons with disabilities belonging to lower castes is more when compared to others.

Table 9.1 Absolute number of persons with disabilities by residence, gender, and social groups in India (2001 and 2011)

Group	Census 2001			Census 2011			Decadal growth		
	Person	Male	Female	Person	Male	Female	Person	Male	Female
Total	21,906,769	12,605,635	9,301,134	26,814,994	14,988,593	11,826,401	22.41	18.90	27.15
SC	3,711,110	2,127,877	1,583,233	4,927,433	2,770,592	2,156,841	32.78	30.20	36.23
ST	1,618,166	903,899	714,267	2,140,763	1,144,627	996,136	32.30	26.63	39.46
Rural									
Total	16,388,382	9,410,185	6,978,197	18,636,358	10,410,559	8,225,799	13.72	10.63	17.88
SC	3,036,356	1,738,176	1,298,180	3,798,625	2,136,872	1,661,753	25.10	22.94	28.01
ST	1,500,725	835,854	664,871	1,927,319	1,028,040	899,279	28.43	22.99	35.26
Urban									
Total	5,518,387	3,195,450	2,322,937	8,178,636	4,578,034	3,600,602	48.21	43.27	55.00
SC	674,754	389,701	285,053	1,128,808	633,720	495,088	67.29	62.62	73.68
ST	117,441	68,045	49,396	213,444	116,587	96,857	81.75	71.34	96.08

Source Calculated from Census of India (2001 and 2011)

Table 9.2 Proportion of population with disability in respective social groups, genders, and regions in India (2001 and 2011)

Group	Census 2001			Census 2011		
	Person	Male	Female	Person	Male	Female
Total	2.12	2.37	1.87	2.21	2.40	2.01
SC	2.23	2.47	1.97	2.45	2.68	2.20
ST	1.92	2.12	1.71	2.05	2.18	1.92
Rural						
Total	2.21	2.47	1.93	2.24	2.43	2.03
SC	2.28	2.53	2.02	2.47	2.70	2.22
ST	1.94	2.14	1.74	2.05	2.18	1.92
Urban						
Total	1.93	2.12	1.71	2.17	2.34	1.98
SC	2.01	2.23	1.77	2.38	2.60	2.14
ST	1.68	1.89	1.46	2.04	2.21	1.87

Source Calculated from Census of India (2001 and 2011)

As per Census 2011, there was 2.68 crore PWDs in India, out of which 0.5 crores were a member of SCs (18.38%) and almost 0.2 crores (7.97%) belong to ST communities. The proportion of the population who belong to SC and ST communities of PWDs is about 2.45% and 2.05%, respectively. Table 9.3 presents the proportion of disabled persons by type of disability and social groups in India in Census 2001 and 2011. It is noticeable that the incidence of disability drastically declined for persons with seeing, movement, and mental disability. However, these declines are offset by the considerable rise in the decadal growth of individuals with hearing and speech

Table 9.3 Proportion of disabled persons by type of disability and social groups in India in census 2001 and 2011

Type of disability	Census 2001			Census 2011		
	Total (%)	SC (%)	ST (%)	Total (%)	SC (%)	ST (%)
Seeing	48.55	48.60	48.38	18.77	19.11	20.00
Speech	7.49	7.30	7.50	7.45	5.19	5.27
Hearing	5.76	5.75	7.72	18.92	17.45	19.36
Movement	27.87	29.13	27.95	20.28	20.50	22.42
Mental	10.33	9.22	8.45	8.31	7.49	7.55
Mental retardation	–	–	–	5.62	5.11	4.92
Mental illness	–	–	–	2.70	2.38	2.63
Multiple	–	–	–	7.89	7.31	8.93
Any other	–	–	–	18.38	22.94	16.47

Source Calculated from Census of India (2001 and 2011)

disabilities. A number of factors can be attributed to these empirical inconsistencies, which mainly emanate from the change in methods and definition of disability data collection in the 2011 Census. In 2001, close to half of the PWDs had "in seeing" disability as compared to 18.77% in 2011. Again, SCs had the highest share in 2001, but in 2011, STs had the highest share. In 2011, the persons who had seeing impairment experienced a drastic decline. It may be partly because of changes in the methods of estimation. One-eyed individuals were considered as disabled in Census 2001, whereas, as per the Census 2011 such persons have not been treated as disabled in seeing. In addition, in the Census 2011 enumerators were asked to apply a simple test to ascertain blurred vision. In Census 2001, no such instructions were given and were mostly based on self-reporting.

The number of speech disabilities has also slightly declined in 2011 from 7.49 to 7.45%. A person was recognized as having "*speech disability*", if he/she is dumb or whose speech is not understood by a listener of normal comprehension and hearing. The definition was made clearer in Census 2011 to record persons with a speech disability. For instance, "*persons who speak in single words and are not able to speak in sentences*" were specifically mentioned to be treated as disabled. There is a very high increase in "*hearing*" disability from 5.76 to 18.92%. STs have the highest in hearing disability at 19.36%. Persons using hearing aid have been treated as disabled in Census 2011. They were not treated as disabled in the Census 2001. Further, persons having problems in hearing through one ear, although the other ear is functioning normally, were considered having a hearing disability in Census 2001. However, in Census 2011, such persons were not considered as disabled. It is interesting to note that movement disability has significantly declined from 27.87 to 20.28% in 2011. There was also a major revision in the category of movement disability. The highest number of persons are rerecorded in the category of movement disability.

Out of the total PWDs, there were 10.33% mentally disabled people in 2001, which has slightly declined to 8.31% in 2011. There were 5.62% mentally retarded people and 2.70% mentally ill persons in 2011. Even after combining these two categories, there is little decline in the proportion of persons with mental disability. Multiple disabled persons constitute 7.89% of the total PWDs, and the highest multiple disabilities are observed for STs. Multiple disabilities cover as many as three types of disabilities. It is important to highlight that "any other" category of PWD has a very high share of the total PWDs, it is 18.38%. SCs are the highest in "any other" PWD category. This category was included for the first time in 2011 to overcome the problem of estimation or counting the numbers of persons with disabilities. In other words, for those PWDs which are not listed in the Census question, the informants were required to report in this category. Census data in the "any other" category is significantly large (around 18%). A very high number of any other disability also shows lacunae in the enumeration process of the Census. It may be possible that the enumerators may not have properly explained the definition of any other disability to the informants. Inaccurate numbers pose a challenge to the development of policy framing.

9.4 Disability Prevalence Across the States of India

A cursory glance at the state-wise distribution of persons with disabilities suggests that PWDs are scattered across every nook and corner in India. Table 9.4 presents the share of PWD males and females out of the total population in the corresponding states and makes a comparative study between 2001 and 2011 Census data. The Table is self-explanatory; however, an attempt is being made to explain in brief. According to the 2001 Census, Sikkim experienced the highest prevalence of disability at 3.77%, followed by Arunachal Pradesh, Jammu and Kashmir, and Odisha. It is important to mention that southernmost states, Kerala, Puducherry, and Tamil Nadu, which are economically better off as compared to most states, had higher disability prevalence. Karnataka and Andhra Pradesh have a lower incidence of disability; also, both of them have almost the same prevalence of disability, around 1.80%. Some north-eastern states like Nagaland, Manipur, and Meghalaya have the least prevalence of disability. Goa had a minimum disability (1.17%). Bihar, Uttar Pradesh, Madhya Pradesh, Rajasthan, West Bengal, and Gujarat are adjacent in terms of disability prevalence; it ranged from 2 to 2.5%. Among the bigger states, Maharashtra had the least number of PWDs.

It can be observed from 2011 Census data that many states have reduced the incidence of disability significantly, some have worsened, and some are on the same ladder. Disability prevalence in the state of Sikkim declined but was still the highest in 2011, followed by Odisha and Jammu and Kashmir. Disability prevalence declined in some of the southern states. However, the disability condition worsened for Andhra Pradesh and Maharashtra. Except for Sikkim, Lakshadweep, and Puducherry, disability condition has improved and are among the lowest for most of the north-eastern states and union territories.

Thus, it is evident that the southern states reported more number of persons with disabilities than their northern counterparts. In this context, Kulkarni et al. (2019) pointed out that the southern states have provided welfare measures to the PWDs in a better way and more comprehensively than the northern states. Therefore, the northern states may be disinclined in reporting the prevalence of disability, as it will add policy liability on the states.

9.5 Disability Prevalence Across the Districts of India in 2011

This section illustrates disability prevalence across diverse regions of India with the help of the Geographical Information System (GIS Mapping). It examines disability distribution at the district level using 2011 Census data.

Spatial images of incidence of male and female disability across districts of India is depicted in Figs. 9.1 and 9.2. It is clear from the figures that male and female topographical examples are fundamentally the same. The prevalence rates

Table 9.4 Proportion of total disabled to the corresponding population in census 2001 and 2011

States	Census 2001			Census 2011		
	Person	Male	Female	Person	Male	Female
Jammu & Kashmir	2.98	3.20	2.74	2.88	3.08	2.65
Himachal Pradesh	2.57	2.93	2.19	2.26	2.48	2.04
Punjab	1.74	1.95	1.51	2.36	2.59	2.09
Chandigarh	1.73	1.88	1.52	1.40	1.51	1.27
Uttarakhand	2.29	2.62	1.96	1.84	2.00	1.67
Haryana	2.15	2.41	1.85	2.16	2.34	1.95
NCT of Delhi	1.70	1.90	1.46	1.40	1.54	1.24
Rajasthan	2.50	2.86	2.11	2.28	2.39	2.17
Uttar Pradesh	2.08	2.37	1.75	2.08	2.26	1.88
Bihar	2.27	2.62	1.90	2.24	2.47	1.98
Sikkim	3.77	3.95	3.55	2.98	3.03	2.92
Arunachal Pradesh	3.03	3.82	2.15	1.93	2.00	1.86
Nagaland	1.33	1.39	1.27	1.50	1.58	1.41
Manipur	1.31	1.41	1.21	2.05	2.17	1.93
Mizoram	1.80	1.91	1.69	1.38	1.48	1.28
Tripura	1.84	2.04	1.64	1.75	1.89	1.60
Meghalaya	1.24	1.30	1.18	1.49	1.56	1.42
Assam	1.99	2.16	1.81	1.54	1.61	1.46
West Bengal	2.30	2.55	2.04	2.21	2.41	2.00
Jharkhand	1.66	1.90	1.41	2.33	2.52	2.14
Odisha	2.78	3.05	2.49	2.96	3.18	2.74
Chhattisgarh	2.02	2.21	1.82	2.45	2.60	2.29
Madhya Pradesh	2.33	2.62	2.02	2.14	2.36	1.89
Gujarat	2.06	2.29	1.81	1.81	1.95	1.66
Daman & Diu	2.00	1.92	2.12	0.90	0.86	0.96
Dadra & Nagar H	1.84	1.91	1.74	0.96	0.98	0.93
Maharashtra	1.62	1.85	1.37	2.64	2.91	2.35
Andhra Pradesh	1.79	2.01	1.57	2.68	2.89	2.47
Karnataka	1.78	2.00	1.55	2.17	2.35	1.98
Goa	1.17	1.29	1.04	2.26	2.30	2.22
Lakshadweep	2.77	2.89	2.63	2.50	2.53	2.48
Kerala	2.70	2.96	2.46	2.28	2.46	2.11
Tamil Nadu	2.63	2.52	2.74	1.64	1.82	1.45
Puducherry	2.65	3.03	2.28	2.42	2.67	2.17
Andaman & Nicobar I	1.98	2.19	1.73	1.75	1.90	1.58

Source Calculated from Census of India (2001 and 2011)

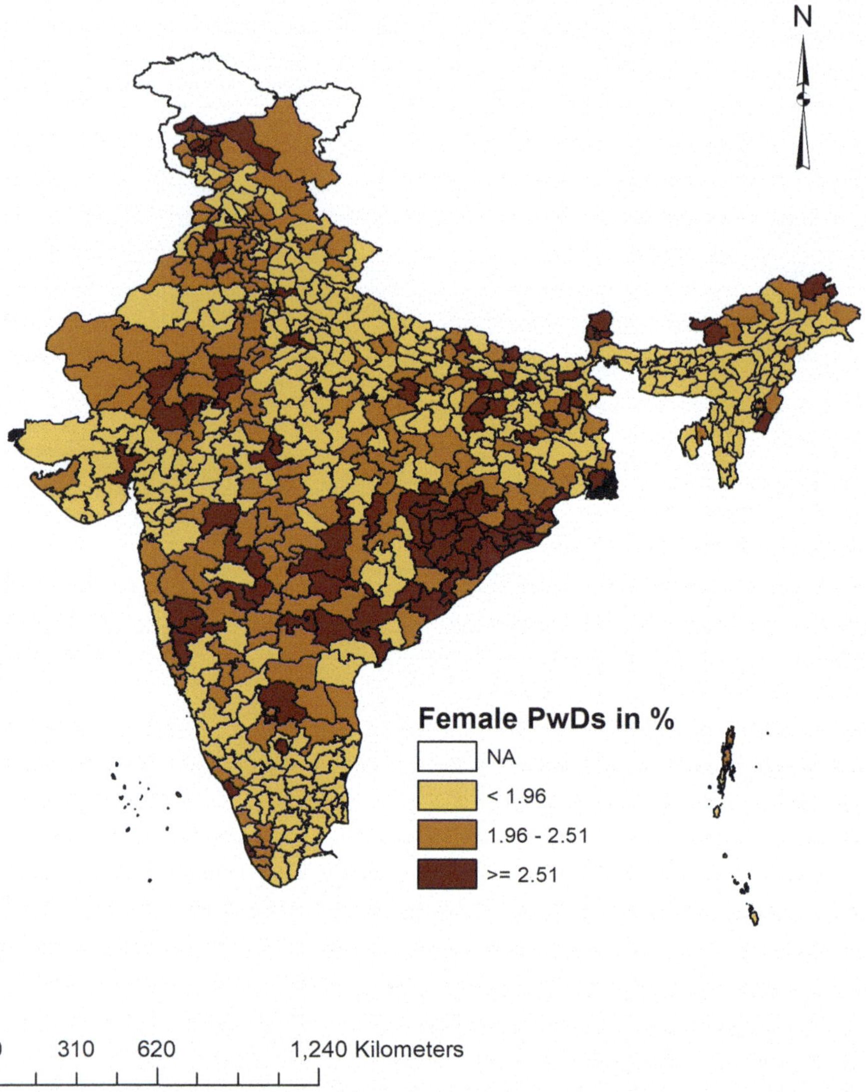

Fig. 9.1 Prevalence of disability across districts of India for males in 2011 (*Source* Prepared by the author, based on Census of India 2011)

are calculated as a proportion of persons with male and female disabilities out of their respective population groups in particular districts. The following GIS maps are self-explanatory. The geographical patterns of males and females with disabilities are found to be very similar to each other. There are many spatial zones where there are higher concentrations of disability prevalence. On the other hand, it can also be observed that there are several discontinuous spatial zones where there is

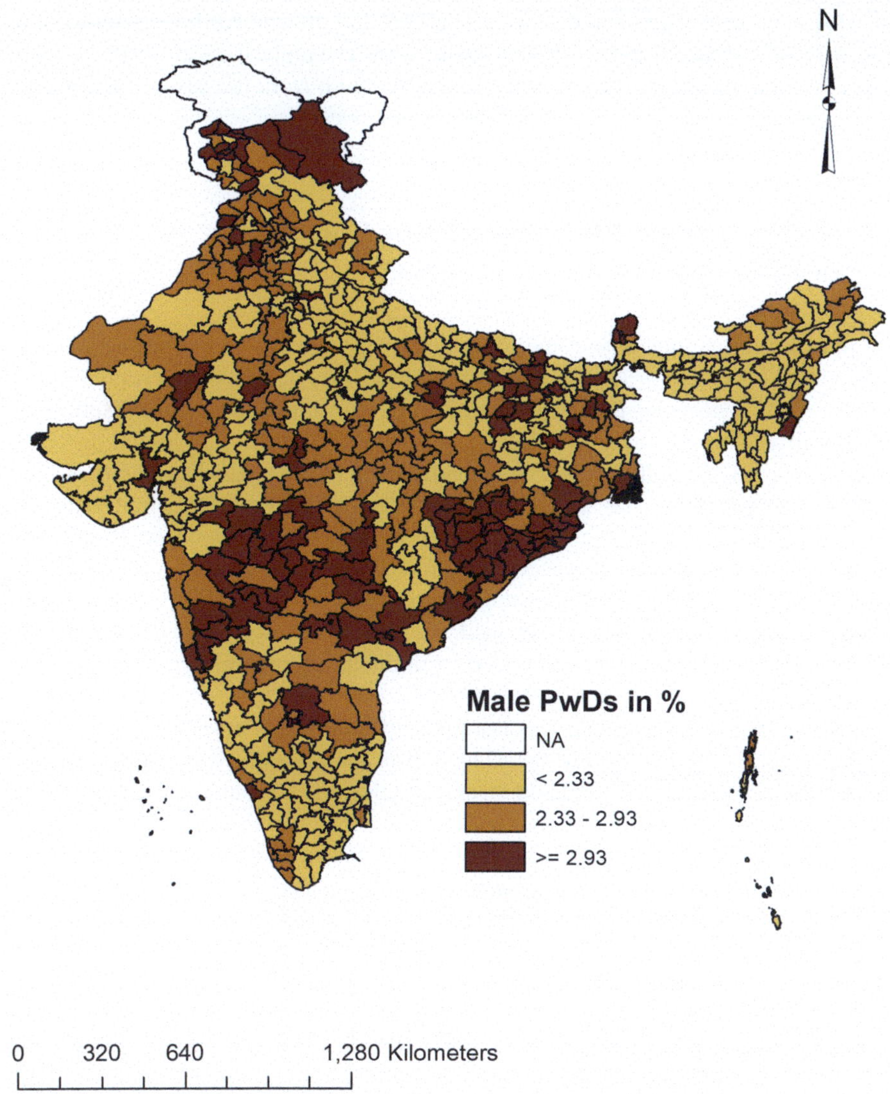

Fig. 9.2 Prevalence of disability across districts of India for females in 2011 (*Source* Prepared by the author, based on Census of India 2011)

lower disability prevalence. Therefore, a cursory glance across the districts of India suggests that disability is prevalent in every nook and corner of India. However, there are some regions (gender-wise), which are more disadvantageously placed.

A closer examination of disability prevalence among males and females provides that although the incidence of disability is lower among women, the prevalence is evenly distributed, and the pattern is largely similar to males. However, female disability is relatively lower in the eastern parts of Jammu and Kashmir, followed by some parts of Karnataka and southern Maharashtra. The highest disability prevalence (about 3% and above) can be observed in continuation from the eastern districts of Odisha to the southern parts of Maharashtra in the west.

9.6 Socioeconomic and Demographic Correlates of Disability Prevalence in India

The term disability involves complex interactions between individual and social environment (Topliss 1979). The United Nations (1982) in the World Programme of Action (WPA) concerning persons with disabilities noted hosts of factors responsible for disability. This includes social conflicts, natural disasters, accidents, poverty, resource constraints, unhygienic living conditions, physical and social barriers, and psycho-social problems. A disability may exist from birth or may be acquired during a person's lifetime. Coppin et al. (2006) found that a higher prevalence of disability is associated with decreasing socioeconomic status. Poor living conditions, particularly among lower caste groups, can make them more vulnerable to disability. Persons with disabilities tend to suffer from a physical, mental, and sensory impairment; hence, these people systematically get excluded from the development narrative (Appunni and Deshpande 2009; Meyer and Mok 2019).

Some of the other important literature provides that socioeconomic inequalities and prevalence of disability are correlates (Braithwaite and Mont 2009; Yeo and Moore 2003; and Elwan 1999). Gannon and Munley (2009) examined that people of older age suffer from multiple disabilities. It means that people tend to more disabled while they start aging. This pattern holds for all types of disabilities, that is, disability is an increasing function of age. Kulkarni et al. (2019) also found that disability has risen among the elderly population of India. In this context, Mishra and Mohanty (2017) noted that hearing and intellectual disability are positively associated with increasing age. Besides, lack of access to health services is a significant cause of disability (WHO 2002).

Table 9.5 shows the descriptive statistics for the outcome and exposure variables for all the 640 districts as per Census 2011. All of these socioeconomic and demographic characteristics examined vary significantly across districts of India. For example, the average value of the percentage of persons with disabilities is 2.15%, and the values vary between 0.76% (minimum) and 4.51% (maximum). The proportion of the population who belong to SC communities range from a minimum of 0% to a maximum of 50.17%. The proportion of the population who are members of ST communities ranges from 0 to 98.58%. The district-level female literacy rate is around 55% on an average; it varies from 24.25 to 88.62%. The proportion of persons living in urban areas is 27%. The proportion of the workers who are main workers constitute about 73% and ranges from 12 to 96%. As far as the profile of the average living conditions of the households is concerned, about 66% of households

Table 9.5 Descriptive statistics for the outcome and explanatory variables across 640 districts of India in 2011 (in %)

Variables	Districts	Mean	Std. Dev	Min	Max
Persons with disabilities (PWDs)	**640**	2.15	0.57	0.76	4.51
Socioeconomic					
Proportion of SC	640	14.86	9.13	0.00	50.17
Proportion of ST	640	17.71	27.00	0.00	98.58
Proportion of female literacy	640	55.23	12.41	24.25	88.62
Proportion of urban population	640	26.40	21.12	0.00	100
Proportion of main workers	640	73.29	12.65	30.65	96.4
Proportion of households in dilapidated buildings	640	4.99	3.11	0.23	17.74
Proportion of households with safe drinking Water	640	66.24	23.30	0.74	99.96
Proportion of households with 2 or more dwelling rooms	640	61.99	15.08	11.57	93.65
Proportion of households with clean fuel for cooking	640	27.15	21.68	1.12	98.64
Proportion of households accessing banking services	640	57.27	16.48	10.27	97.77
Proportion of households with no toilet facility within the premises	640	53.22	26.33	1.13	95.22
Demographic					
Proportion of females	640	48.55	1.64	34.79	54.22
Proportion of persons aged 60 plus	640	8.60	2.08	2.63	17.92

Source Calculated from Census of India (2011)

have access to safe drinking water. However, the condition of households is worse in accessing clean fuel for cooking, for which the mean value is only about 27%. Around 53% of households did not have toilet facilities within the premises. Finally, about 57% of households have access to banking services, which is considerably low.

An analysis of demographic profile shows that the proportion of females on an average is around 49% across districts. It ranges from 34.79 to 54.22%. Last but not the least, the proportion of elderly persons (60 plus) is on an average of about 8.60%, and it ranges between 2.63 and 17.92%.

Table 9.6 provides the outcome of the linear regression model, which assesses the relationships between important socioeconomic and demographic characteristics across districts of India and disability prevalence rates at the district level. There is a strong correlation between male disability and female disability; therefore, total persons with disabilities are chosen as the left-hand-side (dependent variable). The outcome of demographic characteristics across districts suggests that the proportion

Table 9.6 Socioeconomic and demographic determinants of disability prevalence variation across districts

Variables	Coefficients	*P*-values
Socioeconomic		
Proportion of SC	-0.003	0.247
Proportion of ST	0.003	0.005
Proportion of female literacy	−0.014	0.000
Proportion of urban population	0.008	0.000
Proportion of main workers	−0.014	0.000
Proportion of households in dilapidated buildings	0.039	0.000
Proportion of households with safe drinking water	−0.003	0.010
Proportion of households with two or more dwelling rooms	0.001	0.784
Proportion of households with clean fuel for cooking	−0.001	0.778
Proportion of households accessing banking services	−0.001	0.649
Proportion of households with no toilet facility within the premises	−0.001	0.790
Demographic		
Proportion of females	−0.072	0.000
Proportion of persons aged 60 plus	0.185	0.000
R-Squared	**0.257**	
Adjusted R-Squared	**0.241**	

Source Estimated from Census of India (2011)

of females and the proportion of persons who are 60 years or above (elderly population) are significantly associated with disability prevalence. A 1% increase in the number of persons who are above 60 years of age is associated with a 0.18% increase in the prevalence of disability. In this context, WHO and World Bank (2011) noted that "higher disability rates among older people reflect an accumulation of health risks across a lifespan of disease, injury, and chronic illness". Since an absolute number of females with a disability is much lower than that of males with a disability, therefore, the increase in the proportion of females is associated with declining disability prevalence rates, at 1% level of significance.

The findings of the socioeconomic and demographic correlates of disability prevalence are as follows: The districts where a large proportion of the population belongs to STs tend to have higher disability rates, the districts dominated by the scheduled tribe population have impoverished health facilities. On the other hand, no statistically significant link is observed concerning SC communities. Such a result may arise some sort of ecological fallacy. As expected, an additional change in female literacy appears to reduce disability prevalence rates at 1% level of significance. Most other developing countries also represent a similar picture (IIPS 2007). Improvements in female literacy may create better job opportunities and enhanced income standards, and thus better social outcome indicators (i.e., health status) which in turn may reduce disability prevalence among women.

An increase in the proportion of the population living in urban areas is linked to the higher level of incidence of disability at 1% level of significance. Although the social determinants of health outcomes (i.e., of disability) are much worse in rural areas than urban areas, however, the higher level of urbanization is linked to the higher level of incidence of disability at 1% level of significance. A higher level of urbanization is also associated with an increase in accidents. Arguably, a considerable number of residents are also pushed to slums due to increasing urbanization and migration. Besides, aging in urban areas could be another important factor. In this context, Huang et al. (2018) noted that aging in the urban sector is a global phenomenon. In India, also, there is a higher level of aging in the urban sphere. It is also evident that there is a higher level of disability among the Indian elderly. Therefore, it can be argued that the increasing proportion of households in the urban sector is also associated with a rising incidence of disability.

Further, the urban population is increasing; however, there is a lack of accessibility in the urban sphere of India, and the PWDs face multiple barriers. World Report on Disability (2011) noted that "persons with disabilities face widespread lack of accessibility to built environments, from roads and housing to public buildings and spaces and basic urban services such as sanitation and water, health, education, transportation, and emergency response and resilience programs. Barriers to information and communications, including relevant technologies and cultural attitudes, including negative stereotyping and stigma, also contribute to the exclusion and marginalization of persons with disabilities in urban environments" (pp. 262–263).

In this context, the United Nations (2017), in its report "*Good Practices of Accessible Urban Development*", suggested that urban environments have served as a barrier to the inclusion and participation of persons with disabilities. It reported that:

> lack of accessibility contributes greatly to the disadvantaged and vulnerable situations faced by persons with disabilities, leading to disproportionate rates of poverty, deprivation, and exclusion among persons with disabilities. For example, if persons with disabilities cannot access urban education facilities due to the lack of accessible transportation options, they are excluded from programs designed to promote universal education.

Thus, urban amenities or infrastructure can impede or enable participation or inclusion of the disabled depending upon whether these services are accessible or not. The World Urbanization Prospects (2014) estimated that by 2050, 66% of the world's population, will be living in urban areas. In this case, growing urbanization may further exacerbate the burden of disability. Therefore, prioritizing and accessible environment in urban infrastructure is crucial.

An increase in the proportion of main workers is associated with a decrease in rates of disability prevalence. It shows gainful employment can have a significant impact in reducing the burden of disability. The proportion of households living in dilapidated buildings is positively and statistically significantly associated with disability prevalence; therefore, dilapidated buildings can exacerbate the burden of disability in the districts. A unit increase in the accessibility or availability of safe drinking water also appears to reduce the incidence of disability at 1% level of significance. Since India is a water-scarce country, as per capita availability of water is very less. Besides, it is expected that there is a growing risk of acute water shortage in the future. Therefore, the lack of access to safe drinking water may pose a threat to disabled communities. R-Squared of 0.25 shows that the included explanatory variables could explain only 25% of the total variations in the prevalence of disability.

However, the study finds no statistically significant associations between overall disability prevalence and proportion of households with two or more dwelling rooms, the proportion of households with clean fuel for cooking, proportion of households accessing banking services, and the proportion of households with no toilet facility within the premises.

9.7 Concluding Remarks

Inclusive policies for disabilities are a challenge to modern states. A concern has been raised in the present study about the enumeration process. Available literature and empirical investigations carried out in the paper provide that there is an underestimation of disability prevalence in India, particularly among women, mostly caused by sociocultural practices and inadequate enumeration process. Proper and accurate counting under specific categories and age-cohorts of disability would better aim the desired group of individuals, ensuring that benefits are not misappropriated.

Further, the study also stresses that the policymakers need to modify and expand the definition of disability used in the Census of India to produce an internationally comparable estimate of disability prevalence. It is essential that the government comprehensively defined who the disabled are and what constitutes disability from both medical and social perspectives. There is an urgent need to improve the quality and reliability of disability data. By applying appropriate methods, significant improvement in data collection is possible. Properly training the enumerators and awareness among the respondents are the initial steps. A more inclusive nature of data is expected if some of the disabled enumerators are involved at the data collection level. In addition, the disabled communities should also be involved at the upper stage of designing and executing public health policies. The upcoming round of Census enumeration to be conducted in 2021 has gained momentum in India. It is evident that there is underreporting in the disability estimates of India. The issues incorporated and examined in the present paper, if noted sincerely, may lead to improvement and robustness in the validity and reliability of India's disability estimates in the coming Census survey under the broader ambit and commitment for the 2030 Agenda for Sustainable Development Goals.

The important literature on the subject provides that there are sharp variations concerning caste, class, religion, region, and gender for the disabled population in India. Social discrimination is one of the most significant problems experienced by people with disabilities. The evidence shows that disability brings in a socioeconomic burden, and the dynamics of life undergo a significant change as disability sets in. It not only has its implications on the individual but also his household. A cursory glance across the district level distribution of persons with disabilities suggests that the PWDs are scattered across every nook and corner in India with significant variations over time and space. The empirical investigation in the paper provides that the incidence of disability is higher among the marginal and vulnerable sections of society.

An increase in life expectancy and faster growth of the elderly disabled population would present a variety of socioeconomic and healthcare policy challenges in the context of changing family relations and severely limited income supports for the older section of the society. Such issues become more relevant as India is already missing the opportunities of demographic dividend. Further, the "disability-determinant regression model" of the paper reveals that several socioeconomic and demographic factors are associated with higher incidence and prevalence of disability at the district level. Thus, it can be inferred from the findings of the present study that understanding temporal and spatial variations and socioeconomic and regional divides in the disability sector are crucial for designing public health policies.

Limited public health care facilities, increasingly unaffordable private health services, low level of income, and also lower coverage of the social security system will lead to a higher burden of disability on society, which calls for urgent provisioning of social protections and affirmative actions on the part of all concerned stakeholders.

Despite efforts of the government, civil society, and disability rights activists, the largest section belonging to them are living in degenerated conditions. It is high time government, disability rights, and advocacy groups, and the concerned stakeholders focused on the holistic wellbeing of this largest hitherto marginalized, stigmatized, and discriminated section of the society, and accordingly made efforts to amalgamate them in mainstream society. There is also a dire need for effective policy interventions to balance regional development and framing disability policies, particularly in the context of local idiosyncrasies. The country has no choice but to pay adequate attention to the overall wellbeing of the disability sector. Although, India has enacted several legislations in line with international conventions and standard practices to improve the lives of persons with disabilities.

Nonetheless, there is a need to move beyond the charity-based approach, medical model, and patronizing attitude. Most importantly, the recent global pandemics like COVID-19 give an opportunity to the government to reorient its health strategies towards the disability sector. Many Disability Rights Organizations and advocacy groups believe that coining terms like "*Divyang*", enacting the PWDs Act of 2016 (not adequate but at least an improvement over the earlier provisions), and initiatives like "*Accessible India Campaign*" and "*Divyang Kumbh Mela*" recently held at Allahabad, holds a silver lining.

Acknowledgements This research paper is based on the author's Ph.D. thesis under the supervision of Professor Santosh Mehrotra at The Centre for Informal Sector and Labour Studies, Jawaharlal Nehru University, New Delhi. The author is grateful to him for providing valuable inputs and guidance.

References

Appunni SS, Deshpande AP (2009) Work participation among disabilities in India. J Soc Sci 21(2):117–122

Braithwaite J, Mont D (2009) Disability and poverty: a survey of world bank poverty assessments and implications. ALTER-Eur J Disabil Res 3(3):219–232

Census of India (2001, 2011) Data on disabled population. Government of India, Office of the Register General, New Delhi, India

Coppin et al (2006) Low socioeconomic status and disability in old age: evidence from the InChianti study for the mediating role of physiological impairments. J Gerontol Ser A Biol Sci Med Sci 61(1):86–91

Department for International Development (2018) DFID'S strategy for disability inclusive development 2018–2023

Elwan A (1999) Poverty and disability: a survey of the literature, vol 9932. Social Protection Advisory Service, Washington, DC

Gannon B, Munley M (2009) Age and disability explaining the wage differential. Social Sci Med 69(1):47–55

Huang N et al (2018) The relationship between urbanization, the built environment, and physical activity among older adults in Taiwan. Int J Environ Res Ad Publ Health 15(5):1–20

International Institute for Population Sciences and Macro International (IIPS 2007) National family health survey (NFHS-3). 2005–06: India, Volume I

Jeffery R, Singal N (2008) Measuring disability in India. Econ Polit Wkly March 22:22–24

Jha M (2016) Recognising differently abled as minority. Econ Polit Wkly 51(36)

Khanday AZ, Akram M (2012) Health status of marginalised groups in India. Int J Appl Sociol 2(6):60–70

Kulkarni V et al (2019) Has disability risen among the elderly in India. Econ Polit Wkly Vol lIV 29:83–91

Mehrotra N (2013) Disability, gender and state policy: exploring margins. Rawat Publications, New Delhi

Meyer BD, Mok WK (2019) Disability, earnings, income and consumption. J Public Econ 171:51–69

Mishra AK, Gupta R (2006) Disability index: a measure of deprivation among disabled. Econ Polit Wkly 4026–4029

Mishra RS, Mohanty SK (2017) Socioeconomic and health correlates of disability in India. Int J Commun Med Publ Health 5(2):600–610

Mont D (2007) Measuring disability prevalence. World Bank Social Protection Discussion Paper, 706

Pal GC (2010) Dalits with disabilities: the neglected dimension of social exclusion. Indian Institute of Dalit Studies, New Delhi

Reddy CR, Sree KP (2015) Situating census data in disability discourse: an analysis of census 2011 and census 2001. Indian Anthropol Assoc 45(2):59–74

Saikia N et al (2016) Disability divides in India: evidence from 2011 census. PloS One J 12(2)

Sen AK (2004) Disability and justice. In: Keynote speech, second international conference on 'disability and inclusive development'. The World Bank

The World Urbanization Prospects (2014) United nations department of economic and social affairs. https://population.un.org/wup/Publications/Files/WUP2014-Report.pdf

The United Nation (2017) Good practices of accessible urban development. Making Urban Environments Inclusive and Fully Accessible to All

Topliss E (1979) Provision for the disabled, 2nd edn. Basil Blackwell

United Nations (1982) World programme of action (WPA) concerning persons with disabilities. Department of Economic and Social Affairs, The UN

United Nations (2018) Disability and development report. Realising the Sustainable Development Goals by, for and with Persons with Disabilities, Department of Economic and Social Affairs

World Bank (2009) People with disabilities in India: from commitments to outcomes. Human Development Unit: New Delhi, South Asia Region
World Health Organisation (2002) Reducing risks, promoting healthy life. The World Health Report
World Health Organization (2011) World report on disability. World Health Organization
World Health Organization (2018) Disability and health report. World Health Organization
Yeo R, Moore K (2003) Including disabled people in poverty reduction work: nothing about us without us. World Dev J 31(3):571–590

Dr. Baikunth Roy is an Assistant Professor of Economics at College of Commerce, Arts and Science, Patliputra University, Patna. Having completed his Masters in Economics from Hyderabad School of Economics, University of Hyderabad, he went for M.Phil and Ph.D. in Economics from Jawaharlal Nehru University, New Delhi. His areas of interest primarily include development economics, agricultural economics, labor economics, and disability studies. He has authored several research papers in national/international journals and edited books. He has also contributed many articles in newspapers and online portals, such as the Hindu and the Wire. He is also associated with many think tanks and disability rights organizations in India.

Chapter 10
How Far Gender Inequality Suppresses Human Development: Evidence from India

Raju Sarkar and C. M. Lakshmana

Abstract In Indian society, the root cause of gender inequality lies in its patriarchal structure. In this paper, an attempt has been made ment, to a certain extent, to link gender relations with the development of the nation. Furthermore, it examines the spatial variations of gender development and analyzes spatial inequality at the level of human development in the different states of India. Finally, this paper has suggested some important policy implications for reducing gender inequality in India. The Gender Parity Index (GPI) of literacy has been calculated to understand the inequality between the two genders. The Human Development Index (HDI) method has been used to analyze spatial variation in the development of human resources. The study reveals that there are considerable disparities in the level of human resource development among the different states of India, and higher-income states do not show the highest HDI values. The measure to reduce gender inequality in rural India must be developed and implemented according to the needs of specific target groups, that is, women and children.

Keywords Human Development Index (HDI) · Sex ratio · Literacy rate · EAG states

10.1 Introduction

Gender inequality remains a major issue of concern in India. It is reflected in India's low ranking in the Gap Index of the World Economic Forum (GGI-2014), with below-average scores on benchmarks, such as achievement in education, economic activities, health, and survival. The process of gender inequality may not be self-assertive and is a very complex issue. Gender inequality depends upon the power over resources and is affected by technological and socio-economic adaptation (Ridgeway 2011). Amartya Sen (2001) provided an example of different types of inequalities, such as inequality in mortality, inequality in births, inequality in basic facilities,

R. Sarkar (✉) · C. M. Lakshmana
Population Research Centre, Institute for Social and Economic Change, Bangalore, India

M. Mishra et al. (eds.), *Regional Development Planning and Practice*,
Advances in Geographical and Environmental Sciences,
https://doi.org/10.1007/978-981-16-5681-1_10

inequality of special opportunities, professional inequality, ownership inequality, and inequality in households. Gender inequality negatively impacts human development. In the Indian context, a large literature examined regional disparities within the country and suggested that the disparities increased since the reforms (Joseph 2004). However, there has been no discussion of gender disparity at the sub-national level and its effect on human development.

As a rapidly developing country, India is plagued by social problems related to socio-cultural hegemony and gender unfairness that leads to disparities of income and opportunities (Comyn et al. 2014). In India, women comprise half of the total population; but, they cannot appropriately exercise educational rights, economic rights, and human rights. Women are also relatively deprived of the same freedoms and rights that men enjoy. Bhaskar and Gupta (2007) found an unbalanced sex ratio in the northern and western states of India, while in the southern and eastern part, data on the proportion of Child Sex Ratio (CSR) compatible with international evidence indicates a slightly biased sex ratio at birth towards boys. The causes of this unfair CSR, as explained through a kinship pattern (Dyson and Moore 1983; Miller 1989) or avoid of the girl children (Mayer 1999), are mainly due to the social value attributed to women due to their economic usefulness.

The main objective of this paper is to study gender inequality in education, sex ratio, and employment, and to some extent, to link gender relations to the development of the nation. In addition, it attempts to examine the spatial variations of gender development and analyze the spatial inequality in the level of human development of different states of India. Finally, this paper has suggested some important policy implications to reduce gender inequality in India.

10.2 Role of Gender Equality in Human Development

Human development and gender equality are determined by socio-economic and political factors. Human development not only depends on financial achievement, however; it is also determined by three necessary factors, such as life expectancy, literacy, and level of income. Higher human development, at the same time, does not mean an equal advantage for men and women because of the link between income and human development on the one hand and gender equity on the other hand (Bayeh 2016).

According to Guha (2002), a nation cannot build up without sufficient provision for public health, education, food, clothing, shelter, and a decent standard of living. Morrison et al. (2007) argued that gender equality occurs through women's greater access to land and credit markets and their greater decision-making power within households. This leads to improved empowerment of women, poverty reduction, and better economic growth.

In a democratic country, human development enables one to lead a meaningful life, which can help the society to develop economically, educationally, socially, culturally, and so on. Human development also means that the people of a society must be educated and be able to contribute to economic development with good health. Although it is not possible to have a perfect quantitative measure of human development, the United Nations Development Program (UNDP) has developed a composite index, now recognized as the Human Development Index (HDI).

Sen (2001) defined gender inequality as "not one homogeneous phenomenon, but a collection of disparate and interlinked problems." Gender inequality is one of the main barriers to guarantee the socio-economic mobilization of a country. When the value or range of inequality increases, the loss of human development also increases. Once again, countries with high gender inequality also experience an unequal distribution of human development. The gender inequality index measures the inequality among men and women in three dimensions: reproductive health in terms of maternal mortality, empowerment that includes the participation of women in the national parliament and higher education, and the participation of women in the labor market. If we calculate the progress of these three dimensions, as well as their development indicators, we see that there is still a high maternal mortality rate in India compared to several developing countries in the world.

10.3 Gender Inequality in India and Women's Development

Gender inequality means disparity between men and women in various social, economic, political, cultural, and legal aspects. Gender inequality may be defined as discrimination against women based on their sex. Women are exploited, degraded, violated, and discriminated against, both in our homes and outside. Discrimination against women is everywhere in the world and more so in Indian society. Gender discrimination is varying from one community to another. The main reasons for discrimination against women in Indian context are educational backwardness, caste, religious beliefs, culture, low income, unemployment, society, and family situation. Gender-induced disparities are of particular concern in India, discrimination against the girls and the resulting excess female child mortality and adverse sex ratios for females in India.

The Government of India, alongside different states, started various strategies, projects, and plans proposed to decrease the gender gap and to increase the empowerment of women during the period 1989–2015 (Singh 2017). A portion of these remarkable activities is the Rashtriya Mahila Kosh (RMK), Sampoorna Gramin Rozgar Yojana (SGRY), Awareness Generation Projects for Rural and Poor Women, Rajiv Gandhi Scheme for Empowerment of Adolescent Girls-Sable (RGSEAG), Support to Training and Employment Programme for Women, Integrated Child Development Services (ICDS), Balika Samriddhi Yojana (BSY), National Programme of Nutritional Support to Primary Education, National Programme for Education of Girls at Elementary Level, Apni Beti-Apna Dhan (ABAD), and so on (Singh 2017).

10.4 Gender Equality and Human Development in India

The empowerment of women is a standout among the essential issues in India today. The 'empowerment of women', 'gender equity', and 'gender equality' are isolated ideas, but they are firmly linked. Gender equity implies the equivalence in the results of the lives of women and men, recognizing their different desires and interests and requiring a redistribution of intensity and assets. Gender equity recognizes that women and men have different requirements, preferences, and interests (Sarkar 2017).

The fundamental principle of human development is that people expand their own options, realize their potentiality, and enjoy the freedom to lead lives that they value or have reasons to value. Equal opportunities in all areas, for all people, women and men alike, are central to human development. However, these options are still unevenly distributed within the societies, and, unfortunately, many of them are not yet available to women. Disadvantages and discrimination, lack of information about what mainly affects women's lives, including health issues such as physical and emotional, reinforced by social norms and values, institutions, and public policies, exist in all spheres of women's lives and manifest themselves differently in the different environments of the country.

They limit women's capabilities, opportunities, and options, resulting in less progress than men in terms of health, education, the standard of living, empowerment, personal security, and other essential dimensions of human development. Therefore, the full life potential of many women cannot reach an optimal level, and humanity as a whole does not take advantage of its potential.

The Human Development Report (2016) published by the United Nations Development Program (UNDP) reiterates the inferior status of India's human development, with the country ranked 131 out of 188 countries. One of the important factors that are dragging down the development indicators of India is enormous of gender disparities. In Asia, India has a disadvantage over some of the neighbours in terms of gender inequality (as India ranked 125). Countries such as China (37), Myanmar (80), Sri Lanka (87), Bhutan (110), Nepal (115), and Bangladesh (119) have a lower rank of gender inequality than India; however, Pakistan (130) has a higher rank of gender inequality (Jahan 2016).

10.5 Data and Methodology

The relevant data regarding sex ratio, literacy, and occupation have been collected from the Census of India (2011). The sex ratio of children in the 0–6 age group is defined as the number of females for every 1,000 males. In previous censuses up to 1981, it was customary to conclude the literacy rate, taking into account the overall population, and it was called crude or gross literacy rate. If the literacy rate is determined by excluding the population of the 0–6 age group from the overall population, the literacy rate is sufficient since the population seven years and above is only counted as literate and therefore has accepted since 1991, and it is called as effective literacy rate (Census of India, 2011). In this study, on the basis of census definition, the effective literacy rate has been considered as a literacy rate.

The formula for calculating the effective literacy rate (Lee 2008; RGI 2011b) is as follows:

$$\text{Effective literacy rate} = \frac{\text{Number of literate persons aged 7 and above}}{\text{Population aged 7 and above}} * 100 \tag{10.1}$$

The Gender Parity Index (GPI) of literacy has been calculated to understand the inequality between the two sexes (Singh 2017).

$$\text{GPI of literacy} = \frac{\text{female effective literacy rate}}{\text{male effective literacy rate}} \tag{10.2}$$

The most encouraging instrument and weapon or significant indicator of social development is undoubtedly female literacy. Female literacy and sex ratio are two important demographic characteristics of the population. In this context, there is a need to identify the correlation between female literacy and sex ratio in India. Spearman's rank correlation method has been used to understand the relationship between the female literacy rate and sex ratio.

The following formula has been applied (Patrick et al. 2018):

$$r_s = 1 - \frac{6\sum_d 2}{n(n^2 - 1)} \tag{10.3}$$

Human Development Index (HDI) method has been used to analyze spatial variation in human development. The HDI is calculated for women from six selected indicators, which are as follows: (a) average life expectancy rate, (b) literacy rate, (c) percentage of urban population, (d) sex ratio, (e) percentage of female literacy, and (g) per capita income. The HDI was first used by noble laureate Professor Amartya Sen in the year 1990 and Mahbub ul Haq, an economist, developed the idea of human development (Baru 1998). The HDI is an aggregate index, calculated on the basis of United Nations Development Programme (UNDP), and published in its Human Development Reports. In the report, the yield in each dimension is expressed as a value between 0 and 1 by applying the universal formula.

$$\text{Index for each indicator} = \frac{(\text{actual value - minimum value})}{(\text{maximum value - minimum value})} \tag{10.4}$$

$$\text{Human Development Index (HDI)} = \frac{\text{Sum of indicators for all the indicators}}{\text{Number of indicators used}} \tag{10.5}$$

10.6 Results and Discussion

10.6.1 Role of Education and Regional Variation of Literacy Rate in India

Education is significant in the development procedure for two reasons: First, because education can be seen as an end in itself since it improves individuals' observation and personal satisfaction; and, second, education prompts the arrangement of human capital and is a significant imperative for the development process (George 2015; Sarkar 2017). Although the level and quality of education can be considered in a number of ways, literacy figures are necessary for any dimension of educational attainment. The literacy level is an important and the most basic index of the educational achievements of an economy (Shukla and Mishra 2017).

In addition, overall education, female education has a unique role in the development process (Shukla and Mishra 2017; Sarkar 2017); hence, we also separately analyze female literacy as an important determinant of development.

According to the Census of India (2011), literacy is defined as the ability to read and write a simple message with understanding in any language (George 2015; Shukla and Mishra 2017). It is one of the sensitive indicators of development for any society. It is also an accepted fact that the literacy rate is a crucial indicator of social and economic development in any society (George 2015). Literacy in India grew very slowly until Independence in 1947. In the period 1991–2011, there was an acceleration in the growth rate of literacy (see Fig. 10.1). There has been a continuous rise in literacy rates in India. The overall literacy rate increases from 16.6 percent in 1951 to 73.04 percent in 2011.

Gender-based inequality resulted in women's literacy rates being much lower, at 65.46 percent, than their male counterparts, at 82.14 percent in 2011. Figure 10.2 shows the literacy rates by sex and residence. It is evident from Fig. 10.2 that, relative to rural and urban areas, gains in female literacy are improved compared to male literacy. In addition, the gender gap in literacy has dropped from 24.6 percent in 2001 to 19.3 percent in 2011 in rural areas and from 13.4 percent in 2001 to 9.7 percent in 2011 in urban areas.

Inequality in female literacy is observed in states such as Rajasthan, Jharkhand, Chhattisgarh, Madhya Pradesh, Haryana, Bihar, Uttar Pradesh, and Odisha. The 2011 Census notes an eight percent increase in the overall literacy level compared to the 2001 census. It is also noted that the literacy gap between men and women had decreased to 16 percent in 2011 (see Table 10.1). Progress in women's literacy is also evident, as reflected in a decrease in the absolute number of illiterate women between 2001 and 2011. Southern states of India show a higher female literacy rate than most of the North-Indian states, and this is also expressed in the sex proportions as well.

As per the 2001 census, Kerala had the highest literacy rate of 90.92, followed by Mizoram, which in 2001 had a literacy rate of 88.49 percent. Pondicherry, Goa, and Delhi had literacy rates of more than 80 percent. In recent times, particularly in

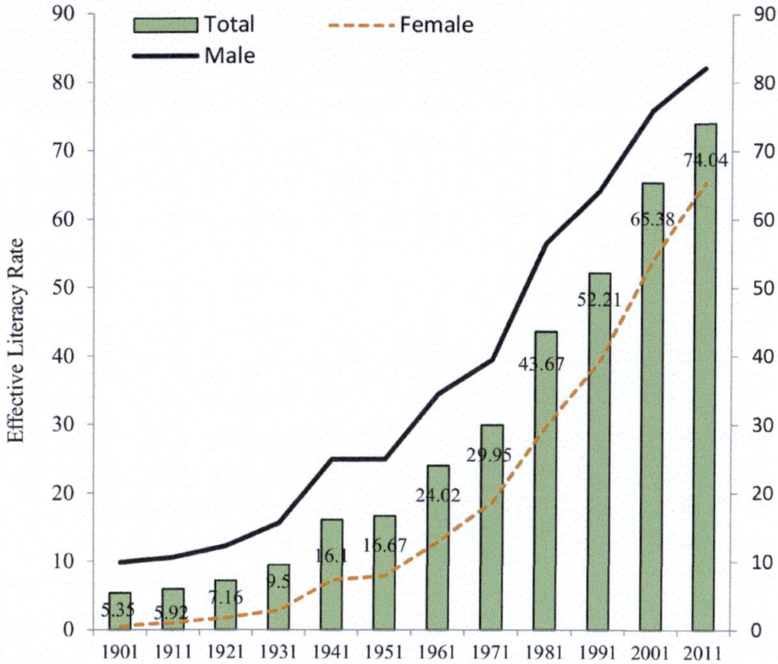

Fig. 10.1 Male–female literacy rate in India, 2011 (*Source* Computed from Registrar General of India 2011a)

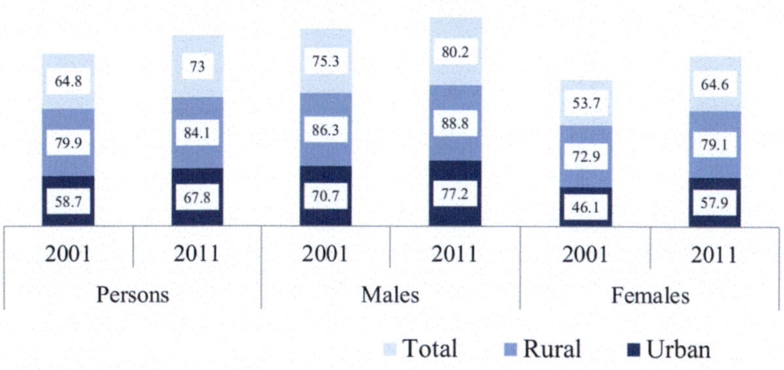

Fig. 10.2 Literacy rates by sex and residence in India (*Source* Registrar General of India (2011b)

the last decade, some of the conventional states with educational delays have also shown significant improvements. Empowered Action Group (EAG) states, such as Madhya Pradesh, recorded a rise from 44.2 percent in 1991 to 64.11 percent in 2001. Similarly, the literacy rate grew from 38.50 percent in Rajasthan in 1991 to 61.03 percent in 2001. In a single decade, literacy in Rajasthan and Madhya Pradesh grew by about 20 percentage points. Bihar's literacy, which was in line with Rajasthan

Table 10.1 Effective literacy rates by sex in EAG and Non-EAG states of India, 2011

Sl. No	State/UTs	Literacy Rate	Female Literacy (fl)	Male literacy (ml)	Difference = (fl-ml)
Non-Empowered Action Group (Non-EAG) States					
1	Andhra Pradesh	67.66	59.74	75.56	15.82
2	Arunachal Pradesh	66.95	59.57	73.69	14.12
3	Assam	73.18	67.27	78.81	11.54
4	Bihar	63.82	51.5	73.39	21.89
5	NCT of Delhi	86.34	80.93	91.03	10.1
6	Goa	87.4	81.84	92.81	10.97
7	Gujarat	79.31	70.73	87.23	16.5
8	Haryana	76.64	66.77	85.38	18.61
9	Himachal Pradesh	83.78	76.6	90.83	14.23
10	Jammu and Kashmir	68.74	65.46	78.26	12.8
11	Karnataka	75.6	68.13	82.85	14.72
12	Kerala	93.91	91.98	96.02	4.04
13	Maharashtra	82.91	75.48	89.82	14.34
14	Manipur	79.85	73.17	86.49	13.32
15	Meghalaya	75.48	73.78	77.17	3.39
16	Mizoram	91.58	89.4	93.72	4.32
17	Nagaland	80.11	76.69	83.29	6.6
18	Punjab	76.68	71.34	81.48	10.14
19	Sikkim	82.2	76.43	87.29	10.86
20	Tamil Nadu	80.93	73.86	86.81	12.95
21	Tripura	87.75	83.15	92.18	9.03
22	West Bengal	77.08	71.16	82.67	11.51
Empowered Action Group (EAG) states					
1	Bihar	63.82	51.5	73.39	21.89
2	Chhattisgarh	71.04	60.59	81.45	20.86
3	Jharkhand	67.63	56.21	78.45	2.24
4	Madhya Pradesh	70.63	60.02	80.53	20.51
5	Odisha	73.45	64.36	82.4	18.04
6	Rajasthan	67.08	52.66	80.51	27.85
7	Uttar Pradesh	69.72	59.26	79.24	19.98

(continued)

Table 10.1 (continued)

Sl. No	State/UTs	Literacy Rate	Female Literacy (fl)	Male literacy (ml)	Difference = (fl-ml)
8	Uttarakhand	79.63	70.7	88.33	17.63
Union Territories					
1	Andaman and Nicobar Islands	86.27	81.84	90.11	8.27
2	Chandigarh	86.43	81.38	90.54	9.16
3	Dadra and Nagar Haveli	77.65	65.93	86.46	20.53
4	Daman and Diu	87.07	79.59	91.48	11.89
5	Lakshadweep	92.28	8.25	96.11	7.86
6	Puducherry	86.55	81.22	92.12	10.9
7	India	74.04	65.46	82.14	16.68

Source Computed from RGI (2011a)

in 1991, has fallen far behind. The literacy rate in Bihar rose to just 47.53 percent compared to 61 percent in Rajasthan in 2001.

According to the 2011 census, among the states of India, Kerala tops the list with a total literacy rate of 94.0 percent, followed by Lakshadweep (91.85%) and Mizoram (91.33%). Bihar represents the lowest total literacy rate and lowest female literacy rate in India in the last decade. States that have literacy rates below the national average are Assam, Arunachal Pradesh, Andhra Pradesh, Jammu and Kashmir, and all Empowered Action Group (EAG) states, excluding Uttarakhand. Historically, various factors have contributed to women's low literacy rate, namely, gender-based inequality, social discrimination, and low enrolment of girls in the schools (George 2015).

10.6.2 Impact of Gender Disparities and the Gender Gap in Literacy Rates

Gender disparity in literacy is the product of one group of people being systematically discriminated against based on gender. The low female literacy rate in India has had a significantly negative effect on family planning and the attempts to stabilize the population. An NSSO report confirms that about 15 percent (about 90 million) of girls never attend school compared to 11 percent of boys (Singh 2017). As a result, lack of education usually creates a set of problems for them. Uneducated mothers do not know the best nutritional options for their children.

A study by Lawrence et al. (2009) recognized the fact that, compared to women with a higher educational level, women with a lower educational level had less control over their families' food options, showed less support for attempts to eat in a way healthy, fewer opportunities to observe and learn good food-related practices, more perceived environmental limitations, and more ambiguous beliefs about the consequences of eating a nutritious diet.

Table 10.1 also shows the gender disparities in literacy across India, as well as within the EAG States. There are ten states, namely, Bihar, Jammu and Kashmir, Haryana, Uttar Pradesh, Uttaranchal, Orissa, Madhya Pradesh, Chhattisgarh, Jharkhand, and Rajasthan, and one union territory, Dadra and Nagar Haveli, have revealed the gender gap in literacy rates is greater than the national average, that is, 16.68 percent. Among the states, Rajasthan had the largest gender disparity with 27.8 percent, followed by Jharkhand 22.3 percent, Chhattisgarh 20.9 percent, Madhya Pradesh 20.5 percent, and so on. In Bihar, 73.39 percent of males were literate compared with 51.5 percent of females in 2011.

The gap was narrower in Mizoram, where male literacy was registered at 90.69 percent and female literacy at 86.13 percent in 2001. But in 2011, for the state of Meghalaya, this changes slightly with the increase in male literacy rate of 77.17 percent and female literacy rate 73.78 percent, confirming that the gap between the male and female literacy rate is very narrow. For Kerala, the two rates were 96.02 and 91.98 percent, respectively, in 2011.

In India, male literacy is higher than female literacy. To minimize the inequality between the two sexes, it is required to minimize the difference between the literacy rates of men and women. The difference between the literacy rates of men and women in India in 2011 is 16.3 percentage points compared to 21.6 percentage points in 2001. The difference between the two sexes has been reduced over the decade by five percentage points. In 2001, the Gender Disparity Index (GPI) of the literacy rate was 0.59 or less in five states, while it was equal to or less than 0.69 with respect to 10 states, that is, Arunachal Pradesh, Bihar, Jammu and Kashmir, Uttar Pradesh, Jharkhand, Madhya Pradesh, Odisha, Chhattisgarh, Rajasthan, and Dadra and Nagar Haveli.

In 2011, the literacy GPI was more than 0.70 in all states and union territories (except Rajasthan). In 2001, the highest gender literacy gap state—Bihar—improved its position to the third in 2011. Rajasthan, which ranked third in 2001 in terms of GPI in literacy, moved to the top spot in 2011, which further expanded the gender disparity (see Fig. 10.3). The performance of states such as Arunachal Pradesh, Uttar Pradesh, and Odisha has been very encouraging in terms of reducing the gender gap in literacy. In general, the gender gap in the literacy rate has been substantially reduced in most of the states of India during the last decade, which is undoubtedly a significant achievement.

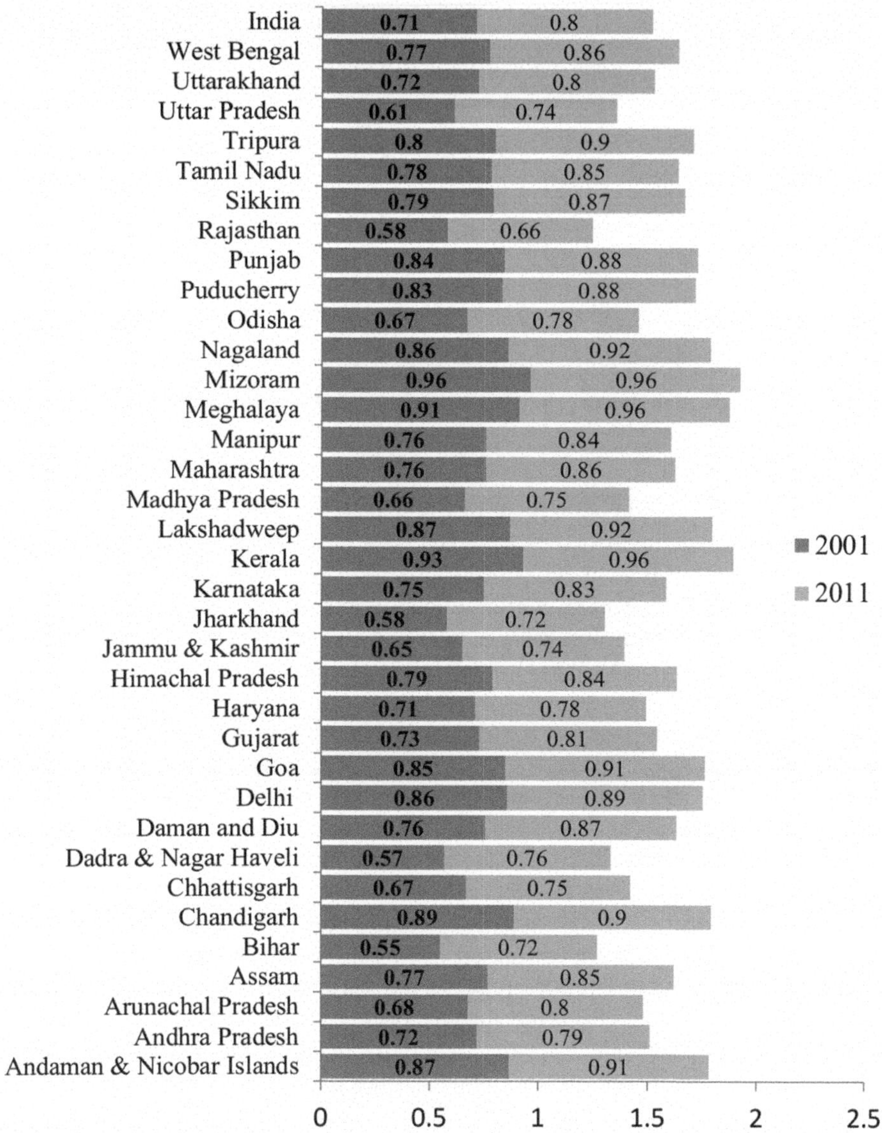

Fig. 10.3 Gender Parity Index (GPI) of literacy across states (*Source* Computed from Registrar General of India 2011a)

10.6.3 Imbalance Sex Ratio and Declining Child Sex Ratio in India

The sex ratio is representative of gender inequality in India (measured in terms of the number of women per 1000 men). Biologically, in almost all countries of the

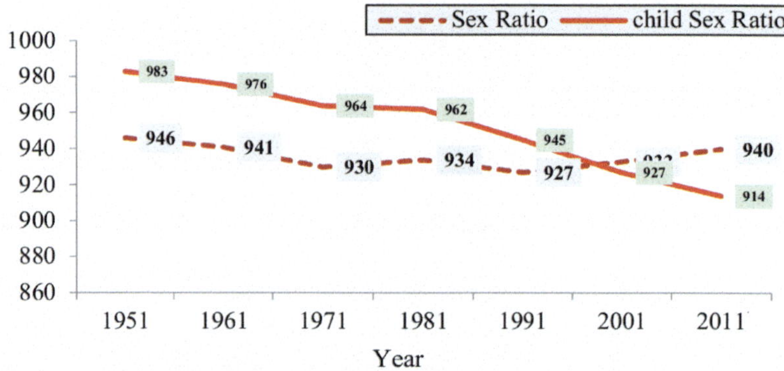

Fig. 10.4 Sex ratio and declining child sex ratio (age group 0 to 6) in India (*Source* Computed from Registrar General of India 2011c)

world, the sex ratio should be in favor of women, and so it is. However, most states in India have been distinguished by a marked skew in sex ratios in favor of men. The gender ratio in India has been unbalanced and of 971 women per 1000 men in 1901, dropped to 940 in 2011. The proportion of children by sex in India has experienced a downward trend, and the proportion of boys and girls in India has decreased to 914 girls against 1,000 boys, one of the lowest since Independence according to the 2011 Census (see Fig. 10.4).

According to the 2011 census, Kerala has the highest sex ratio in India, with 1084 women for every 1,000 men; on the other hand, Haryana has the most unfavorable sex ratio with 877 women for every 1,000 men. This is largely attributed to lower female status in Indian society, which contributes to early marriage, lower levels of literacy, and higher levels of fertility and mortality, thereby resulting in unfavorable progress toward human development.

The ratio also decreased significantly in some of the best-located states in 1951, for example, Orissa, where it decreased from 1022 in 1951 to 978 in 2011, and Tamil Nadu, which recorded a decrease in the sex ratio from 1007 to 995 in the same period. Bihar experienced the sharpest decline, from 1000 in 1951 to 916 in 2011. During the last decade, Rajasthan reported a marginal improvement from 910 to 926, which is noteworthy given its low starting point. The decline over the last decade is of concern, particularly in the relatively better-off states such as Haryana and Punjab (see Tables 10.2 and 10.3). In 1951, India had as many as ten states and Union territories that had sex ratios in favor of women. However, now, in 2011, only Kerala and Pondicherry have a sex ratio in favor of women (see Table 10.4).

Table 10.2 Sex ratio by residence in Non-EAG states of India (1991, 2001, and 2011)

Sl. no	States	No. of Females per 1000 Males								
		1991			2001			2011		
		Total	Rural	Urban	Total	Rural	Urban	Total	Rural	Urban
1	Jammu and Kashmir	–	–	–	892	917	819	883	899	840
3	Himachal Pradesh	976	990	831	968	989	795	974	988	853
2	Punjab	882	888	868	876	890	849	893	906	872
4	Chandigarh	790	632	810	777	621	796	818	691	821
5	Haryana	865	864	868	861	866	847	877	880	871
6	NCT of Delhi	827	807	830	821	810	822	866	847	867
7	Sikkim	878	892	750	875	880	830	889	883	908
8	Arunachal Pradesh	859	880	728	893	914	819	920	929	889
9	Nagaland	886	917	749	900	916	829	931	942	905
10	Manipur	958	951	975	974	963	1009	987	966	1038
11	Mizoram	921	912	932	935	923	948	975	950	1000
12	Tripura	945	942	958	948	946	959	961	956	976
13	Meghalaya	955	966	910	972	969	982	986	983	997
14	Assam	923	934	838	935	944	872	954	956	937
15	West Bengal	917	940	858	934	950	893	947	950	939
16	Gujarat	934	949	907	920	945	880	918	947	880
17	Goa	967	993	930	961	988	934	968	997	951
18	Maharashtra	934	972	875	922	960	873	925	948	899

Source Census of India 1991 and Office of Registrar General
Note In India, excluding Jammu & Kashmir where the 1991 Census was not held

Table 10.3 Sex ratio by residence in EAG states of India (1991, 2001, and 2011)

Sl. no	EAG States	No. of Females per 1000 Males								
		1991			2001			2011		
		Total	Rural	Urban	Total	Rural	Urban	Total	Rural	Urban
1	Rajasthan	910	919	879	921	930	890	926	932	911
2	Uttar Pradesh	879	884	860	898	904	876	908	914	888
3	Uttarakhand	–	–	–	962	1007	845	963	1000	883
4	Bihar	911	921	844	919	926	868	916	919	891
5	Jharkhand	–	–	–	941	962	870	947	960	908
6	Orissa	971	988	866	972	987	895	978	988	934
7	Chhattisgarh	–	–	–	989	1004	932	991	1002	956
8	Madhya Pradesh	931	943	893	919	927	898	930	936	916

Source Census of India 1991 and Office of Registrar General

Table 10.4 Sex ratio by residence in South Indian States and UTs (1991, 2001, and 2011)

Sl. No	States/Uts	No. of females per 1000 males								
		1991			2001			2011		
		Total	Rural	Urban	Total	Rural	Urban	Total	Rural	Urban
1	Andhra Pradesh	972	977	959	978	983	965	992	995	984
2	Karnataka	960	973	930	965	977	942	968	975	957
3	Kerala	1036	1037	1034	1058	1059	1058	1084	1077	1091
4	Tamil Nadu	974	981	960	987	992	982	995	993	998
5	Daman and Diu	969	922	1024	710	586	984	618	867	550
6	Dadra and Nagar Haveli	952	965	817	812	852	691	775	863	684
7	Lakshadweep	943	959	930	948	959	935	946	954	944
8	Puducherry	979	970	985	1001	990	1007	1038	1029	1043
9	Andaman and Nicobar Islands	818	837	769	846	861	815	878	871	891
	India	894	897	855	933	946	900	940	947	926

Source Census of India 1991 and Office of Registrar General

Table 10.5 Correlation between female literacy and sex ratio in India, 2011

Correlations				
			Female literacy	Sex ratio
Spearman's rank method	Female literacy	Correlation coefficient	1.000	0.110 (*)
		Sig. (2-tailed)		0.531
		N	35	35
	Sex ratio	Correlation Coefficient	0.110 (*)	1.000
		Sig. (2-tailed)	0.531	
		N	35	35

Source Calculated by authors' using census data 2011
*Note** Correlation is significant at the 0.05 level (2-tailed).

10.6.4 Relation Between Sex Ratio and Female Literacy in India

Female literacy and sex proportion are two significant demographic characteristics of the population. Female literacy plays a very crucial role in raising the sex ratio (Rao and Jain 2001). Literacy is considered as one of the imperative social aspects of the development of any society, especially in rural areas, and denotes a necessary criterion of human development (Mayer 1999; Bayeh 2016; Jahan 2016). However, the change in the sex ratio and female literacy rates in individuals in Indian states is not uniform. Table 10.5 confirms the correlation between the female literacy rate and the sex ratio in India. Spearman's method is measure of the correlation between two variables, that is, female literacy and sex ratio. The sex ratio and female literacy have moved in the same direction in India. It is observed that there is a positive correlation between ($r = 0.11$) the female literacy rate and sex ratio in India. However, the correlation is not strong. The pattern of spatial variation in gender development is not uniform in India.

10.6.5 Work Participation Rate of Women in India

Female Work Participation (FWP) refers to the proportion of total female workers with respect to the total female population. Labor participation implies contributing to the production of goods and services for the marketplace or for own utilization (Sarkar 2017). The participation rate for women in the work also indicates women's economic empowerment (GoI 2009; Singh 2017). It is believed that parity in the WPR occurs when 100 women work against 100 men. Any deviation from there creates a gender gap.

In the last decades, significant changes have been observed in the case of employment and the educational status of women. However, the change has been very slow. The economy of India is mainly agro-based. The population of India is 1.21 billion, of which 586.5 million are female. It was observed that the population growth of women was greater than that of men. According to the 2011 census, there was an increase of 90.97 million men and an increase of 90.99 million women.

India occupies the second lowest place in the group of twenty (G20) economies when the participation of women in the workforce is considered. According to the World Economic Forum, India obtained a score of 101 as an overall score and 124 of 136 nations in labor force participation (Khattar 2016). In India, 39.10% population was economically active in 2011. In the participation rate in the labor force, there is also a significant gap, with men occupying 53.26% and women 25.51% (of the age group of 15 to 49 years), where the gap is 27.75%. The workforce participation rate is less equal for rural and urban men, but there is a large gap in the labor force participation of rural and urban women. Of the total working men, 43.8% were main principal workers, and 9.4 percent were marginal workers, while of the total workers, 15.2% were principal workers, and 10.3% were marginal workers.

The level of education, work participation rate, and earnings are some of the areas that show the lower status of women compared to that of men. Women generally receive lower wages than men for the same work. Comparatively, they also suffer from a poor state of health, a low educational level, and a lower skill compared to men. In India, of the total of 30 percent who are below the poverty line, 70 percent are women. The poverty of women in India is directly interconnected to the lack of economic opportunities and autonomy, lack of access to economic resources, land ownership, lack of access to education and support services, and their minimal participation in the decision-making process (Sarkar 2017).

10.6.6 Gender Equality and Development of Human Resources in India

Human development and social development indices can be used as instruments to reallocate resources to initiatives and schemes designed to resolve social inequalities at all levels of governance by frequently monitoring and measuring progress and implementation; provide women with access to assets and income opportunities, such as the right to work for all citizens; provide access to decent jobs to enable people to rise out of poverty and thereby decrease gender inequalities in employment and living standards; provide access to clean water to minimize the burden of disease; and provide access to better health facilities and timely access to medical facilities to reduce gender differences in morbidity and death (GOI 2009).

The Human Development paradigm defines development as a process of expansion of people's options. Gender equality and human development are interrelated and are based on three essential components. (a) Equality: an opportunity for all. (b) Sustainability: such opportunities from one generation to another. (c) Empowerment: people to participate in the development process and benefit from it.

The development of human resources is also an important creation to measure the level of development. The demographic composition of any region is closely related to the level of development. Developed regions are often characterized by a favorable level of population, density, sex ratio, high degree of urbanization, and literacy rate. When applying the human development index method, the state is divided into high, moderately low, and very low development categories based on combined scores of six indicators. The broad category of development disparities is shown in Fig. 10.5. The value of 0.1 in the index represents the highest total of literates. It can be safely argued that the total number of literates and development is positively related, which means that the higher the literate, the higher the development.

10.6.7 Human Development Index (HDI) and the recent scenario of Indian states

The process of Human development is an expansion of people's decisions and increasing their living standards (Jahan 2016). Unlike earlier approaches to development, the human development approach analyzes all social problems viz. economy, trade, education, political opportunity, and culture from the viewpoint of the citizens (Varatharajan 2006). India's first 50 years of Independence achieved a substantial rise in human development. However, as per the latest global Human Development Report (HDR) in 2011, India ranks 134 out of 187 countries with a GDI value of 0.547.

Table 10.6 shows the distribution of the overall human development index in the bigger states of India. The index is the aggregate of the indices of the six indicators selected to calculate the human development index in the present study. Among

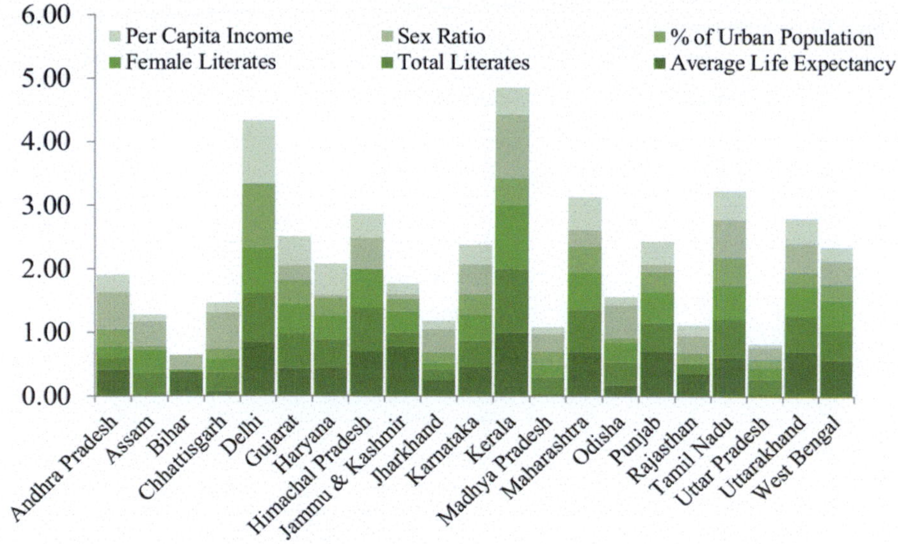

Fig. 10.5 Human Development Index for each dimension, 2011 (*Source* Calculated by author's using HDI)

the region, Southern states have generally done better. Some states fall into the highest development category, such as Kerala and Delhi. These two states have several facilities, that is, educational, medical, transportation, etc. are the responsible factor for the high human development index.

Moderate development is found in two states—Tamil Nadu (0.54) and Maharashtra (0.52). However, the most economically wealthy states like Haryana (0.35) and Punjab (0.41) come under the low development category. It is observed that the states with the highest per capita income do not show the highest HDI values; economic growth over time was generally followed by an increase in the human development index.

Table 10.6 reveals that Madhya Pradesh, Rajasthan, Jharkhand, Uttar Pradesh, Jammu and Kashmir, and Bihar fall into the very low development category (0–0.30). In 2011, the HDI for Bihar recorded the lowest value of 0.11, which is an improvement over the previous years. However, it has been observed that lack of awareness, perceptions of people, transportation, and educational facilities are factors responsible for the low rate of human development in these states. The development of human resources can be attributed to more developed infrastructure facilities, the availability of resources, and the mental perception of people.

Table 10.6 Human Development Index, 2011

Category	States	HDI	Rank
Higher Development (0.70–1)	Kerala	0.808	1
	Delhi	0.722	2
Moderate Development (0.50–0.70)	Tamil Nadu	0.538	3
	Maharashtra	0.521	4
Low Development (0.30–0.50)	Himachal Pradesh	0.478	5
	Uttarakhand	0.466	6
	Gujarat	0.417	7
	Punjab	0.405	8
	Karnataka	0.396	9
	West Bengal	0.391	10
	Haryana	0.346	11
	Andhra Pradesh	0.318	12
Very Low Development (0–0.30)	Jammu & Kashmir	0.295	13
	Odisha	0.260	14
	Chhattisgarh	0.244	15
	Assam	0.213	16
	Jharkhand	0.197	17
	Rajasthan	0.186	18
	Madhya Pradesh	0.181	19
	Uttar Pradesh	0.138	20
	Bihar	0.107	21

Source Calculated by authors' using HDI

10.7 Policy Implication and Conclusions

India is plagued by social problems related to socio-cultural hegemony and gender inequality that result in income and opportunity disparities. One of the reasons for the enormous gender discrimination is the lack of empowerment of women, which is also reflected in the representation of women in the Indian parliament. In the National Population Policy (NPP-2000), the Government of India identified the low status of women in India, characterized by factors such as discrimination against girls and adolescents, the minimum age for marriage, and high rates of maternal mortality, as an important barrier to the achievement of population and maternal and child welfare goals (Rao and Jain 2001).

In the last decades, there have been important changes in the labor and educational scenario of women; however, the change has been terribly low. There is a need to increase investment toward women issues. The government makes a significant contribution to educational grants for women and health grants for women, which have the ability to change the actions of society toward women. We suggest allowing full participation of women in personal and family decisions, particularly those related to motherhood. Guha (2002) expressed that the Central Government has to play an energetic role in removing interstate disparities in Human Development.

To reduce the gender gap and, furthermore, achieve greater socio-economic development in India, suggestions and subsequent policy initiatives are recommended:

1. Create an enabling atmosphere through positive economic and social policies for the development of women and the awareness of their complete potential.
2. Equalize women's access to public services and decision-making in the social, political, and economic spheres.
3. Change the attitudes of social groups and community practices through the energetic contribution and participation of each man and woman.
4. Eliminate discrimination and every type of violence against women and girls.
5. To ensure that all girls receive a good quality education; and
6. The government should promote the Beti Bachao-Beti Padhao (BBBP) scheme at the grassroots level.

The study reveals that there are huge disparities in the level of human resource development among the different states. Gender equality is an urgent need to ensure the positive social transformation of a nation like India. It can be concluded that countries like India have a long way to go itself to become a gender unbiased country. The measures to reduce gender inequality in the rural population must be developed and implemented according to the needs of the specific target groups, i.e., women.

Acknowledgements The authors would like to express their gratitude to the reviewers for their insightful comments and suggestions, as well as Ms. Amrin Kamaluddin Fakih for her moral support and motivation.

References

Bayeh E (2016) The role of empowering women and achieving gender equality to the sustainable development of Ethiopia. Pacific Sci Rev B Humanities Soc Sci 2(1):37–42

Baru S (1998) Mahbub ul Haq and Human Development: A Tribute. Econ Political Weekly 33(35):2275–2279

Bhaskar V, Gupta B (2007) India's missing girls: biology, customs and economic development. Oxford Rev Econ Policy 23(2):221–238

Comyn P, Kemmis RB, Smith E (2014) How can the expansion of the apprenticeship system in India create conditions for greater equity and social justice? Australian J Adult Learn 54(3):369–387

Dyson T, Moore M (1983) On kinship structure, female autonomy, and demographic behavior in India. Popul Dev Rev 9(1):35–60

George A (2015) Literacy Rate in India: A Census Based Analysis. Indian J Adult Educ 76(2):70–86

GOI (2009) Gendering Human Development Indices: Recasting the Gender Development Index and Gender Empowerment Measure for India. New Delhi: Ministry of Women and Child Development, Govt of India with UNDP India. https://www.undp.org/content/dam/india/docs/gendering_human_development_indices_summary_report.pdf

Guha B (2002) Human development in India: a study of interstate disparity. In: Datt R (ed) Human development and economic development, Deep and Deep Publication Pvt. Ltd, New Delhi, p 211–229

Jahan S (2016) Human development report 2016: human development for everyone. New York: United Nations Publications (UNDP)

Joseph M (2004) Performance of the Northern States: a comparative analysis. Econ Political Weekly 39(6):564–579

Khattar P (2016) Demographic dividend—Focusing on Women in India. Indian J Adult Educ 77(2):94–101

Lawrence W, Skinner C, Haslam C et al (2009) Why women of lower educational attainment struggle to make healthier food choices: The importance of psychological and social factors. Psychol Health 24:1003–1020. https://doi.org/10.1080/08870440802460426

Lee T (2008) Benchmarking the effective literacy rate. Math Soc Sci 56(2):233–239. https://doi.org/10.1016/j.mathsocsci.2008.03.002

Lindsey LL (1995) Gender roles: a sociological perspective. Englewood Cliffs, New York

Mayer P (1999) India's falling sex ratios. Popul Develop Rev 25(2):323–343

Miller BD (1989) Changing patterns of juvenile sex ratios in rural India, 1961 to 1971. Econ Pol Wkly 24(22):1229–1236

Morrison A, Raju D, Sinha N (2007) Gender Equality, Poverty Economic (Growth Policy Research Working Paper No. WPS4349). Washington: World Bank

Patrick et al (2018) Correlation Coefficients: Appropriate Use and. Anesthesia Analgesia 126(5):1763–1768. https://doi.org/10.1213/ANE.0000000000002864

Rao M, Jain D (2001) National Population Policy 2000: Re-Examining Critical Issues. Econ Polit Weekly 36(16):1299–1302

RGI (2011a) Rural urban distribution of literacy (chapter 3), Provisional Population Totals, Paper 2 of census 2011. New Delhi: Office of the Registrar General and Census Commissioner of India.

RGI (2011b) Socio-cultural series of census. New Delhi: Office of the Registrar General and Census Commissioner of India.

Ridgeway CL (2011) Framed by gender: How gender inequality persists in the modern world. Oxford University Press, New York

Sarkar R (2017) Recent Status of Education, Employment and Empowerment of Women in West Bengal. Int J Sci Res Publ 7(1):263–269

Sen A (2001) Many faces of gender inequality. Frontline 18(22):35–40

Shukla V, Mishra US (2017) Literacy achievement in India: A demographic evaluation. Prospects 47(3):257–274

Singh S (2017) The State of Gender Inequality in India. Gender Stud 15(1):139–157. https://doi.org/10.1515/genst-2017-0009

Varatharajan D (2006) India's human development: Gaps and prospects. Health Popul Perspect Issues 29(1):1–20

Raju Sarkar is a Ph.D. Research Scholar at Population Research Centre, Institute for Social and Economic Change, Bangalore, India. He holds two Master's Degree- Master's of Population Studies (MPS) from the International Institute for Population Sciences (IIPS), Mumbai, and Master's in Geography and Applied Geography from the University of North Bengal, Darjeeling, India. He has several national and international publications to his credit in his area of interest, which includes Demography, Urban Geography, Gender Studies, Migration Studies, and Fluvial Geomorphology.

Dr. C. M. Lakshmana is a Professor at Population Research Centre (PRC), ISEC in Bangalore.. He has published various research papers in well-reputed international and national journals on various issues related to Population Geography, Urbanization, Migration, Environment, Gender Development, and Health. He has completed many projects for both the state and federal government, and he is also a member of many important committees set up by ISEC and the state Government of Karnataka.

Part IV
Urban Dimensions of Regional Development

Chapter 11
Association of Morbidity Among Children with Housing, Water and Sanitation in Urban India: A Policy Perspective

Sk. Karim, Kalosona Paul, Mamta Chaudhary, and Somnath Mukherjee

Abstract India is one of the fastest-growing developing nations of the world today. The urban population is expected to rise at least seven percent by 2026. This expansion and explosion of the urban sector will initiate several environmental and health issues which will directly and indirectly promote the morbidity and thrust many challenges upon us Presently, around half a million children below five years of age have been suffering from diarrhoea annually due to unsafe potable water, poor sanitation, and unhygienic condition in India. At this backdrop, this chapter examines the association of urbanization, the condition of drinking water, patterns of sanitation with the health condition (ARI/diarrhoea) of children in urban India. The present study is based on the latest published NFHS-4 dataset. Descriptive analysis and Logistic Regression techniques have been applied to explain an association between ARI & diarrhoea among the children under investigation and water, sanitation, and various socio-economic and demographic characteristics of Indian households. We found that diarrhoea (12.3%) and Acute Respiratory Infection (ARI) (2.3%) are significantly diseased among urban India. Among the children (<1 year) they were more reported than the reference category. A male child is more susceptible than a female to both diseases. Similarly, the Model also proved that ST/SC child is more vulnerable to ARI and diarrhoea. We recommend community engagement in the planning and implementation of water-sanitation programs that may be effective to render better health condition of the children.

Keywords Multivariate analysis · Diarrhoea · Acute respiratory infection

---

Sk. Karim
Department of Migration & Urban Studies, International Institute for Population Sciences, Mumbai, India

K. Paul (✉)
Department of Geography, Sidho-Kanho-Birsha University, Purulia, West Bengal, India

M. Chaudhary
School of Health Systems Studies, Tata Institute of Social Sciences, Mumbai 400088, India

S. Mukherjee
Bankura Christian College, Bankura, West Bengal, India

M. Mishra et al. (eds.), *Regional Development Planning and Practice*,
Advances in Geographical and Environmental Sciences,
https://doi.org/10.1007/978-981-16-5681-1_11

11.1 Introduction

India is one of the leading countries in the world in terms of the growth of the population. This huge population is not only concentrated in rural India but the rapid growth of urbanization now a day shares the burden of a huge amount of population in urban areas as well (Paul and Sharma 2016). India's urban population (31%) (RGI 2011) is expected to rise to 535 million (38%) by 2026 (NPC 2006). The estimated projection of the urban population of India in the year 2031 will reach 654 million, which is nearly 40% of the country's population (NPC 2006). The number of million-plus cities in India has gone up from 16 in Census 1991 to 27 in 2001 to 53 in Census 2011. Similarly, the number of towns in India increased from 5161 in 2001 to 7935 in 2011 (RGI 2011).

The rapid growth of the population in the urban sector creates huge pressure on the quality of living. People with diverse economic conditions exaggerate certain diversity of standards of living. One of the indicators of such living conditions can be determined by the quality of urban housing. The low quality of urban housing is characterized by over crowed population with high room density, poor housing-ventilation-widows problems, safe water, inadequate toilet and bathroom, unscientific sewage system, etc. These characteristics of a typical crowded urban housing generate various health diseases among children. According to the Ministry of Housing and Urban Poverty Alleviation reports, it is estimated that an around 18.78 million shortage of housing units in 2012 (MHUPA 2013; Tiwari and Rao 2016). It wills double (34.1 million) nearly the next ten years, i.e., by 2022. The housing problem has manifested itself and led to a proliferation of slums, and the lack of coherence in the policies of the administration has contributed to the problem. The irony is that the casual, daily cheap wages, temporarily based labourers who contribute to building residential, commercial urban complexes themselves remain outside the purview of any organized housing arrangement (Gandhi 2012).

The Millennium Development Goals (MDGs) on ensuring environmental sustainability, combat HIV/AIDS, malaria, diarrhoea, ARI, and other diseases should be cut off level 6.2 by 2030 (UNICEF/WHO 2012). An around 16% of urban India is using improved drinking water between 1990 and 2015. According to SDG 6 goal: nearly 0.84 billion people were accessing unimproved water sources, and 2.3 billion people were not access to proper sanitation facilities (UNICEF/WHO 2018). Moreover, water-borne diseases, improper sanitation, and hygiene conditions resulted in 4% of all deaths and 5% of the total disease burden (DALYs) occurring worldwide (Prüss et al. 2002; Nath 2003; Joshi et al. 2014). The report also exposed that only 30% of sewage/wastewater is treated before it the felt into the mainstream or river or canal. Approximately four lakhs of children died due to causes of water-borne infectious diseases (cholera, diarrhoea, malaria, and dysentery), etc. Estimated approximately is $53.8 million loss due to an absence of a proper sanitary system (Van and Hung 2011).

Impoverished Water Supply, Sanitation, and Hygiene (WASH) conditions may be associated with nutritional outcomes and infectious diseases such as diarrhea

(World Bank 2008) among children. Diarrhea disease burden is highly concentrated in children and disproportionately affects those living in low and middle-income countries (LMICs) (Paul et al. 2020; Walker et al. 2013). Every two minutes, we lost a child due to symptoms of diarrhoea as a result of using unclean sources of drinking water. Diarrhea accounts for the largest share, causing an estimated 1.4 million deaths annually (Patel et al. 2019; Prüss-Ustun et al. 2014), or 19% of all under-five deaths in low-income settings (Boschi-Pinto et al. 2008). Diarrhoea is the leading cause of death among under-five children, and it is around 760,000 children die every year globally (UNICEF/WHO 2012, 2018). WASH accounts for 60% of the deaths due to diarrhea. Only handwashing with soap can cut the risk of diarrhea by at least 40% and significantly bring down the danger of respiratory infections. WASH interventions show diarrhea risk reductions between 27 and 53% in children 0–5 years old, depending on intervention type, providing ample evidence to support the scale-up of WASH in low and middle-income countries (LMICs) (Curtis and Cairncross 2003; Darvesh et al. 2017).

In India, this had 564 million open defecators in 2015, 61% of the rural and 10% of the urban population (WHO/UNICEF 2015; UN 2015). The ARI is the second leading cause of death among children under age five worldwide after diarrhoea diseases. According to NFHS-3, overall, 6% of children under five years of age suffer (Esrey et al. 1991; Fewtrell et al. 2005; Ashwani and Kalosona 2016). In developing countries, seven out of 10 deaths occurred due to ARI in under 5 year children. Similarly, about 26.3 million cases of ARI were estimated in India (Sharma et al. 2013). ARI contributes to 15–30% of all under-five deaths in India, and most of these deaths are preventable. National Family Health Survey (NFHS-4) revealed that two weeks before the survey, 2.3% of under-five children had symptoms of ARI in urban India (IIPS and ICF 2017). The present study examines the effects of quality of sanitation, carrying capacity of housing, potable water on the health conditions of the juvenile population of urban India. It will also be valuable in understanding the possible relation of these circumstances with selected illness prevalence.

11.2 Materials and Methods

11.2.1 Geographical Coverage

All the components of the National Family Health Survey (NFHS-4) were implemented in 29 States and 7 Union Territories. The first time NFHS-4 has provided indicators at the district level for all 640 districts per the 2011 Census. Given the need to report demographic and health indicators at the district level, the NFHS-4 sample size has been increased to approximately 601,509 households compared to 109,041 households in NFHS-3. In these selected households, information on approximately 259,627 children below age five years will be collected. However, our study is based on only urban areas of India. In this study, we have used only 55,256 samples of urban India.

11.2.2 Statistical Analysis

Prevalence of diarrhoea/ARI was calculated per 1000 population. The following formula was used to calculate morbidity prevalence (Paul and Singh 2017).

$$M_i = \frac{Ai}{Pi} * 1000 \tag{11.1}$$

where, Ai = No. of ailing persons (diarrhoea/ARI) in last 14 days reference period;
Pi = Total number of persons alive in the sample households.

11.2.3 Multivariate Analysis: Associating the Sanitation and Living Conditions with Morbidity

The multivariate binary logistic regression analysis is a systematic model-building procedure. For this study, the adopted Model-I took water to make safe, stool disposal, age, and gender of children. This allowed us to test whether there is a significant difference in the likelihood of diarrhoea and ARI among children. Model-II included all the independent variables to check the consanguineous likelihood status with dependent variables (Kalam et al. 2017).

$$log_e[P(Y_i = 1|X_i)/1 - P(Y_i = 1|X_i)] = log_e[\pi|1 - \pi] = \alpha + \beta_1 X_{i1}, \cdots, \beta_k X_{ik} \tag{11.2}$$

where, Y_i is the binary response variable;
X_i is the set of descriptive variables, such as socio-demographic characteristics as mentioned in case of the multivariate model; and
$\beta_1, \ldots, \beta_k$ are the coefficients of the X_{ii} variables.

The results of multivariate logistic regression are depicted by estimated odds ratio with 95% Confidence Interval (CI). The entire analysis was carried out STATA-14 and QGIS 3.14.

11.2.3.1 Dependent/Response Variables

In the NFHS-4 data collection, there was a question (Q. No. 516) asked to the women for the children '*had diarrhoea and ARI in the last 14 days from the survey date?*'. So the response was categorized as '*Yes*' or '*No*'.

11.2.3.2 Independent/Predictor Variables

Control variables had been used from the same data set, i.e., NFHS-4 data; they were principal sources of the age of child, gender of child, mother's working status, mother's education, religion, social group, wealth quintile, drinking water, toilet facility, bathroom facility, stool disposal, drainage facility, house types, kitchen type (within the household), cooking fuel and regions.

11.2.4 Ethical Statement

The data used by this study are available in the public domain for the researchers upon request.[1] The protocol for the NFHS survey, including the content of the survey questionnaires, was approved by the IIPS Institutional Review Board and the ICF Institutional Review Board. The protocol was also reviewed by the U.S. Centers for Disease Control and Prevention (CDC). Data collection procedures were also monitored and approved by the ORC Macro institutional review board. All individuals selected in the National Family Health Survey were provided with informed voluntary and written consent. Each individual's approval was sought, and only then was the interview conducted (Singh et al. 2012).

11.3 Results

11.3.1 Diarrhoea and ARI Morbidity Scenario in Urban India

The highest prevalence of diarrhoea was observed in Uttaranchal (17.5%), followed by Uttar Pradesh (14.6%), Himachal Pradesh (13.9%), and Chhattisgarh (11.9%). At the same time, the lowest prevalence of diarrhoea was in Sikkim (0.3%) followed by Daman & Diu (2.4%), Kerala (2.8%), Goa (3.0%), Assam (3.3%), and Tripura (3.4%). However, the overall, North-east, West, and South regions were reported the lowest prevalence of diarrhoea than the counterpart of other regions. In the case of symptoms like ARI, the highest in Uttaranchal (5.4%) followed by Punjab (5.3%), Jammu & Kashmir (4.1%), Uttar Pradesh, Meghalaya, and Maharashtra (4.0%), etc. However, the Northern region was reported the highest prevalence of ARI than the other regions. The lowest prevalence's states are Sikkim (0.3%), Kerala (0.5%), Goa (0.7%), and Assam (0.9%) (Fig. 11.1a, b).

[1] See for details: https://dhsprogram.com/data/dataset/India_Standard-DHS_2015.cfm.

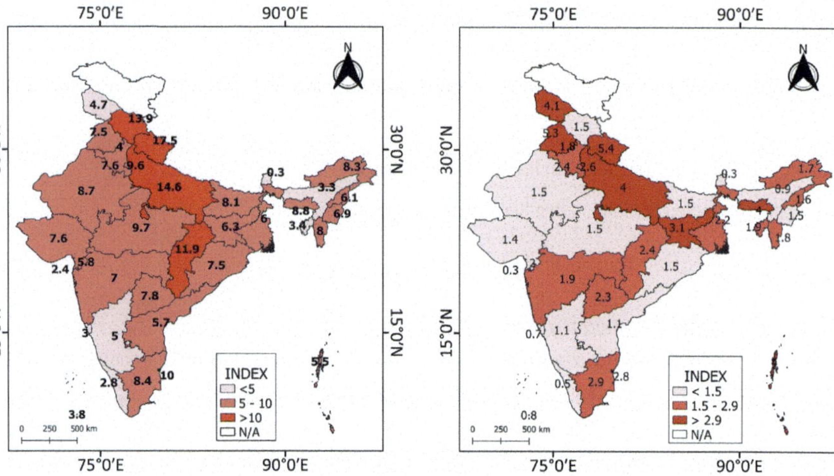

Fig. 11.1 **a** Diarrhoea and **b** ARI scenario in urban India (*Source* Drawn by the authors based on NHFS-4, 2015–16 datasets)

11.3.2 Diarrhoea Among Under-Five Age Children by Selected Socio-economic and Demographic Characteristics

Less than 1 year age children are (12.7%) more affected by diarrhea than the 1–4 year (7.3%) age group (Table 11.2). Mostly male children (8.5%) are exposed by diarrhea than female children (8.1%). Drinking water has played a vital role in the prevalence of diarrhea among under-five children. The children in households with improved drinking sources of water (8.3%) are less affected than those households which are not using improved sources of drinking water (8.9%) in urban India. The table shows that open defecation (9.4%) increased the risk of diarrhoea. HHS, which disposes of the child stool at the right place, has fewer chances of being affected with diarrhoea prevalence (7.7%) than the unsafe disposal (9.1%). Among Muslim children (9.4%) prevalence of diarrhoea is higher than Hindu (8.0%), Christian (5.8%). It should also be noted that poor wealth quintile households reported the highest prevalence of diarrhoea (9.0%), whereas, among the rich quintile, it is 8.1%. The central region reported the highest prevalence of diarrhoea (12.9%), followed by North (8.8%), West (7.1%), East (6.8%), South (6.6%), and North-east (4.9%) in India (Table 11.1).

Table 11.1 Prevalence of diarrhoea and ARI among under-five in Urban India, 2015–16

Background characteristics	Diarrhoea (%)	ARI (%)	Sample (N)
Age of child			
<1 year	12.7	3.3	10,311
1–4 year	7.3	2.2	44,925
Sex of child			
Male	8.5	2.3	28,935
Female	8.1	2.2	26,301
Drinking water			
Improved	8.3	2.4	50,016
Unimproved	8.9	2.7	5,220
Water to make safe			
Yes	7.4	2.1	28,168
No	9.1	2.4	27,068
Toilet facility			
Open defecation	9.4	2.6	7,566
Improved	8.2	2.4	45,326
Unimproved	7.2	3.1	2,150
Other	3.8	1.5	194
Stool disposal			
Safe	7.7	2.3	31,772
Unsafe	9.1	2.3	23,246
House type			
Kachha	10.3	2.2	865
Semi-pucca	9.7	2.7	11,725
Pucca	7.9	2.3	40,996
Kitchen			
Separate	7.9	2.1	32,876
Not separate	9	2.7	16,475
Type of fuels			
Clean fuel	8.1	2.2	39,181
Solid fuel	9.1	2.5	16,027
Age of mother			
15–19	9.3	2.7	655
20–29	8.7	2.5	29,659
30–44	7.5	1.9	20,801
45 and above	9.3	2.4	4,121
Mother education			

(continued)

Table 11.1 (continued)

Background characteristics	Diarrhoea (%)	ARI (%)	Sample (N)
No education	9.4	2.6	9,020
Up to secondary	8.6	2.5	34,350
Higher	6.9	2.2	11,567
Mother's working status			
Not working	8.5	2.3	8,703
Working	7.9	2	1,300
Religion			
Hindu	8	2.2	36,306
Muslim	9.4	3	12,465
Christian	5.8	2.1	4,375
Others	8.7	3.2	2,090
Caste/tribe			
SC	8.9	2.7	9,289
ST	6.5	1.9	6,730
OBC	9.1	2.3	23,525
GEN	7.4	2.6	13,748
Wealth			
Poor	9	2.4	8,364
Middle	8.6	2.6	10,477
Rich	8.1	2.3	36,395
Regions			
North	8.8	3	11,522
Central	12.9	2.9	14,766
East	6.8	2.1	7,050
North-east	4.9	1.7	7,143
West	7.1	1.6	5,658
South	6.6	1.9	9,097
Total	**8.3**	**2.3**	**55,236**

Source National Family Health Survey-4, 2015–16
Note All values in the table represent absolute numbers and percentages unless otherwise stated. Percentages are weighted and numbers are not weighted

Table 11.2 Results of multivariable binary logistic regression showing the risk of diarrhoea in children below 5 years of age, in Urban India, 2015–16

Socio-economic and demographic characteristics	Adjusted odds ratio (95% CI)	
	Model 1	Model 2
Water to make safe		
Yes®		
No	1.37*** (1.29, 1.45)	1.13*** (1.05, 1.21)
Stool disposal		
Safe®		
Unsafe	1.09*** (1.03, 1.17)	1.06 (0.98, 1.13)
Age of child		
<1 year®		
1–4 year	0.57*** (0.53, 0.61)	0.57*** (0.53, 0.61)
Sex of child		
Male®		
Female	0.93** (0.88, 0.99)	0.92*** (0.87, 0.98)
Toilet		
Improved®		
Unimproved		1.00 (0.91, 1.11)
Mother education		
No education®		
Upto secondary		1.05 (0.96, 1.15)
Higher		0.94 (0.83, 1.06)
House type		
Katcha®		
Semi pucca		0.86 (0.68, 1.09)
Pucca		0.77** (0.61, 0.97)
Religion		
Hindu®		
Muslim		1.30*** (1.20, 1.40)
Christian		1.16 (0.95, 1.42)
Others		1.04 (0.87, 1.23)
Caste/tribe		
SC®		
ST		0.80*** (0.69, 0.93)
OBC		0.93 (0.85, 1.02)
Others		0.83*** (0.75, 0.92)
Wealth		

(continued)

Table 11.2 (continued)

Socio-economic and demographic characteristics	Adjusted odds ratio (95% CI)	
	Model 1	Model 2
Poor®		
Middle		1.01 (0.90, 1.14)
Rich		1.00 (0.88, 1.13)
Regions		
North®		
Central		1.23*** (1.13, 1.34)
East		0.69*** (0.61, 0.78)
North-east		0.56*** (0.48, 0.67)
West		0.83*** (0.74, 0.94)
South		0.60*** (0.53, 0.67)

Sources National Family Health Survey-4, 2015–16
®Reference group for the multivariate logistic analysis. *p < 0.1, **p < 0.05; ***p < 0.01

11.3.3 Acute Respiratory Infection (ARI) Among Under-Five Age Children by Selected Socio-economic and Demographic Characteristics

It is revealed from the same table (Table 11.1) that the infants (3.3%) are having the highest ARI prevalence than the 1–4 year (2.2%) age group of children. Similarly, male children are more vulnerable to ARI than female children. The households with separate kitchens (2.1%) have fewer chances to report ARI than their counterparts (2.7%). Furthermore, if the households are using solid fuels (2.5%) that mean HHs children are more prone to ARI symptom. Mother's education in the HH is also an essential factor in increasing/decreasing ARI prevalence. Christian children are also reported with a higher prevalence of ARI (3.0%), followed by Hindus (2.2%) and others. North region was having a maximum prevalence of ARI (3.0%), followed by Central (2.9%), East (2.1%), South (1.9%), North-east (1.7%), and West (1.6%).

11.3.4 Multivariate Analysis

11.3.4.1 Diarrhoea

In Model-I & II, those households which are not using safe water have high chances of diarrhea among the children with respect to the reference category (OR = 1.37 & 1.13; $p < 0.001$). Also, children in the HHS who are using safe stool disposal are less likely with respect to unsafe disposal households (OR = 1.09; $p < 0.001$). As depicted in Model-I, a positive relationship between infant child and diarrhea is highly significant. It means less than one year's children are more reporting diarrhoea than the 1–4 age group (OR = 0.57; $p < 0.001$). However, in Model-I & II, female children (OR = 0.93 & 0.92; $p < 0.001$) have fewer chances of diarrhea than the counterpart of the male child. As expected, children living in Kachha houses have a significantly higher prevalence of ARI (OR = 1.00; $p < 0.001$) than those living in Pucca houses (OR = 0.77; $p < 0.05$). As depicted from the given table Muslim (OR = 1.30; $p < 0.001$) children are more likely to diarrhoea symptoms than the other religious group in urban India. Model-II also exposed that STs children (OR = 1.00; $p < 0.001$) are more likely to face the diarrhea problem than OBC (OR = 0.80; $p < 0.001$) & Other castes (OR = 0.83; $p < 0.001$) in India. The central region (OR = 1.23; $p < 0.001$) children are more affected with diarrhoea than followed by North (OR = 1.00; $p < 0.001$) and West (OR = 0.83; $p < 0.001$) regions, and the less affected region is north-east (OR = 0.56; $p < 0.001$) in India (Table 11.2).

11.3.4.2 Acute Respiratory Infection (ARI)

ARI is more associated with the type of kitchen as either separate or not separate. The children in households which are not having any separate kitchen are more likely to be affected with ARI diseases. At the same time, households using solid cooking fuels have more chances of report ARI among children. Age has a positive impact on the prevalence of ARI in both models. Model-I & II explained infants have a considerably higher risk of having ARI than children aged 1–4 years (OR = 0.73 & 0.77 $p < 0.001$). Similarly, a male child has a higher prevalence of ARI than a female child (OR = 0.87; $p < 0.001$). Children of the Christian community (OR = 1.76; $p < 0.001$) have a very significant association with ARI (OR = 0.57; $p < 0.001$) than followed by others and Muslim religions (OR = 1.19; $p < 0.001$). Children belong to ST/SC (OR = 0.78; $p < 0.01$) have a significantly high risk of ARI than the counterpart of OBC and Others. By socio-economic and demographic region, children in the northern (OR = 1.00; $p < 0.001$) and central regions (OR = 0.91; $p < 0.001$) reported maximum morbidity prevalence of ARI (lowest: NE region: (OR = 0.44; $p < 0.001$)) than the rest of the region (Table 11.3).

11.4 Discussion

India aims to meet targets to reduce the proportion of people without sustainable access to safe drinking water. However, the population is not able to achieve the goals due to rapid urban growth (Lilje and Mosler 2017). A sharp reduction in urban inequities associated with drinking-water coverage in the country is an excellent achievement in this regard. It becomes more critical for a country dominated by a patriarchal society like India. Girls and women have to do household duties, look after their wards, and collect water for cooking, drinking, washing, and other household chores. A significant proportion of Indian urban households have poor sanitation

Table 11.3 Results of multivariable binary logistic regression showing the risk of ARI in children below 5 years of age, in Urban India, 2015–16

Socio-economic and demographic characteristics	Adjusted odds ratio (95% CI)	
	Model 1	Model 2
Kitchen		
Not separate®		
Separate	0.78*** (0.69, 0.88)	0.77*** (0.67, 0.88)
Cooking fuel		
Clean fuel®		
Solid fuel	1.11* (0.98, 1.25)	1.09 (0.93, 1.29)
Age of child		
<1 year®		
1–4 year	0.73*** (0.64, 0.83)	0.77*** (0.67, 0.88)
Sex of child		
Male®		
Female	0.87** (0.78, 0.97)	0.87** (0.78, 0.98)
Mother education		
No education®		
Upto secondary		1.08 (0.92, 1.27)
Higher		1.03 (0.83, 1.28)
House type		
Katcha®		
Semi pucca		1.14 (0.72, 1.80)
Pucca		1.02 (0.64, 1.62)
Religion		
Hindu®		
Muslim		1.19** (1.03, 1.38)
Christian		1.76*** (1.28, 2.43)
Others		1.65*** (1.27, 2.13)
Caste/tribe		
SC®		
ST		0.78* (0.60, 1.03)
OBC		0.94 (0.80, 1.11)
Others		0.91 (0.76, 1.10)
Wealth		
Poor®		
Middle		1.03 (0.84, 1.28)
Rich		1.02 (0.80, 1.29)

(continued)

Table 11.3 (continued)

Socio-economic and demographic characteristics	Adjusted odds ratio (95% CI)	
	Model 1	Model 2
Regions		
North®		
Central		0.91** (0.78, 1.07)
East		0.68*** (0.56, 0.84)
North-east		0.44*** (0.33, 0.59)
West		0.54*** (0.43, 0.68)
South		0.61*** (0.50, 0.74)

Sources National Family Health Survey-4, 2015–16
®Reference group for the multivariate logistic analysis. *p < 0.1, **p < 0.05; ***p < 0.01

facilities. The target on sanitation will not be met unless progress is much accelerated, and if it is not. Less developed states like Bihar, Uttar Pradesh, and Chhattisgarh, with around 90% of households with availability to an improved source of drinking water, have already achieved the MDG target in this regard (UNICEF/WHO 2012). However, concerning sanitation, all the Northern and central regions (except Uttarakhand) lack all sanitation components. According to NFHS-4, around 18% reduced practicing open defecation at the urban empowered action group (EAG) of states in India (Clasen et al. 2014; Kumar and Das 2014). The paper also suggested that households with unimproved toilet facilities within the premises are more prone to diarrhoea among under-five children.

Considering the models that have been applied for estimation, findings would suggest that indoor air pollution from unclean fuel combustion is a public health problem in significant parts of India. Results from this study indicate that exposure to solid cooking smoke is significantly associated with the prevalence of ARI among children under age five (Ashwani and Kalosona 2016). The result also reflects that switch to clean fuels, and separate kitchens in India can be reduced the prevalence of ARI among children under age five by 42%. A significant effect of mother's smoking reflects the greater vulnerability of child's health because of their compromised respiratory system from mother's smoke. Similarly, cooking smoke is greater for infants than children aged 1–4 years, consistent with expectation; because infants are more exposed to cooking smoke as spending more time with mothers for breastfeeding and care and cooking is considered to be the primary duty of women in India (Acharya et al. 2003). However, we also explored that male children are more vibrant with ARI than female children. It has due to female children are biologically stronger than males. Another interesting result is that the prevalence of diarrhoea/ARI among Muslim children quite common may be due to their poor standard of living in urban slum areas.

11.5 Limitations

The target population itself is one of the significant limitations of the study. The dearth of data on children below five years of age is an evident example of ignorance from all concerns. This outshines section is actually the future of the nation. But understanding the truth of life and living of this youngest population is quite difficult as this section of the population is restricted only within the ambit of Birth Rate, Infant Mortality Rate, and child abuse now a day and that's all. There is a massive scope of study on them, which is unseen till now. Thus, surmising the little ones will be done on the overall socio-economic condition, and it will have some limitations to explore the truth. Besides this, detrimental factors which lead to even death of the children are also not well determined like the nature of potable water. There is no universal definition of pure and safe drinking water. It can't be universalized either. Water is a natural object, and it is available with different pure and impure constituents depending upon location and other factors. So, one cannot draw a single line to decide the effect of water on the population. The limitation of data also is found in such cases. Although this study provides a snapshot of the emerging patterns of selected self-reported morbidity, covering a census and the latest national-based sample, the findings need to be taken in light of a few limitations (Paul and Singh 2017). In general, the definition of sources of drinking water, all types of sanitation, drainage of Census data, whereas NFHS-4 data did not cover drainage, garbage disposal related questions. Another major drawback is that self-related morbidity among under-five likes of malaria, stomach, and skin diseases is also not covered. On the other hand, there has been a slight mismatch in the classification of every component of water sanitation categories from 1991 to 2016. Therefore, it is not getting good pictures of the other important child health-related prevalence of self-reported morbidity. Despite these limitations, the levels trend of urbanization, housing, sanitation, and associated with the best health outcomes in this study is useful to understand the specific morbidity conditions in the states of urban India.

11.6 Conclusion

Unfortunately, over the years, the level of urbanization is significantly increased in India. Simultaneously, reporting of prevalence's like diarrhoea and ARI of morbidities also increased day by day in urban India. Although it is derived that urban areas have sources of potable water, toilet facilities, and garbage disposal systems, require improved and safe sources of these necessary facilities of healthy life and living in the urban world. The key finding of the paper is practicing sources of drinking water, toilet facilities, and garbage disposal has been tremendously improved. It is important to note that among the states, Manipur has the worse source of drinking water, way below the national average, i.e., only 47%. Besides this, states like Karnataka (70%) followed by Tamil Nadu, Telangana, etc.

Similarly, regions like North Eastern and Northern (except Rajasthan) are less practicing open defecation. In contrast, Central and Eastern regions (except West Bengal) are maximum practicing open defecation in urban India. The situation of Odisha was very poor for practicing open defecation (28.4%) and bathroom facility (64.6%). On the contrary, the poor economic states having indicated a lower prevalence rate in most of the morbidities examined in the study. Similarly, North-Eastern states like Manipur (65.3%), Chhattisgarh (67.3%), and Arunachal Pradesh (70%) are less using bathroom facilities as comparing the national average 87.0%, a very low prevalence of self-reported morbidity from the very first round of the survey. Practicing of open drainages found maximum in all north-eastern states except Tripura. An infant male child is more vulnerable than a female child. Here, another positive association was observed between Kachha housing structures with diarrhoea. However, Muslim children are more vibrant diarrhoea as well as ARI symptoms. Central and northern states are more reporting of both diarrhoea and ARI diseases among under-five children in urban India. Sanitation Index suggests that Odisha (Rank-35) was in the worst condition, followed by Jharkhand, Chhattisgarh, and Bihar. On the opposite, Punjab (Rank-1) was occupied the top position, then Uttarakhand and all Union Territories of India.

11.7 Policy Prescriptions

Sanitation is in a state of crisis that needs to be addressed with due urgency. Safe drinking water, supply, and sanitation are a State responsibility under the Indian Constitution. States may give the responsibility to the municipalities in urban areas, called Urban Local Bodies (ULB), Ministry of Housing and Urban Poverty Alleviation, Ministry of Urban Development, and Ministry of Water and Sanitation at the Federal Level are also responsible administrative bodies for policy setting and their proper implementation. The promotion of well infrastructural development will require eradicating the practice of open defecation in Urban Municipal Corporation (UMC). Health care provisions for both diarrhoea and ARI at the primary level need to be ensured for early screening and treatment, which is almost non-existent at present. Notably, primary health care for both diarrhoea and ARI needs to be made available in urban areas. Appropriate policies aimed at under-five care are the need of the hour. At the basic level of education, like Primary education, specialized health care provisions for the children need to be synchronized. There is a more important requirement for promoting care education of both diarrhoea and ARI disease among the said targeted population. Additionally, health promotion measures may be taken to inform people to inculcate healthy habits to prevent particular diseases.

Acknowledgements The authors are grateful to the Demographic and Health Survey (DHS) for granting access to the National Family Health Survey (NHFS) datasets of India.

References

Acharya D, Prasanna KS, Nair S, Rao RS (2003) Acute respiratory infections in children: a community based longitudinal study in south India. Indian J Public Health 47(1):7–13

Ashwani K, Kalosona P (2016) Effect of indoor air pollution on acute respiratory infection among children in India. Soc Nat Sci J 10(2)

Boschi-Pinto C, Velebit L, Shibuya K (2008) Estimating child mortality due to diarrhoea in developing countries. Bull World Health Organ 86:710–779

Clasen T, Boisson S, Routray P, Torondel B, Bell M, Cumming O, Ray S (2014) Effectiveness of a rural sanitation programme on diarrhoea, soil-transmitted helminth infection, and child malnutrition in Odisha, India: a cluster-randomised trial. Lancet Glob Health 2(11):e645–e653

Curtis V, Cairncross S (2003) Effect of washing hands with soap on diarrhoea risk in the community: a systematic review. Lancet Infect Dis 3(5):275–281

Darvesh N, Das JK, Vaivada T, Gaffey MF, Rasanathan K, Bhutta ZA (2017) Water, sanitation and hygiene interventions for acute childhood diarrhea: a systematic review to provide estimates for the Lives Saved Tool. BMC Public Health 17(4):101–111

Esrey SA, Potash JB, Roberts L, Shiff C (1991) Effects of improved water supply and sanitation on ascariasis, diarrhoea, dracunculiasis, hookworm infection, schistosomiasis, and trachoma. Bull World Health Organ 69(5):609

Fewtrell L, Kaufmann RB, Kay D, Enanoria W, Haller L, Colford JM Jr (2005) Water, sanitation, and hygiene interventions to reduce diarrhoea in less developed countries: a systematic review and meta-analysis. Lancet Infect Dis 5(1):42–52

Gandhi S (2012) Economics of affordable housing in Indian cities: the case of Mumbai. Environ Urban Asia 3(1):221–235

International Institute for Population Sciences (IIPS) and ICF (2017) National Family Health Survey (NFHS-4), 2015–2016, India

Joshi A, Prasad S, Kasav JB, Segan M, Singh AK (2014) Water and sanitation hygiene knowledge attitude practice in urban slum settings. Global J Health Sci 6(2):23

Kalam MA, Sharma SK, Ghos S, Ro S (2017) Change in the prevalence and determinants of consanguineous marriages in India between national family and health surveys of 1992–1993 and 2015–2016. Hum Biol

Kumar A, Das KC (2014) Drinking water and sanitation facility in India and its linkages with diarrhoea among children under five: evidences from recent data. Int J Humanit Soc Sci Invent 3(4):50–60

Lilje J, Mosler HJ (2017) Socio-psychological determinants for safe drinking water consumption behaviors: a multi-country review. J Water Sanit Hyg Dev 7(1):13–24

Ministry of Housing and Urban Poverty Alleviation (2013) Affordable housing in partnership: scheme guidelines. MHUPA, New Delhi

Nath KJ (2003) Home hygiene and environmental sanitation: a country situation analysis for India. Int J Environ Health Res 13(sup1):S19–S28

National Population Commission (2006) National population census. New Delhi, India

Nations U (2015) World population prospects: the 2015 revision. United Nations Econ Soc Aff 33(2):1–66

Patel SK, Patel S, Sharma SK, Paul KS (2019) Water and sanitation conditions and its negative impact as disease prevalence in India: a state level analysis. J Adv Soc Sci Humanit 5(2):606–615

Paul K, Singh J (2017) Emerging trends and patterns of self-reported morbidity in India: evidence from three rounds of national sample survey. J Health Popul Nutr 36(1):1–13

Paul K, Chaudhary M, Chowdhary R, Sengupta R (2020) Changes in levels of morbidity and hospitalisation in Kerala: a district level analysis (1995–2014). Clin Epidemiol Glob Health 8(1):21–28

Paul K, Sharma D (2016) Post facto, trends, pattern of urbanity in Sikkim. Geogr Malays J Soc Space 12(13)

Prüss A, Kay D, Fewtrell L, Bartram J (2002) Estimating the burden of disease from water, sanitation, and hygiene at a global level. Environ Health Perspect 110(5):537–542

Prüss-Ustün A, Bartram J, Clasen T, Colford JM Jr, Cumming O, Curtis V, Freeman MC (2014) Burden of disease from inadequate water, sanitation and hygiene in low-and middle-income settings: a retrospective analysis of data from 145 countries. Tropical Med Int Health 19(8):894–905

Registrar General of India (2011) Household data (H-series). Registrar General of India, New Delhi, India

Sharma D, Kuppusamy K, Bhoorasamy A (2013) Prevalence of acute respiratory infections (ARI) and their determinants in under five children in urban and rural areas of Kancheepuram district, South India. Ann Trop Med Public Health 6(5):513

Singh PK, Rai RK, Alagarajan M, Singh L (2012) Determinants of maternity care services utilization among married adolescents in rural India. PloS One 7(2):e31666

Tiwari P, Rao J (2016) Housing markets and housing policies in India

UNICEF, World Health Organization (2012) Progress on drinking water and sanitation: 2012 update. WHO/UNICEF Joint Monitoring Programme for Water Supply and Sanitation, United States

UNICEF, World Health Organization (2018) Progress on drinking water and sanitation: 2018 update. WHO/UNICEF Joint Monitoring Programme for Water Supply and Sanitation, United States

Van Minh H, Hung NV (2011) Economic aspects of sanitation in developing countries. Environ Health Insights 5:EHI-S8199

Walker CLF, Rudan I, Liu L, Nair H, Theodoratou E, Bhutta ZA, Black RE (2013) Global burden of childhood pneumonia and diarrhoea. Lancet 381(9875):1405–1416

WHO/UNICEF Joint Water Supply, Sanitation Monitoring Programme, World Health Organization (2015) Progress on sanitation and drinking water: 2015 update and MDG assessment. World Health Organization

World Bank (2008) Environmental health and child survival: epidemiology, economics, experiences. World Bank

Sk. Karim is a Doctoral Fellow at the Department of Migration and Urban Studies at the International Institute for Population Sciences, Mumbai. His area of interest is Migration and Urbanization, Population Health, and Gender Issues. He worked on several national and international projects (i.e., Kala-azar project, WHO-Study on global ageing and adult health, Comprehensive National Nutrition Survey (CNNS), and WHO-Global youth tobacco survey. He has published multiple research papers and having expertise in using statistical software.

Dr. Kalosona Paul is an Assistant Professor at the Department of Geography in Sidho-Kanho-Birsha University, Purulia, West Bengal. His area of interest is Migration, health, and development. He earned a second Masters & M.Phil. in Population Studies from IIPS, Mumbai and his PhD awarded in School of Development Studies from TISS, Mumbai. He worked on different large-scale International (e.g., NFHS-4) and National (e.g., Maharashtra State Livelihood Rural Mission) projects. He has written research papers using various sophisticated statistical and GIS software.

Mamta Chaudhary is a Doctoral Fellow at the School of Health Systems at Tata Institute of Social Sciences, Mumbai. Her area of interest is Population Health and Ageing. She has written research papers using various sophisticated statistical and GIS software.

Dr. Somnath Mukherjee is an Assistant Professor, Department of Geography in Bankura Christian College, Bankura, West Bengal. His area of interest of Tribal Health and Development. He has published various research papers using statistical and geographical software.

Chapter 12
Urban Heat Island Formation in Relation to Land Transformation: A Study on a Mining Industrial Region of West Bengal

Soumen Chatterjee and Krishnendu Gupta

Abstract The Earth's surface has been transformed a lot by various deliberate and inadvertent anthropogenic activities; most profound among them is perhaps the changes in the land-use and land cover (LULC) pattern of any area. It is considered to be responsible for many irreversible changes in the regional heat budget and water balance. Expansion of the built-up area has modified the surface interaction of solar radiation, thereby altering the surface temperature pattern of any region. The present study has been carried out in the Jamuria community development (C.D.) block of West Bengal to examine the relationship between the LULC change and its impact on the rising land surface temperature (LST), using multi-temporal satellite data. Urbanization, industrialization, and mining activity in the study area alter the LULCsignificantly. It seems to be responsible for the general rise in the ambient temperature and microclimatic variation. A definite association has been found to exist between the LST and LULC pattern of the study area. The rise in the LST pattern is considered to be spatially associated with those areas where the intensity of land-use change is maximum. Correlation between them, land use indices like normalized difference vegetation index (NDVI), normalized difference built index (NDBI), etc., and transect profile also confirmed these facts.

Keywords Land surface temperature · Land use land cover · Urbanization · Industrialization · Mining · Urban heat island

12.1 Introduction

Urbanization, coupled with industrialization, seems to be the significant and perhaps the most pervasive force shaping rural areas' land-use patterns, especially those lie adjacent to major urban centres in recent times. They are considered to be the most important determinants and agents in the land transformation process whereby land use of an area is replaced profoundly by built-up area composed chiefly of impervious

---

S. Chatterjee (✉) · K. Gupta
Department of Geography, Visva Bharati University, Santiniketan, India

M. Mishra et al. (eds.), *Regional Development Planning and Practice*, Advances in Geographical and Environmental Sciences,
https://doi.org/10.1007/978-981-16-5681-1_12

surface that is opaque to solar radiation (Mallick et al. 2008; 2012 Asmiwyati 2016). They are also responsible for the gradual decrease in vegetative cover, modification of wetland, increase in the bare and wasteland, quicker runoff generation, slow rate of infiltration, environmental pollution, etc. This modification leads to the alteration of a host of ecosystem process, which can be manifested by the generation of urban heat island, greater amount of absorption and storage of solar radiation, a higher incident of surface albedo, modification of evapotranspiration rate, etc. (Mallick et al. 2008, Omran 2012, Jalan and Sharma 2014; Ali et al. 2017). Briefly, they alter the water and heat budget of any area to a great extent. Urbanization often leads to accelerate the pace of industrialization, especially in the developing world like India, where economic development always remains an important issue (Chatterjee 2011). Hence, the faster the process of land transformation faster will be the modification of ambient environmental conditions (Singh et al. 2014). The current land use and land cover pattern of any region do not only indicate the current anthropogenic process operating over the land surface but also focuses on the interaction between the human and with their environment, especially in any urban and suburban landscape.

Keeping this in view, different studies have been attempted to explain various aspects of land surface temperature change and its linkages with the land use land cover change in different parts of the globe. Howard (1833) was probably the first person who explains the thermal anomaly between the cities and countryside, which was explained under the term 'Urban Heat Island' (UHI) by Manley (1958). Consequently, different types of UHI have been identified depends on the atmospheric mechanism and urban environment like Canopy Layer Heat Island (CLHI), Boundary-Layer Heat Island (BLHI), and Surface Urban Heat Island (SUHI). Changes in LST may be associated with the interruption of a number of meteorological processes like cloud, precipitation, and thunderstorm activity; it is also found to modify the intensity of diurnal and seasonal temperature (Bornstein and Lin 2000). This modification may trigger several changes in the environmental process and impacts on several organisms and their ecosystem, which have significant consequences on human health and economic well-being. The distribution of LST and their association with various geographical features have been studied by Lo et al. (1997), Liu and Zhang (2011), Wang et al. (2017) in different parts of China. The thermal band obtained from various sensors (Landsat, MODIS, etc.) has been widely used to retrieve LST in different parts of the world (Sobrino et al. 2012) for estimate LST.

More specifically, an increase in the impervious surface associated with urbanization, industrialization, and mining activity undoubtedly modify the surface interaction process of solar radiation that eventually resulted in urban heat island and microclimatic variation over any territory. With the rapid pace of urbanization in India, the UHI problems seem to extend over the medium and smaller urban area; hence to monitor the problem and formulating policies studies over such regions are becoming increasingly important. In India, several such studies have been carried out mainly in urban areas. Jalan and Sharma (2014) conducted a study in Jaipur using satellite data and established that area with sparse vegetative cover and increased built-up cover has experienced a positive change in LST. Grover and Singh (2015) carried out a comparative study on the changing pattern of LST in Delhi and Mumbai. They found that the UHI, due to the lesser vegetation cover effect, has been more prominent over Mumbai. Similar studies have been carried out by Mallick et al. 2008; 2012, over the Delhi region; Joshi et al. 2015, over Ahmedabad city; Ali et al. in 2017 in Bhopal, etc. All the study confirms the increasing incidence of UHI formation is related to the increase in the built-up area. Govind et al. (2019) have established the LST of Bengaluru that has increased by 6°C, and the maximum LST has increased over half of the city region. In West Bengal, various studies have also been undertaken to assess the extent of LST change that emanates from LULC modification. Pal et al. (2016) examined the issue of LST with reference to LULC change in English bazaar and found that the urban area is comparatively warmer than the surrounding landscape. They used various geospatial tools and found a significantly strong correlation between LST and other land use indices like NDVI, NDBI, etc. In a similar study over Dhaka Bangladesh, Ahmed et al. (2013) also established that LST has a positive correlation with NDBI and negative correlations with NDVI and NDBaI. Dhar et al. (2019) attempted such a study over the Rajarhat block near Kolkata and found that urbanization has converted 44% of the LULC into built-up, which is responsible for the intensification of UHI. They also found a significant negative correlation between LST and NDVI. Most of the studies discussed so far are undertaken over the large cities and considered only the impacts of urbanization. As mentioned earlier, the small and medium cities, where there is no such strict land regulation, are also facing these kinds of problems. They need to be addressed appropriately to mitigate. Besides, the unplanned and rapid urbanization instigate several other activities like industrialization and mining, which need to be assessed for their potential to form UHI. The present study has been undertaken as an effort to fill such a research gap.

So far as the general objective of the study is concerned, this study has been intended to examine the formation of UHI in the far-flung area of the urban landscape along with the urban centre in relation to land-use change. Specific objectives of the study are: to identify the LULC change of the study area; to calculate the LST pattern of the study area over different periods; to investigate the association between LST and LULC pattern by using various tools.

12.2 Study Area

The present study has been carried out in the Jamuria CD block of Paschim Bardhaman district, West Bengal, India. It is one of the oldest mining regions of the country, located in the Raniganj Coalfield area, where coal has been extracted for centuries (Peterson 1910). Besides, the study area located in the Asansol Durgapur industrial area, which is the second-largest industrial zone of the state, next to the capital city of Kolkata. The study region has an extent of 158 km^2 constituting Jamuria municipality and 48 villages with a population of almost 1.25 lakh (Census of India 2011). The northern and southern limits of the study area have been marked

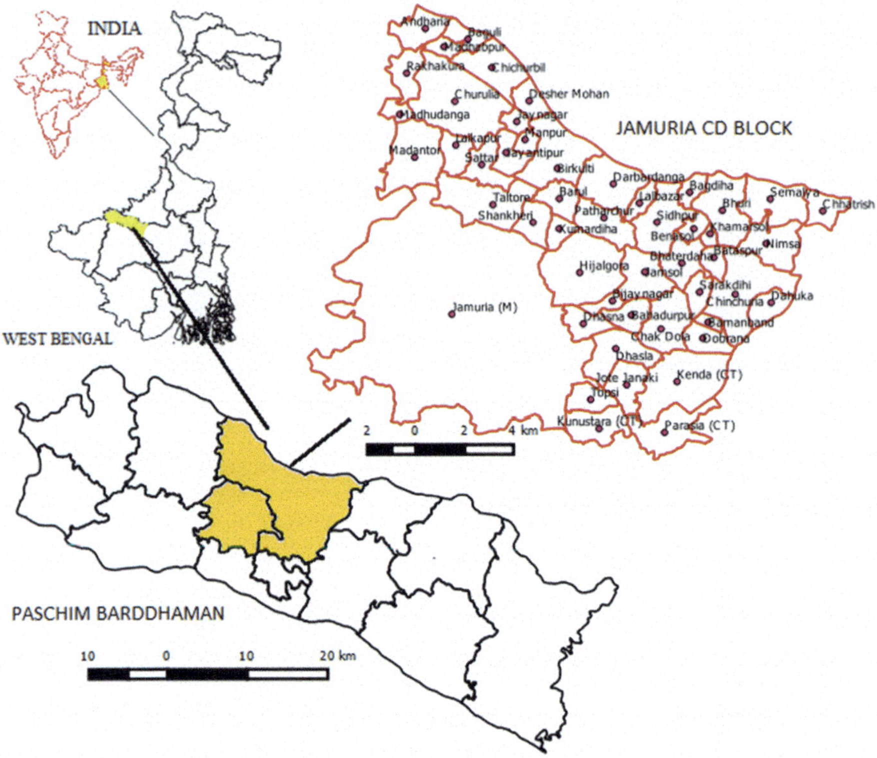

Fig. 12.1 Location of the study area (*Source* Prepared by the authors)

by the River Ajoy and Raniganj CD block, respectively. The Barabani and Andal CD block marked its eastern and western boundary, respectively (Fig. 12.1). Geologically the area is composed of a soft and medium type of sandstone, shale, and coal seam of Raniganj and Barakar formation, which lay down on the hard foliated granitic gneissic complex of Chhotanagpur (Chatterjee 2016). Geomorphologically, it constitutes part of the rolling upland of the Chhotonagpur plateau fringe region. The area experiences a subtropical monsoonal type of climate characterized by seasonal reversal of temperature and rainfall patterns (Chatterjee 2016).

Economically, the area depends on agriculture and mining activities. Summer paddy has been cultivated widely all over the region along with the rain-fed uplands. Wheat, pulses, and oilseed are cultivated during the winter in low-lying areas. Several medium and small-scale industries, mainly sponge iron and cement, have come up in the area during the past two decades. It actually forms the hinterland of the Asansol Raniganj coal belt, which has a long history of mining-industrial and urban development. However, the area lies under the jurisdiction of the Asansol Durgapur Development Authority.

Table 12.1 Data set used in the study of UHI in the mining industrial region of West Bengal

Satellite	Sensor	Path/Row	Date of acquisition
Landsat5	TM	139/044	1990-04-11
Landsat5	TM	139/044	1990-12-23
Landsat5	TM	139/044	2010-12-14
Landsat5	TM	139/044	2011-03-20
Landsat7	ETM +	139/044	2018-12-12
Landsat7	ETM +	139/044	2019-05-21

(*Source* Compiled by the authors)

12.3 Database

Landsat 5 and 7 satellite images of summer and winter time (1990, 2011, and 2019) have been selected database for the present study (Table 12.1). They were downloaded from the official website (*earthexplorer.usgs.gov*) of the United States Geological Survey (USGS). The images have spatial resolution of 30 m and are located in the Landsat path of 139 and row 44. The level-one terrain-corrected (L1T) product has been selected in the present study. It is noteworthy to mention that scan line correction has been performed for the Landsat 7 images. Survey of India topographical map bearing the numbers of 73 M/1 and 73 M/2 and Google earth image of the area has been used to obtained land use information. QGIS 3.6 software has been used in this study.

12.4 Methods

The methodological framework adopted in this study can be presented with a flowchart (Fig. 12.2) which is self-explanatory. Details about the procedure have been discussed in the following sections.

12.4.1 Atmospheric Correction

There is every possibility that the atmospheric condition might differ between the above-mentioned time span as the images belong to three different periods. Hence, the omission of atmospheric effects is an essential part while working with the thermal band. The reflected energy can be attenuated and enhanced by the atmosphere condition, which needs to be adjusted by using a radiative transfer model (Barsi et al. 2003). Following formula commonly used for atmospheric correction.

$$L_{TOA} = \mathrm{L_T} + \mathrm{L_u} + (1-)\mathrm{L_d} \quad (12.1)$$

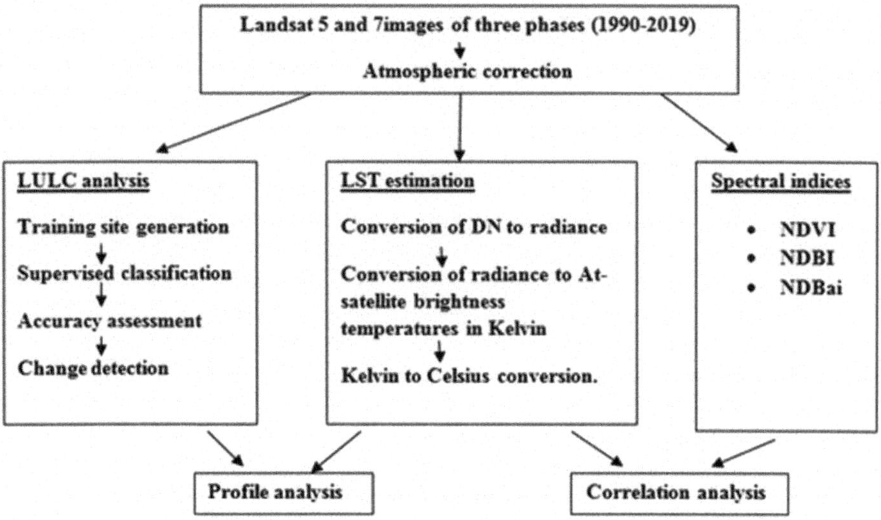

Fig. 12.2 Flow chart for methodological framework (*Source* Authors' perception)

Where τ is the atmospheric transmission;

ε is the emissivity of the surface;

LT is the radiance of a body;

Lu is the upwelling or atmospheric path radiance;

Ld is the downwelling or sky radiance; and

LTOA is the space-reaching or TOA radiance

The values of Lu, Ld, and τ were evaluated using the Atmospheric Correction Parameter Calculator online tool.[1]

The input values of the parameters are available at the metadata of the respective images and Landsat handbook (*landsathandbook.gsfc.nasa.gov/ pdfs-landsat7_handbook.pdf*). Chavez (1996) explained that within the image, there might be some pixels that are in shadow and their radiance received at the satellite by the process of scattering, which is related to the concept of a dark object that also needs atmospheric correction. The Semi-Automatic Classification Plugin (SCP) tool of QGIS software has been used for this correction.

[1] See http://atmcorr.gsfc.nasa.gov for details.

12.4.2 Computation of LST using the Thermal Band

12.4.2.1 Conversion of Digital Number to Spectral Radiance

Four Landsat images of various period belongings to two particular seasons, namely summer and winter, have been employed to explain the land surface temperature of the study area. They show both seasonal variations of temperature and temporal changes over the past three decades. Following the principle of remote sensing, the reflected solar energy from terrestrial objects had been captured by the sensor at the space-borne platform. The thermal sensor of Landsat has the capability of sensing thermal electromagnetic energy emitted by any terrestrial objects and converts them into the sensor radiance value, which is then scaled into an 8-bit digital number (DN). It is now possible to convert these DNs to ToA (Top of the atmosphere) reflectance using the following formula obtained from the Landsat handbook.[2]

$$L\lambda = \left(\frac{LMAX\lambda - LMIN\lambda}{QCALMAX - QCALMIN}\right) * (QCAL - QCALMIN) + LMIN\lambda \quad (12.2)$$

where, Lλ stands for Spectral Radiance;

LMAXλ and LMINλ indicates the spectral radiance of the thermal band;

QCAL stands for the DN value of an individual pixel and; CALMAX and QCALMIN indicate the maximum and minimum pixel value, which is 255 and 0, respectively.

12.4.2.2 Conversion of Spectral Radiance to At-satellite Brightness Temperatures

After getting the spectral radiance value for each pixel, they have been converted into effective at-satellite temperatures in Kelvin by applying the following formula obtained from Landsat handbook mentioned earlier:

$$T = \frac{K2}{\ln\left(\frac{K1}{L\lambda} + 1\right)} \quad (12.3)$$

where, T stands for Effective at-satellite temperature in Kelvin;

K1 and K2 represent Calibration constant with the corresponding value of 607.76 watts/meter^2 and 1260.56 Kelvin, respectively for Landsat 5 products and finally;

Lλ is the spectral reflectance. The rest of the parameters has been collected from the Landsat handbook.

[2] See http://landsathandbook.gsfc.nasa.gov/pdfslandsat7_handbook.pdf for details.

12.4.2.3 Estimation of LST in Celsius

After executing all the above formulas by the raster calculator tool in the GIS environment, the resultant scene generates the surface temperature value in Kelvin (^{0} k). For ease of calculation, the unit has been converted into the degree Celsius by deducting the value by 273.15.[3]

Past climatic data from secondary sources have been employed to cross-check and validate the model. Different land use indices have been calculated to compare with the land surface temperature pattern. Both are indicative of the spatial pattern of the surface temperature.

12.4.3 Computation of Spectral Indices

Different spectral indices using the various bands of the Landsat image have been calculated to understand the relation between the land-use change and land surface temperature as they are thought to be correlated strongly with the LST pattern (Xiao et al. 2008). They are:

NDVI (Normalized Difference Vegetation Index) is a useful index to estimate the health of green vegetation in the study area. NDVI has been widely used for biomass content of any area. Red and IR bands have been employed to derive the NDVI in the following way (Bonfiglio et al. 2002):

$$NDVI = \frac{(NIRBand - RBand)}{(NIRBand + Rband)} \tag{12.4}$$

In Landsat 5 and 7, B4 represents the NIR, and B3 represents the Red band. The corresponding value lies between -1 to $+\ 1$. Atmospheric noise, snow, and water body may yield lower output, while greater output may come from different types of vegetation cover.

NDBI (Normalized differential built-up index) has been used to understand the extent of built-up area. They have been calculated as (Zha et al. 2003):

$$\text{NDBI} = \frac{(SWIRBand - NIRBand)}{(SWIRBand + NIRband)} \tag{12.5}$$

In Landsat 5 and 7, B4 represents the NIR, and B5 represents the SWIR band.

Finally, NDBaI (normalized difference bareness index) has been used to identify the barren land of the study area by using the equation (Zhao and Chen 2005):

$$\text{NDBaI} = (\text{SWIR} - \text{TIR})/(\text{SWIR} + \text{TIR}). \tag{12.6}$$

[3] As 0^0 in Celsius scale represents 273.15 k temperature. Hence the resultant imagery shows the land surface temperature in Celsius unit.

In Landsat 5 and 7, B5 represents the SWIR, and B6 represents the TIR band. It is very useful to identify the barren land in the study area. NDBI is used to show the extent of the built-up area.

12.4.4 Classification of Image

12.4.4.1 Supervised Classification

One of the major advantages of multispectral satellite images is that we can cluster the pixel having a similar spectral signature, which is determined by the relative reflectance in the various wavelength bands. With prior knowledge of the study area, a supervised classification with a maximum likelihood algorithm has been applied to the image with significant land-use classes. These are built-up, water body mining-industrial, forest, barren/waste, sand bar, and agricultural land. In this type of classification, the analyst defines on the image a small area having a similar spectral value, called a training site, which is representative of that particular land use class from the whole image. After the clusters for each training site are defined, the computer then classifies all the remaining pixels in the scene, which yields the final land use land cover map of the study area. The area under the respective classes has been extracted from the post-processing tools of QGIS software.

12.4.4.2 Accuracy Assessment

Accuracy assessment is an integral part of image classification in the post-processing stage. In brief, it is a procedure of quantifying how accurate the land use classification is. It is generally done by comparing the classified product with some reference data. In the present study, 150 and175 random points have been created on the classified map for the years 1990 and 2011 using a stratified sampling method that represents all the classes. These points are actually represented by some pixels over the satellite image. Then the selected pixels have been compared with Google Earth images and some time with field verification to know the actual land use. Thereafter, corresponding classes have been checked to prepare an error matrix. Kappa coefficient, which is a measure of the difference between an automated pixel and reference data, has also been calculated for each year.

12.4.4.3 Change Detection

After the classification of both the images, the change detection analysis has been performed to assess the quantitative aspects of the land-use change using the image processing software. They enable us to compute the land-use change statistics and the relative changes of individual land-use classes.

12.4.5 Profile Analysis

Three longitudinal profiles have been constructed over the study area in different directions for 1990 and 2011 using the q profile tool in QGIS. They were selected in such a way that they cover all the land use classes so that a general idea of the thermal impression of all the land use categories can be obtained. Finally, the profiles for both the year has been presented in a single frame with two different colours.

12.4.6 Correlation Analysis

Eventually, correlation analysis using *Pearson's product-moment correlation coefficient*(r) has been computed to understand the relationship between LST and various spectral indices like NDVI, NDBI, and NDBaI.

12.5 Results

12.5.1 LULC Dynamics

LULC of the study area has seven broad categories, namely agriculture, built up, fallow/ barren, vegetative cover, mining-industrial area, water body, and sand bar. The overall accuracy for 1990, 2011, and 2019 are estimated as 86.19, 93.73, and 94.02 percent with the *kappa coefficient* of 0.83, 0.91, and 0.92, respectively. LULC map of the study area (Fig. 12.3) along with their corresponding statistics (Table 12.2) are given below.

Over time, the forest cover of the study depleting at an alarming rate. In 1990 the forest land was found to be the second largest land use category after agriculture but within the next three decades, significant loss in area has been reported for forest land. Almost 10% of the land surface that had vegetation covers previously has been converted into other land use categories. The northwestern part of the study area witnessed significant loss of vegetation cover due to mining operations, followed by the western part. Here, the expansion of agricultural activity and the built-up area played a significant role in decreasing forest cover. However, the southern part of the study area manages to maintain the vegetation cover to some extent.

Mining and industrial land use, which was practically absent in the study area during 1990, become the second-largest land use category having almost 10% of the area in 2019. The onset of open cast mining through digging pit left large areas unsuitable for other activities besides making them vulnerable for land degradation (Fig. 12.4) The central part of the study area remains an important agent of land transformation as several sponge iron industries and open-pit coal mines comes into the region.

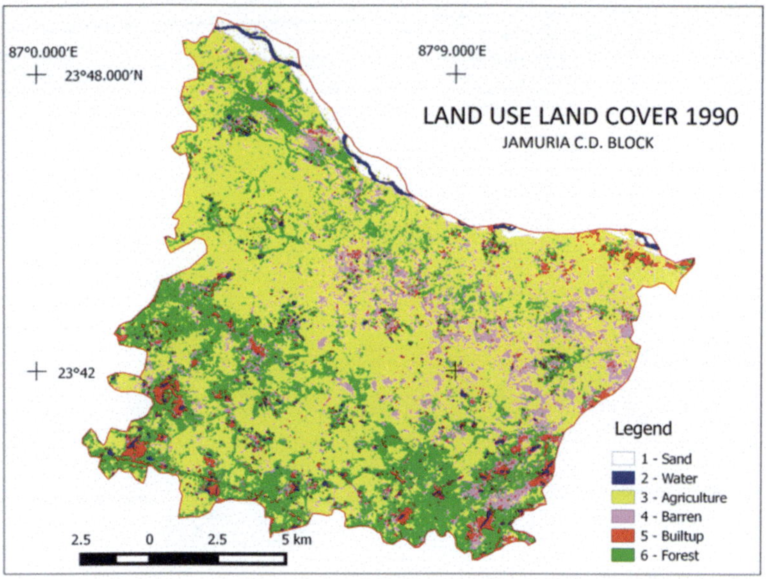

(a)

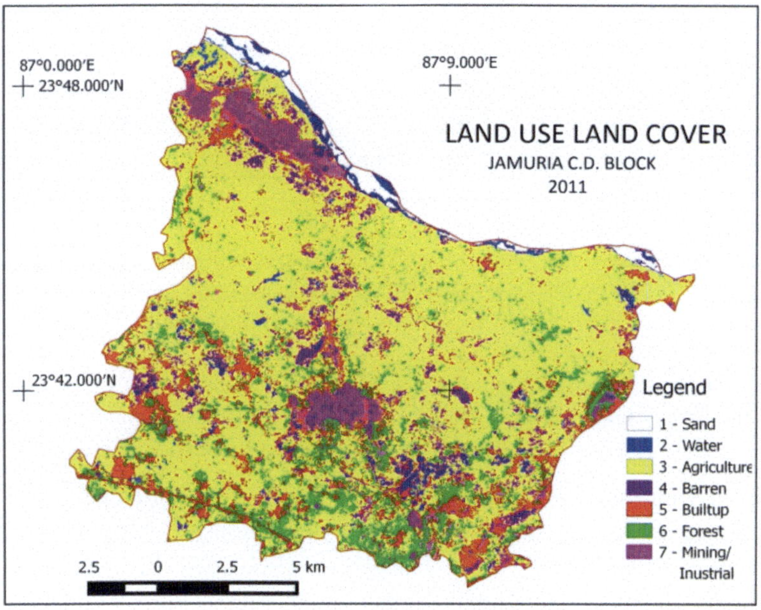

(b)

Fig. 12.3 Land use land cover (LULC) in a1990, b2011 and c 2019 (*Source* Prepared by the authors)

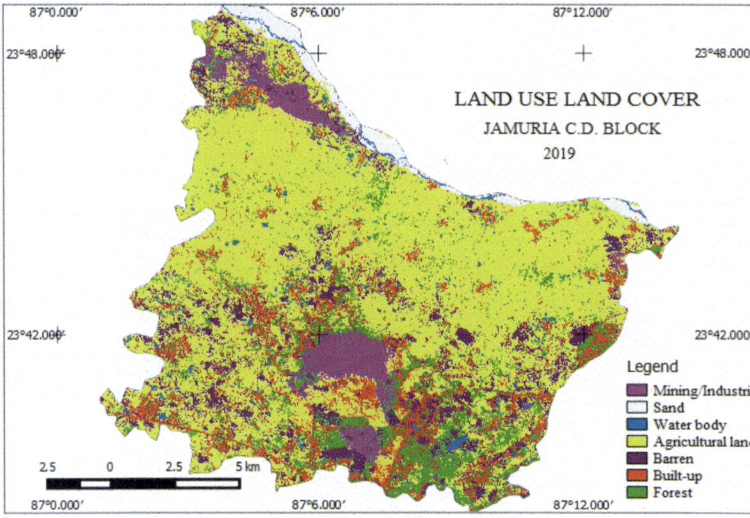

(c)

Fig. 12.3 (continued)

Table 12.2 LULC statistics of Jamuria for the last three decades

Class	1990		2011		2019		Change (1990–2019)	
	Area#	%	Area#	%	Area#	%	Area#	%
Sand	2.85	1.78	2.93	1.83	5.07	3.17	2.22	1.39
Water	2.94	1.84	4.88	3.05	2.13	1.33	−0.81	−0.51
Agriculture	95.23	59.52	89.86	56.16	87.03	54.4	−8.20	−5.12
Forest	32.59	20.37	16.87	10.54	17.97	11.23	−14.62	−9.14
Barren/Fallow	17.66	11.04	19.67	12.30	18.88	11.8	1.22	0.76
Built up	8.72	5.45	11.66	7.29	13.34	8.34	4.62	2.89
Mining/industrial	0	0	14.11	8.82	15.57	9.73	15.57	9.73

#area in km^2 (*Source* Compiled by the authors)

Coupled with industrial mining growth, other built up particularly settlement areas, also experienced positive growth of almost 3%. Considerable growth has been recorded to occur mainly in the southern part of the study area. The growth of Jamuria town in the southwestern part and the growth of census towns like Topsi in the southeastern part can be considered as examples in this regard. Agriculture remains the most important and dominant land use category in the study area. More than half of the study area has been put to raise crops. The fertile riverine tracts situated in the northern part of the study area are mostly suitable for agricultural purposes. Besides, the western part also has a consolidated amount of vast agricultural tract. However, this type of land use has faced an almost 5% decrease during the past three decades.

(a) (b)

Fig. 12.4 Loss of agricultural land by **a** waste dump and **b** by open cast mining at Churulia (Captured by authors from location 23° 46′ 54.45″N, 87° 03′ 56.71″E and 23° 48′15.29″N, 87° 04′ 05.18″E, respectively, dated 26 Oct 2016)

It is interesting to note that despite the encroachment of agricultural land over the barren, fallow, and forested land, they face a negative growth due to the expansion of mining industries.

There is a slight increase in the case of barren and fallow land. So far as 1990 is concerned, fallow land is found to be critical in the eastern and central part of the study area. However, the wasteland originated from mining activity found to be prevalent in the southeastern part of the study area. One thing that should be mentioned that the mines in the study area have a high stripping ratio (WBPDCL 2019). For instance, Churulia open pit, the largest open-pit in the study area, situated in the northwestern part, has a stripping ratio of 1: 10.72. (WBPDCL 2019) It means10.72 million tonnes of overburdened rock will be generated to extract 1 million tonnes of coal. So, a huge amount of overburdened waste rock is found to be accumulated in the study region. This type of wasteland is quite prevalent (Fig. 12.4) in post-2011. Conversion of fallow land into cropland is common in the eastern part, where a sizable amount of fallow land has been found to be brought under the plough. However, the amount of wasteland is found to be increased due to mining activity.

Finally, the amount of land remains almost static so far as the case of sand bar is concerned. Although the amount of water body found to be increased slightly, this is somewhat unusual and might be associated with abandoned open cast pit which left as water-filled. They might cause the amount of surface water body to register positive growth. To understand the nature of land-use change, a cross operation has been performed (Fig. 12.5) to detect the magnitude of change of individual land-use class. It appears that the vegetation and agricultural land are more susceptible to change in the face of the expansion of built-up areas. Significant changes have been shown in Fig. 12.5.

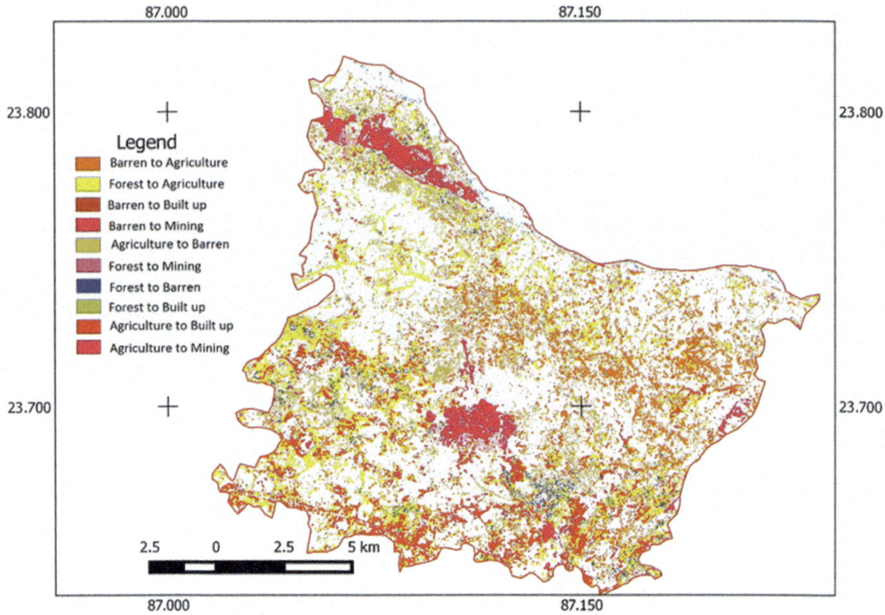

Fig. 12.5 Change detection between 1990 and 2011 (*Source* Prepared by the authors)

12.5.2 Changing Pattern of LST

12.5.2.1 Spatial Pattern

Every land use land cover object has its own thermal expression; hence land use of any area always responds differently to the thermal character (Weng et al. 2007; Weng 2009; Buyantuyev et al. 2010; Li et al. 2014). In the present study, three sets of image belongings to summer and winter season has been used to compute LST. Images captured during 1990, 2010, and 2019 of which the first marks the pre-industrial-mining phase of the region. The latter two mark the post-mining—industrial and current phases. General surface temperature conditions on April and December 1990 (Fig. 12.6a, b). Three distinct patches of the *high-temperature zone* can be identified from April 1990 image where surface temperature is found to prevail near about 35°C. They comprised chiefly the built-up area of Jamuria Municipality, the barren land tract of Churulia, and the mining tract of Topsi situated in the northern and eastern fringe of the study area, respectively. Agricultural land is mostly confined in the northern part of the study area in Shyammala, Chinchuria, and Hijalgora village, and some pockets at the south form the area of the *low-temperature zone*.

In the subsequent two phases, there are no significant changes found in the basic pattern of LST except intensification of the earlier patches and addition of a more recent patch of high temperature in the central part of the study area. In the first instance, the opening of the mining area at Churulia might have played a significant

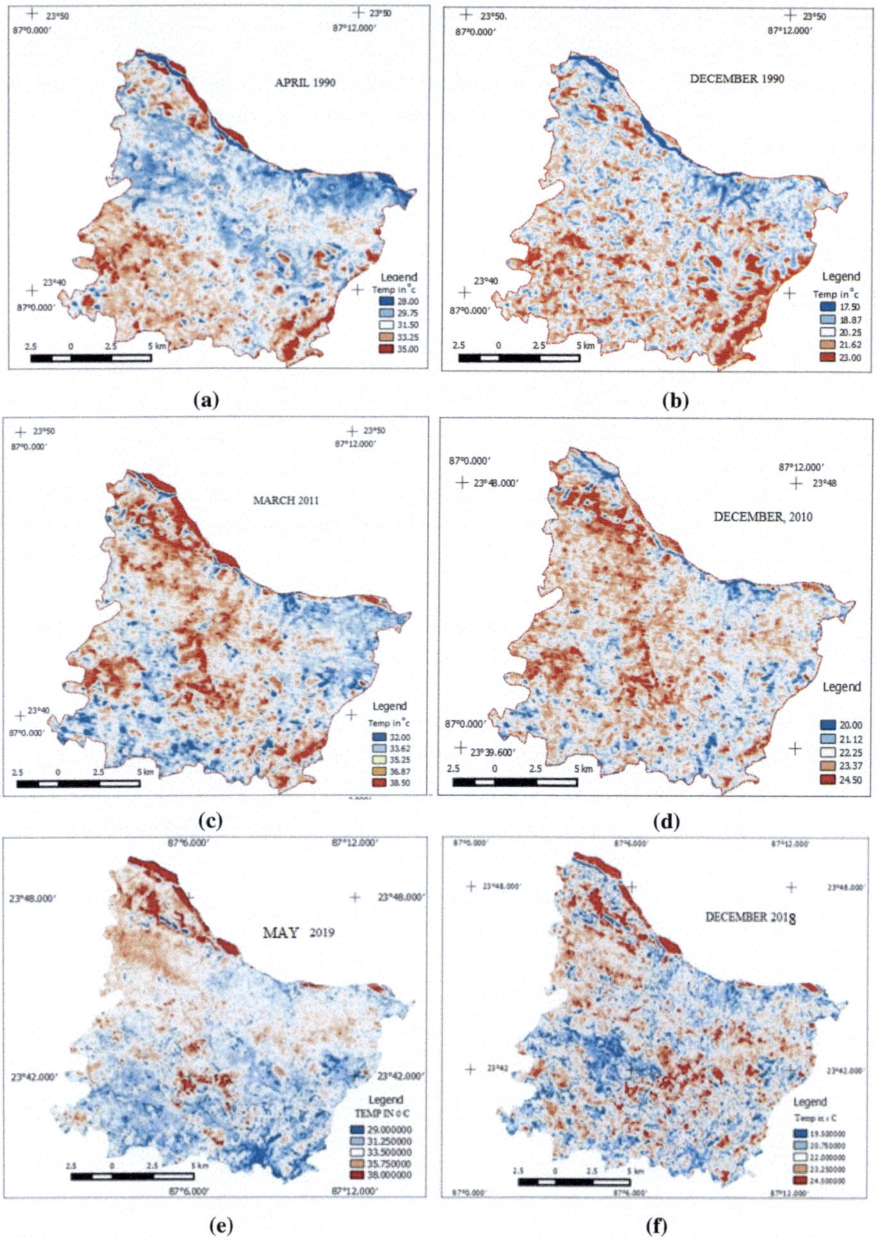

Fig. 12.6 Spatial pattern of land surface temperature: **a** Apr 1990 & **b** Mar 2011in the top; **c** Dec 1990 & **d** Dec 2010 in the middle; and **e** May 2019 & **f** Dec 2018 in the bottom (*Source* Prepared by the authors)

role in increase the LST drastically. The LST of the Municipal area and mining area at the eastern end was also found to rise abruptly. Setting up of industrial cluster at the central part of the study area is responsible for increasing LST there. So, the changing pattern of LULC seems to have modified the LST distribution pattern over the past decades.

12.5.2.2 Seasonal Pattern

A seasonal comparison between the summer and winter temperatures is now necessary to assess the seasonal variation of land surface temperature distribution. Figure 12.6a, c and e represent the summer months. It appears that the northern and central cropland forms the zone of the *low temperature* area. The built-up land, barren, and sand bar forms the pockets of *high temperature.* Like the summer temperature pattern, winter surface temperature (Fig. 12.6b, d and f) also tends to be high in the urban, industrial, and mining areas compared to the agricultural region. However, the range is quite low from the summer months. The maximum temperature prevailed over the region during the summer of 1990, 2011, and 2019 has been 35.0 °C, 38.5 °C, and 38.0°C, respectively. While the winter months of 1990 and 2010 have LST of 23 °C in and 24.5 °C, respectively. Therefore, the maximum temperature in summer and winter is increased by 2 to 3°C over the past three decades (Fig. 12.7).

As a corollary of the above discussion, it is now essential to describe the changes in the areal distribution of various microthermal zones in the study area. Considering the highest and lowest temperature value from each set (3) of raster images, a range of

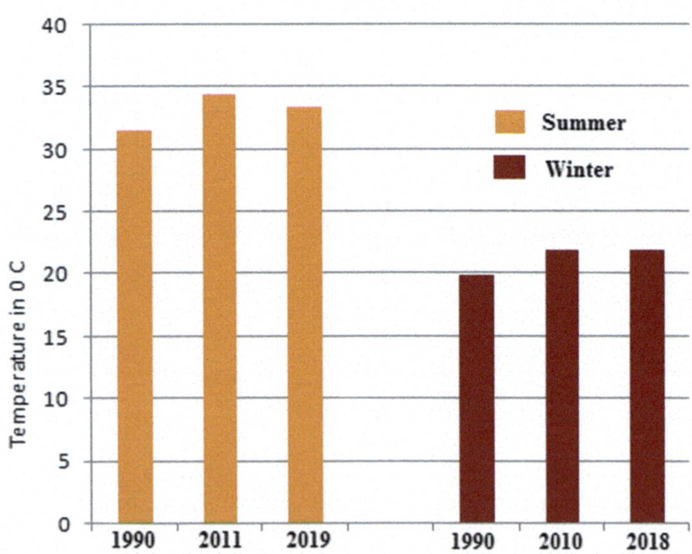

Fig. 12.7 Maximum land surface temperature: summer and winter (*Source* Prepared by the authors)

Table 12.3 Spatial and seasonal pattern of LST distribution

	Area (%)				Area (%)		
Temperature(c)	Dec-90	Dec-10	Dec-18	Temperature(c)	Apr-90	Mar-11	Apr-19
below18	9.92	0	0.09	below30	15.22	0.42	3.14
18–20	24.8	9.84	11.4	30–33	74.38	5.59	31.73
20–22	58.57	30.08	28.43	33–36	9.37	75.66	52.54
22–24	6.71	56.42	56.65	36–39	1.03	17.66	11.91
24 +	0	3.66	3.43	39 +	0	0.67	0.68

(*Source* Prepared by the authors)

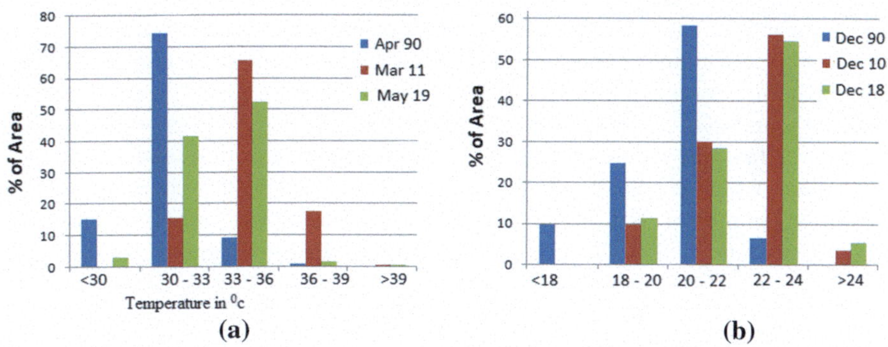

Fig. 12.8 **a** Spatial and **b** seasonal pattern of LST distribution (*Source* Prepared by the authors)

6° for winter and 9° for summer is determined. The range has been divided into five equal classes to prepare the following table (Table 12.3 and Fig. 12.8). Almost 60% of the area experienced surface temperature between 20 and 22°C during the month of December 1990. In the extreme, only 10% and 7% of the study area experienced a surface-temperature of below 18 °C and above 22 °C, respectively, during the said time. There has been a general rise in temperature in almost all parts of the area and in the later phases when almost 90% of the study area experienced above 20 °C temperature. Of them, almost 4% have surface-temperature above 24 °C. The surface-temperature during the summer reveals the same pattern. During 1990, there is only 10% area that has a surface temperature above 33 °C, whereas almost 84% area experiencing that temperature ranges during 2011, and there are some pockets that have more than 39 °C temperature.

12.5.3 Relation Between LST and LULC

In order to measure the relation between LST and LULC of the study area, two approaches have been adopted. Firstly, the LST of various land cover units has been examined by drawing a series of profiles. Secondly, the relation between LST and LULC has been measured by calculating different land-use indices like NDVI, NDBI, and NDBaI. Moreover, the linear relation between the LST and these indices has been calculated by *Pearson's correlation coefficients* to understand the direction and degree of association between the two variables in the following section.

12.5.3.1 Profile Analysis

To understand the LULC wise LST pattern, three cross-sections (Fig. 12.9) have been drawn over the map that runs across various land use land cover types. As raster data contains individual radiance values for each pixel, it is easier to determine the particular LST value for them. Three section lines have been constructed using the LST value. The first section line is selected on the southern central part where major land use classes are mainly agricultural land and built up both settlement and industrial found to occur.

However, from the profile, it is found that built-up areas are associated with higher temperature, which is near about 38 °C while agricultural land is characterized by below a temperature of 36 °C. The second profile has been chosen over the major land use of forest, mining, agricultural land, and barren land. Mining land is characterized

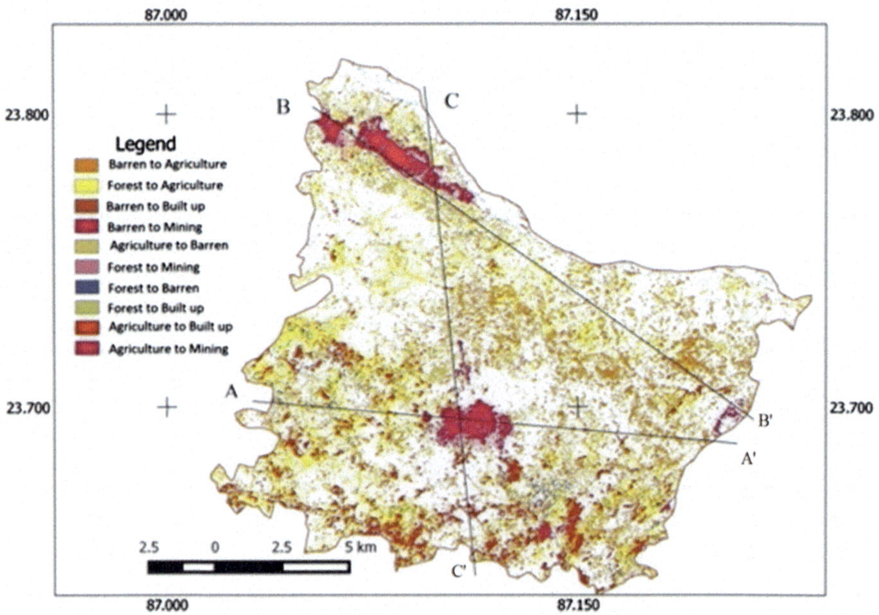

(a)

Fig. 12.9 **a** Sections lines and Transect profile of temperature over various LULC along **b** Line AA', **c** Line BB', and **d** Line CC' (*Source* Prepared by the authors)

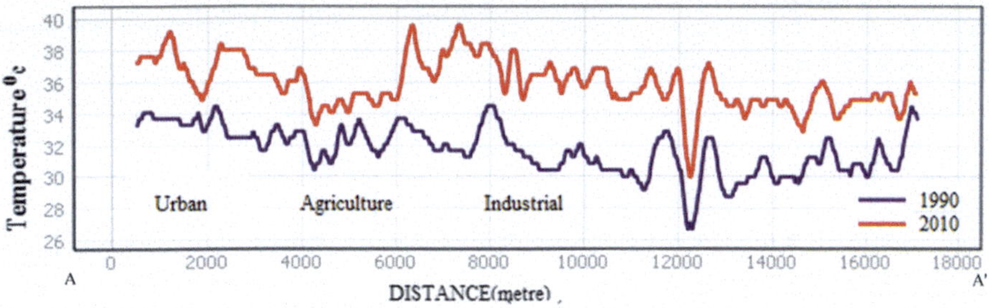

(b)

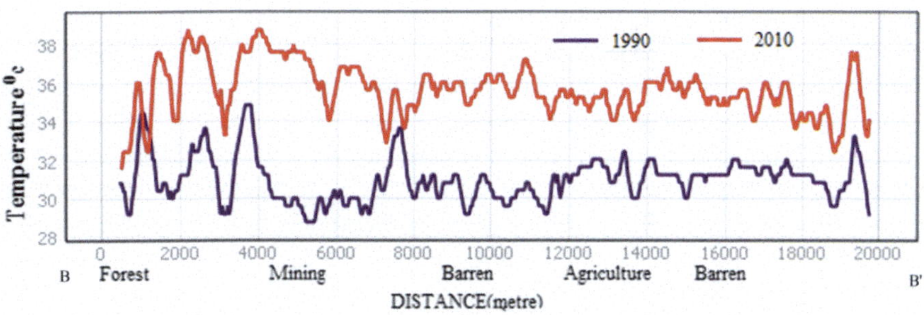

(c)

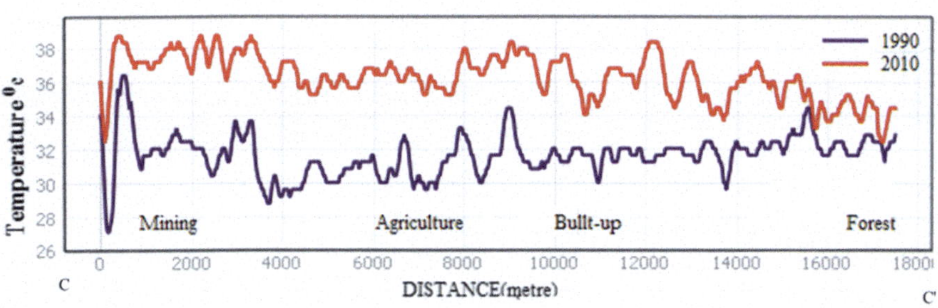

(d)

Fig. 12.9 (continued)

by higher temperature, in this case, followed by barren land. Low temperature is found to prevail over the forest and agricultural land.

The longest profile has been taken over the central part of the study area running across the north–south direction, which is mainly composed of built-up (mining and industrial), river and sand bar, forest, and agricultural land.

This profile also confirms that the land use types as the higher temperature is associated with the built-up that running like a continuous line around 37 °C, and the rest forms two isolated patches of the microthermal zone. Thus, the thermal association of LST and LULC can be empirically explained; however, for more precision quantification approach has been adopted in the next section of the study that deals with various land use indices.

12.5.3.2 Indices

As mention earlier the study incorporate various land use indices for assessing the relation between various land use category and LST. Keeping in mind the major LULC categories, three indices have been chosen—NDVI, NDBI, and NDBaI. The vegetation cover of any area is supposed to regulate the LST pattern as there lies a negative relation between NDVI and LST (Yue et al. 2007). The study area's NDVI map shows a gradual weakening of the index value over the last three decades.

Depletion of vegetation is quite marked in the northern part of the study area (marked by the red circle in Fig. 12.10) and the central part as well as a consequence of the mining industry's growth. Consequently, these are the areas where the LST pattern is modified profoundly over the last three decades. Hence, it can be assumed logically that the gradual decrease of vegetative cover leads to the general increase in the LST pattern of the study area as a whole. Moderate NDVI is commonly associated with barren land which is found to lie in the moderate to high temperate zone in the study area. Agricultural land and village orchards of the northeastern belt (marked by a yellow circle in Fig. 12.10) are associated with low temperature and high NDVI index zone. The correlation between them also cites the negative relation in the study area.

The built-up area is commonly composed of impervious and concert masses that are heat absorbents. Hence, in most cases, the urban built-up areas are comparatively hotter than the surrounding countryside (Ahmed et al. 2013). So there lies a positive relation between the LST and the amount of built-up area; the greater the amount of built-up area greater will be the expected LST. Black-topped roads, building materials, urban design and plan, presence or absence of industrial structure are all responsible for to increase in the LST in urban areas. In the present study, it is observed that the formation of urban heat island is directly associated with the formation and expansion of built-up areas.

The NDBI map also conforms to the LST pattern of the study area as the higher NDBI value in the central and southwestern part of the study area mark the setting up of industrial unit and expansion of the built-up area, respectively, which are usually shown in the area with higher LST.

12.5.4 Correlation Analysis

Land use indices always show some sort of relation with the surface temperature pattern. In the present section, an effort has been made to measure the relation between NDVI, NDBI, NDBaI, which are considered as independent variables (x), and LST, which is considered as a dependant variable(y). A bi-variate linear regression is

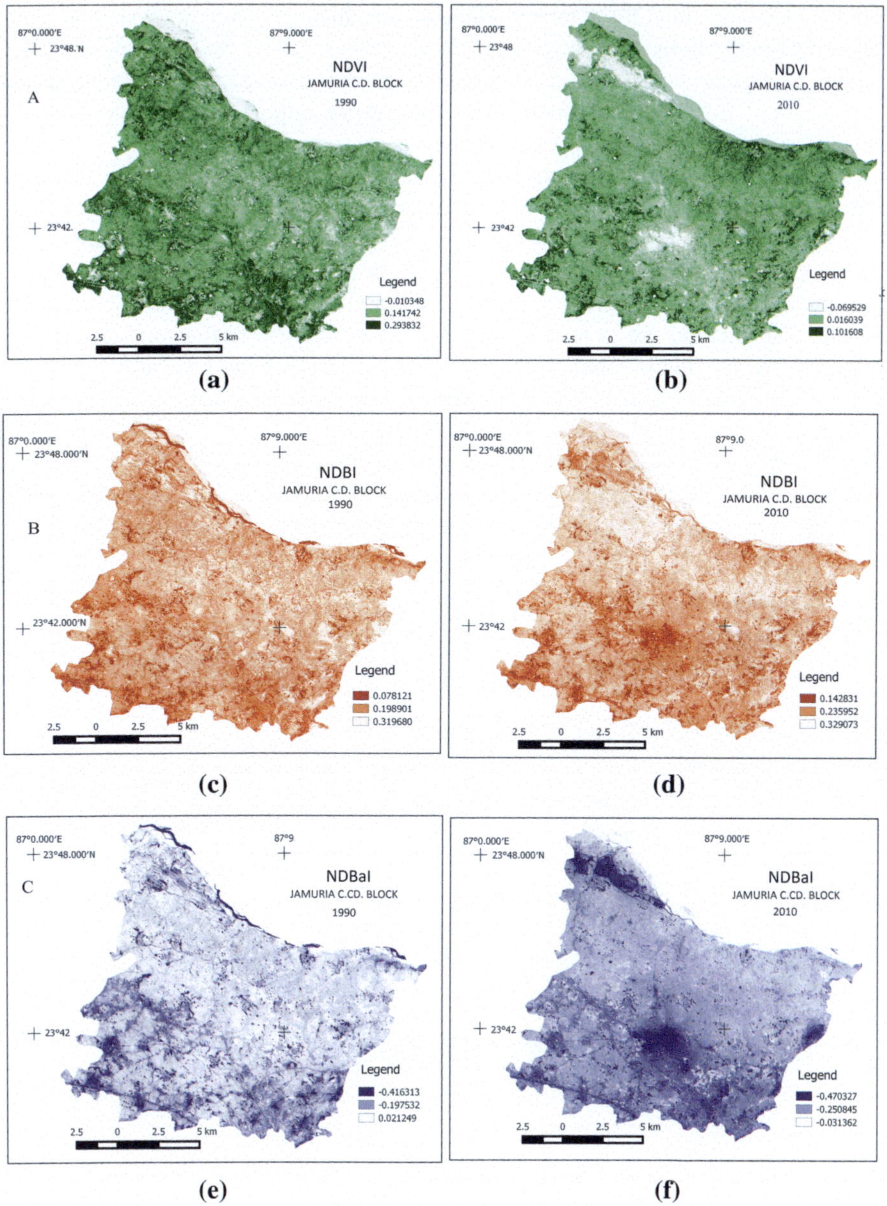

Fig. 12.10 Spatio-temporal pattern of various land use indices namely **a** NDVI, **b** NDBI and **c** NDBaI for 1990 and 2010, respectively (*Source* Prepared by the authors)

employed to understand the relation. The data for this regression analysis has been prepared from the individual raster file of LST, NDVI, NDBI, NDBaI, respectively. A series of regular sample points is taken on the raster map.,and information has been extracted from the respective raster files. After importing the whole data in Microsoft excel, regression analysis has been performed (Fig. 12.11).

NDVI creates a single band data set that contains a value between + 1 and – 1. In the present study NDVI for 1990 ranges between 0.29 to -0.01. It decreases to 0.17—0.01 in 2019, which represents a considerable loss in vegetal density. It should be noted that the NDVI value of 0.29 of 1990 indicates moderate density arises out mainly from the grass and shrubland, and the negative value indicates the rocky and barren land. So, it may be stated that the bareness of the soil increases to a great extent. Correlation and regression analysis between the LST and NDVI for 2011 with R^2 value of almost 50% confirms the negative relation indicates the lower the NDVI higher the LST pattern in the study area.

NDBaI is a useful index to analyze the extent of the bareness of any area. For the given period, the NDBaI value was found to decrease from 0.02 to −0.41 to −0.09 to −0.56. They also exhibit a positive relationship with the LST pattern with an R^2

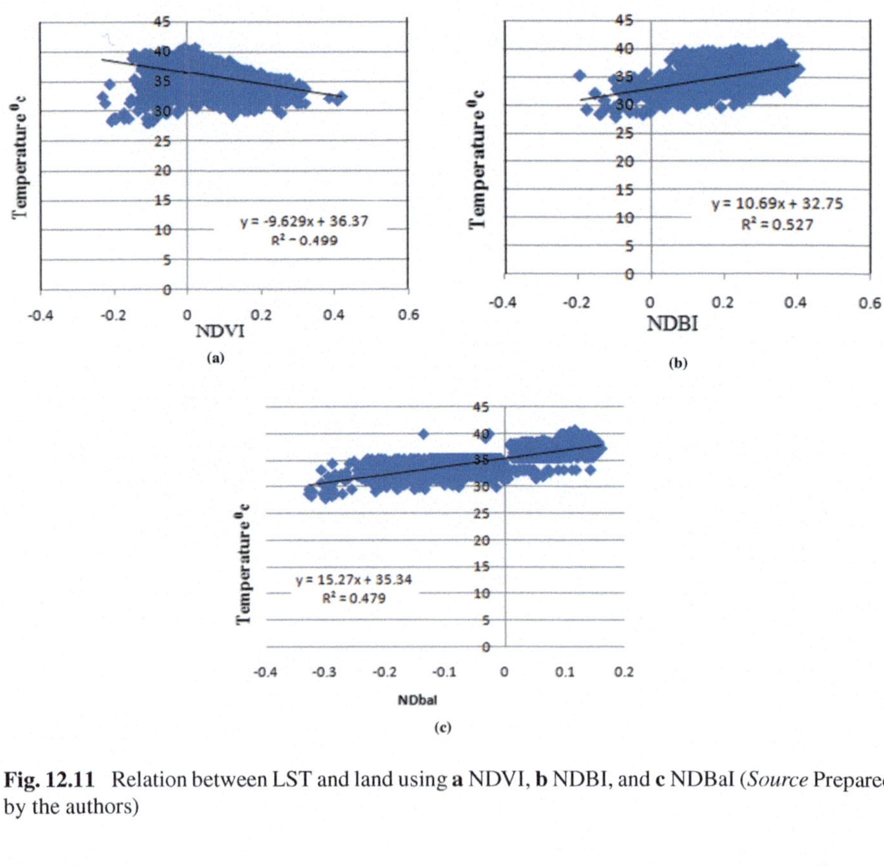

Fig. 12.11 Relation between LST and land using **a** NDVI, **b** NDBI, and **c** NDBaI (*Source* Prepared by the authors)

value of 48%. As barren land practically has absence of vegetation cover and water holding capacity is comparatively less, they tend to possess much higher surface temperature than the surrounding land.

NDBI can use to map the built-up effectively. In the present study, the NDBI value is found to increase slightly as the corresponding value is found to increase from 0.07 to 0.31 in 1990 to −0.19 to 0.17. They are positively associated with the LST pattern of the study area with R^2 value of almost 48%. So, it may be inferred from the result that the higher is the NDBI greater the chances of UHI formation due to its impervious nature.

12.6 Discussion

This study is conducted with three specific objectives. The first one is related to the examination of the existing LULC of the study area. Three Landsat images of 1990, 2011, and the most recent of 2019 have been employed to understand the dynamic of LULC. Agriculture and forest are found to be the most dominant land use category in the study area for all three phases. However, simultaneously, a considerable area of these two categories has been transformed into built-up areas. Within three decades, agriculture and forest have lost almost 14% of LULC area.

Contrary to this, urban, mining, and industrial areas are increased to a great extent. Hectares of land are found to be converted into open mining pit from forest and agriculture land in the northwestern part of Churulia village. Cropland has been found to be transformed into industrial land at the central part, and the urban area in the western part itself is found to be expanded.

The second objective of the study is to assess the LST pattern of the study area in light of the changing LULC pattern. Like the LULC, LST has been calculated for the same years with summer and winter separately. A series of calculations have been performed using the raster calculator in QGIS by following the Landsat handbook guideline to estimate the LST. On average, 1990 was found to be cooler than the other two years. Initially, the urban area in the west and few pockets of barren land in the north are found to be warmer than the surrounding agricultural tract. The LST pattern has been modified to some extent by the formation of the central industrial region and open pits at the north and eastern part in the next phase, which is quite clear from Fig. 12.6. Overall the maximum temperature in the summer and winter has been marked by the general rise of 2 to 3 °C. The pockets of higher temperature patches found to be located in the area experienced built-up growth. Hence LST pattern is not only marked by a temporal increase, but their spatial anomalies are also of special interest.

The final objective of the study is related to establish a relation between them. To fulfill these two approaches has been taken. Firstly, a series of three transect profiles have been drawn in such a way that they cover almost all types of LULC classes. Following the profile, the LST has been plotted for both the year. It has been found that built-up areas are warmer places than the surrounding agricultural area. These profiles also reveal that LST patter also modified in the area where agricultural and forest land has been converted into industrial or mining sites. Now it is necessary to examine this empirical relation with some statistical tool. Hence a second approach has been adopted to set a correlation between the LST and the various spectral indices, which indirectly reflect the LULC class. Accordingly, three spectral indices, i.e., NDVI, NDBI, and NDBal have been calculated, which reflects the vegetation cover, built-up area, and barren land, respectively. As expected, NDVI is found to be lower in the central and northern parts where industrial mining land use exists. Contrary to this, NDBI and NDBal is found to be greater in those areas. LST is found to be negatively associated with NDVI and positively associated with NDBI and NDBal. So from this outcome, it can be inferred that the LULC of the study area has been transformed to a great extent by the expansion of the built-up area in the form of settlement and mining-industrial growth, which in terms responsible for the microclimatic variation and UHI formation in the study area.

12.7 Conclusion

The study analyzed the major LULC change in the study area over the past two decades, which is mostly brought about by anthropogenic factors. This irreversible change in the land use pattern modifies the surface heat budget of the study area manifests in the form of UHI formation general rise in the LST etc. Over time agricultural land and forested area decrease alarmingly and are mostly replaced by the built-up area in the form of industrial development and opening of the mining pits. They contribute significantly to the general rise of the LST, which is measured to be an almost 3 °C rise in the average surface temperature. The longitudinal profile created over the various land use classes also shows that built-up and mining areas are comparatively warmer than the surrounding agricultural or forest patch. Even the LST pattern is found to be modified to a great extent over the areas where original forest or agriculture has been replaced by the built-up area. Considering the dominant type of land use, three major spectral indices, i.e., NDVI, NDBI, and NDBal have been used to explore the relationship between these indices and the LST. The former is found to be negatively associated with the LST, and the latter two indices have been found to be positively related to the LST. So it is quite clear that decrease in the forest cover and increase in the built-up area and barren land are aggravating the problem of UHI formation. Eventually, there is a general consensus between the rising pace of land transformation and LST pattern, which can be proved from the fact that the area that suffered from medium to large scale land transformation forms the worst patches where the LST has been found to be significantly changed.

To conclude, this study represents an unique example of UHI formation in relation to land transformation. Assemblage of several categories of land use has been found to be exists within this study and their propensity of changes is quite fast. Various agents of land transformation including expansion of urban area, formation of industrial cluster, increasing number of open pits coal mines has made the area vulnerable to environmental degradation. Degradation of forest landscape shrining of agricultural land increase in the mine derilict land in the study area may have serious environmental consequences along with the formation of UHI. Any type of management practice to nullify these impacts is extremely necessary. In this context, the present study area may be useful for undertaking any action research plan for restoration of the sustainability of the whole landscape.

References

Ahmed B, Kamruzzaman Md, Zhu X, Rahman S, Choi K (2013) Simulating land cover changes and their impacts on land surface temperature in Dhaka Bangladesh. Remote Sens 5:5969–5998

Ali SB, Patnaik S, Madguni O (2017) Microclimate land surface temperatures across urban land use/ land cover forms. Global J Environ Sci Manage 3(3):231–242

Asmiwyati AR (2016) Impact of land use change on urban surface temperature and urban green space planning; Case Study of the Island of Bali, Indonesia. https://espace.curtin.edu.au/bitstream/handle/20.../242463_Asmiwyati%202016.pdf?. Retrieved 26 Jan 2018

Barsi JA., Barker JL, Schott JR (2003) An Atmospheric correction parameter calculator for a single thermal band earth-Ssensing instrument. Retrieve from https://atmcorr.gsfc.nasa.gov/Barsi_IGARSS03.PDF

Bonfiglio A, Cuomo V, Lanfredi M, Macchiato M (2002) Interfacing NOAA/ANHRR NDVI and soil truth maps for monitoring vegetation phenology at a local scale in a heterogeneous landscape of Southern Italy. Int J Remote Sens 23(20):4181–4195

Bornstein R, Lin Q (2000) Urban heat islands and summertime convective thunderstorms inAtlanta: Three case studies. Atmos Environ 34(3):507–516

Buyantuyev A, Wu J (2010) Urban heat islands and landscape heterogeneity: linking spatiotemporal variations in surface temperatures to land-cover and socioeconomic patterns. Landscape Ecol 25:17–33

Census of India (2011) District census handbook Barddhaman.

Chatterjee S (2011) Source, Dispersal and Impacts of Airborne Pollutants: A Case Study of Mangalpur Industrial Complex. Raniganj. J Hum Ecol 35(3):195–201

Chatterjee S (2016) Spatio temporal transformation of land and its implication on the economy of asansol Durgapur planning region. Unpublished PhD thesis submitted to Visva Bharati university.

Chavez PS (1996) Image-Based Atmospheric Corrections - Revisited and Improved Photogrammetric Engineering and Remote Sensing, [Falls Church, Va.] American Society of Photogrammetry, 62, 1025–1036

Dhar RB, Chakraborty S, Chattopadhyay R, Sikdar PK (2019) Impact of Land-Use/Land-Cover Change on Land Surface Temperature Using Satellite Data: A Case Study of Rajarhat Block, North 24-Parganas District, West Bengal. J Indian Soc Remote Sens https://doi.org/10.1007/s12524-019-00939-1

Grover A, Singh RB (2015) Analysis of Urban Heat Island (UHI) in Relation to Normalized Difference Vegetation Index (NDVI): a Comparative Study of Delhi and Mumbai. Environments 2:125–138

Govind NR, Ramesh H (2019) The impact of spatiotemporal patterns of land use land cover and land surface temperature on an urban cool island: a case study of Bengaluru. Environ Monit Assess 191:283. https://doi.org/10.1007/s10661-019-7440-1

Jalan S, Sharma K (2014) Spatio-temporal assessment of land use/land cover dynamics and urban heat island of Jaipur City using satellite data. Int Arch Photogramm Remote Sens Spatial Inform Sci XL-8: 767–772.

Joshi R, Raval H, Pathak M, Prajapati S, Patel A, Singh V, Kalubarme M (2015) Urban heat island characterization and isotherm mapping using geo-informatics technology in Ahmedabad City, Gujarat State India. Int J Geosci 6(3):274–285

Liu L, Zhang Y (2011) Urban heat island analysis using the Landsat TM data and ASTER data: A case study in Hong Kong. Remote Sens 3:1534–1552

Li L, Tan Y, Ying S, Yu Z, Li Z, Lan H (2014) Impact of land cover and population density on land surface temperature: case study in Wuhan China. J Appl Remote Sens 8. 1117/1.JRS.8.084993

Lo CP, Quattrochi DA, Luvall JC (1997) Application of high-resolution thermal infrared remote sensing and GIS to assess the urban heat island effect. Int J Remote Sens 18:287–304

Mallick J, Kant Y, Bharath BD (2008) Estimation of land surface temperature over Delhi using Landsat-7 ETM+. J Ind Geophys Union 12(3):131–140

Mallick J, Singh CK, Shashtri S, Rahman A, Mukherjee S (2012) Land surface emissivity retrieval based on moisture index from Landsat TM satellite data over heterogeneous surfaces of Delhi city Int. J Appl Earth Obs Geoinf 19:348–358

Omran EE (2012) Detection of land-use and surface temperature change at different resolutions. J Geogr Inf Syst 2012(4):189–203

Pal S, Ziaul S (2016) Detection of land use and land cover change and land surface temperature in English Bazar urban centre. Egypt J Remote Sens Space Sci. https://www.researchgate.net/publication/ Accessed 12 Sept 2017

Peterson JCK (1910) Bengal District Gazetteer Burdwan. Bengal Secretariat Book Depot, Calcutta

Singh RB, Grover A, Zhan J (2014) Inter-seasonal variations of surface temperature in the urbanized environment of Delhi using landsat thermal data. Energies 7:1811–1828

Sobrino JA, Oltra-Carrió R, Jiménez-Muñoz JC, Julien Y, Sòria G, Franch B, Mattar C (2012) Emissivity mapping over urban areas using a classification-based approach: Application to the Dual-use European Security IR Experiment (DESIREX). Int J Appl Earth Obs Geoinf 18:141–147

Wang X, Guo W, Qiu B, Liu Y, Sun J, & Ding A (2017) Quantifying the contribution of land use change to surface temperature in the lower reaches of the Yangtze River. Atmos Chem Phys 17:4989–4996. http//www.atmos-chem-phys.net/17/4989/2017/ accessed on 12/2/2018

WBPDCL (2019).EIA&EMP for Tara (East) & Tara (West) Opencast Coal Mine Project. Kolkata

Weng Q, Liu H, Lu D (2007) Assessing the effects of land use and land cover patterns on thermal conditions using landscape metrics in city of Indianapolis, United States. Urban Ecosyst 10:203–219

Weng Q (2009) Thermal infrared remote sensing for urban climate and environmental studies: methods, applications, and trends (Review article).ISPRS. J Photogramm Remote Sens 64:335–344

Yue W, Xu J, Tan W, Xu L (2007) The relationship between land surface temperature and NDVI with remote sensing: application to Shanghai Landsat 7 ETM+ data. Int J Remote Sens 28(15):3205–3226

Xiao R, Weng Q, Ouyang Z, Li W, Schienke EW, Zhang W (2008) Land surface temperature variation and major factors in Beijing. China. Photogram Eng Remote Sensing 74(4):451–461

Zha Y, Gao J, Ni S (2003) Use of normalized difference built-up index in automatically mapping urban areas from TM imagery. Int J Remote Sens 24(3):583–594

Zhao HM, Chen XL (2005) Use of normalized difference bareness index in quickly mapping bare areas from TM/ ETM. In: Proceedings of 2005 IEEE international geosciences and remote sensing symposium, Vol 3, pp. 1666–1668. Seoul, Korea

Soumen Chatterjee is an independent researcher from Durgapur, West Bengal, India. He has been awarded a Ph.D degree from Visva Bharati University, India in 2016 for completing doctoral thesis on land transformation over Asansol Durgapur planning region. He mainly works in the field of urban and environmental issues for the past one decade and has published a number of research articles in several national and international journals.

Dr. Krishnendu Gupta is Assistant Professor of Department of Geography, Visva Bharati University, India and formerly senior project Scientist & group Head of State Remote sensing Centre, Department of Science and Technology, Government of West Bengal. He has vast academic and research excellence. His areas of interest are urban and regional planning, land degradation, applied geomorphology, and application of remote sensing. He has supervised and coordinated several research and technical projects in different time. More than 30 research articles have been published by him.

Chapter 13
Housing Conditions with Reference to Seepage in Different Residential Localities in Mumbai

Vidya Yadav, Arun Kumar Yadav, Rahul Mishra, and Ram Babu Bhagat

Abstract The monsoon winds bring humid weather and heavy torrential rain to Mumbai every year, causing considerable damage to buildings from water leaks. Meanwhile, loss of life is also experienced due to the collapse of buildings. These incidents are seen in the city center and suburbs. In Mumbai, people are forced to live in these dilapidated buildings, so their lives are always full of risk due to this. Thus, this study examines issues related to dilapidation and leakage in buildings using primary data collected from 450 households located in five different housing typologies in Greater Mumbai. Appropriate statistical techniques have been used to analyse the quantitative data, and qualitative data to explore deeper insights. Results indicate that seepage is a very common problem that is widespread across all seasons and housing typologies. However, it takes a serious form in the monsoon season. People living in dilapidated houses do not want to shift to transit tenants, which increases the risk of their lives. People with strong social ties in their area (especially those living in slums and the slum rehabilitation houses) do not want to leave these dilapidated buildings and relocate. The risk of their lives also increases. These residential structures experience heavy live loads. Providing emergency services in unplanned urbanization that includes slums and chawls also becomes a major challenge. Therefore, considering the meteorological aspects of the region, effective engineering practices of building construction and renovation work are urgently needed to ensure the safety of life and wealth.

Keywords Dilapidated buildings · Monsoon · Slums · Risk · Chawal

---

V. Yadav (✉)
P.G. Department of Geography, College of Commerce Arts & Science, Patna, Bihar, India

A. . K. Yadav
Evidence Action's Technical partner in India, Patna, Bihar, India

R. Mishra
International Institute for Population Sciences (IIPS), Mumbai, India

R. B. Bhagat
Department of Migration and Urban Studies, International Institute for Population Sciences (IIPS), Mumbai, India

M. Mishra et al. (eds.), *Regional Development Planning and Practice*,
Advances in Geographical and Environmental Sciences,
https://doi.org/10.1007/978-981-16-5681-1_13

13.1 Introduction

The art of constructing buildings has developed and enriched with the growth of human civilization (Sweansen et al. 2020). The housing design and the use of construction material change and modify over time with the alignment of the geographical factors (Housing and Urban Development 2017). However, irrespective of the type of building materials, any construction may get dilapidated in the absence of proper maintenance over the period by the effect of weathering agents (Dias 1994; Dias and Jayanandana 2003; Dias and Sivasubramaniyam 1989). Many times we come across such buildings which need major repairing before time (within the age of less than ten years. These are mainly due to poorly designed or constructed buildings and neglect of timely repair (Shinde and Arwikar 2018). The premature deterioration of structures creates a heavy burden not only on owners but also on the municipalities and the nation as a whole. The structures having a long life enables better utilization of natural resources.

Moreover, the buildings/structures should be adequately designed for resistance to natural calamities like earthquakes, cyclones, floods, etc. The geographical variation of landscape heavily influences the housing design and settlement typology in India (Bowen and Gleeson 2018). Moreover, the socio-economic status of the dweller directly affects the used materials of house construction (Yadav et al. 2018).

If one looks and recalls from a historical perspective, one can demarcate the year 1945, which shows the worldwide transition in building materials used (Dias 2013). Most commonly, structural steels used for beams and columns were substituted with reinforced concrete. The year 1975 is also the changeover time to recognize when there was a change in the cement manufacturing process, which results in the greater strength in concrete (Dias 2013). Over time, increasing awareness and importance were given to the durability of the structures. As the quality of building materials improved through the inclusion of waterproofing materials, the performance of modern structure was subsequently enhanced. Although, despite the improvement in construction material and technology, building longevity has decreased over time in a few areas due to large profit-making by the contractor and other stakeholders involved with it (Dias 2013).

The housing unit is not merely a material or physical object. However, also a setting for the work of the society, and indeed the activity of living itself. Moreover, dwelling constitutes the more significant part of the built environment, as that are the spaces where most people spend most of their time, and where probably the real work of society is done. 'Goal 11' of the sustainable development (SDGs) is also to 'make cities and human settlements inclusive, safe, resilient and sustainable' (United Nations 2020). Considering the housing specific target, this goal seeks to ensure access for all to adequate, safe, and affordable housing and basic services and upgrade slums. The 'Goal 11' represents an expansion in thinking about housing from millennium to sustainable development goal. In this context, dilapidated dwelling becomes an expensive metaphor for urban life in general (Thompson 2017). While looking into the statistics, a considerable number of households in India are in a congested and dilapidated situation, which necessitates un-congestion, up-gradation, and renovations (Tiwari and Rao 2016).

Empirically, a close association between housing, health, and wellbeing of people has been recognized (Garner 2006; Firdaus and Ahmad 2013) and it is evident that the miserable condition of housing in the city has always been in the center of attraction from planning and policies perspectives (Krieger and Higgins 2002). The census of India (2011) indicates that approximately 2 million households in India are in dilapidated condition. The Census of India definition measures dilapidation condition based on a self-rated technique which is influenced by the element of subjectivity. In India, Mumbai (37,995) has the second-highest number of dilapidated buildings after Delhi (62,317), which has developed in the various phases of history. Hence, along with the finest buildings, there are many old and dilapidated buildings (Tiwari and Rao 2016). These housing developments are a function of cities changing landscape. Therefore, it has been categorized into different housing typologies (Van Doorn et al. 2019).

There is scanty literature available on the subject of the expected service life of structures. The lifespan of RCC (Reinforced Cement Concrete) generally is taken as 100 years (Penttala 2009). RCC is a process when plain cement concrete, which is a mixture of cement, coarse aggregate, and fine aggregate, is reinforced with steel bars. RCC is used in structures to provide the tensile strength that the concrete is ineffective to handle. The service life of structures depends on a variety of factors, such as (i) their purpose, (ii) socio-economic considerations, (iii) materials of construction, (iv) surrounding environment, and (v) degree of maintenance (Dias 2003). However, there are some expected as well as prevalent conventions about life span. Monumental structures (temples, mosques, or churches) have long service life even up to 500 years or more, steel bridges, steel buildings, or similar structures bears up to 100–150 years, concrete bridges or high-rise buildings or stones bridges stand up to 100 years. In contrast, residential houses or general offices/commercial buildings lifespan is up to 50–60 years (Ahmad 2003).

The factors affecting the service life of structures can vary, not only from building to building but even within a given building. E.g., (i) The quality of the substructure and roof may vary (ii) the environment to which the building is subjected may also vary from seaward to the leeward side (Dias 2013). Moreover, various buildings receive different degrees of maintenance, depending on the accessibility and inspectability (Dias 2003). However, moisture, excessive heat, leakage of water in settlement, other chemical and biological factors cause cracks and seepage of water (Penttala 2009).

Every year with the onset of monsoon in June month, Brihanmumbai Municipal Corporation (BMC) or Maharashtra Housing and Development Authority (MHADA) declares the name of the most dangerous building in their purview area. The BMC considers 30 years as the shelf life for a building to be surveyed or audited for redevelopment (Services 1888). Generally, it is stipulated that if the age of a building is 15–30 years, the structural audit must be carried out once in 5 years and for 3 years for building older than 30 years (Suryawanshi et al. 2018).

Most buildings that were constructed between 1969 and 1996 in Mumbai are in dilapidated condition now. Due to a series of building collapses incident, more than 800 people have died in the city since 1971. Between 1971 and 2018, close to 3528 building accidents were reported, which is the highest in India (MHADA 2019). The dilapidated buildings are under the following categories by BMC: First, the C1 category is extremely dangerous, needing immediate demolition. C2A category needing evacuation and structural repairs/demolished. C2B category buildings require structural repairs when they are occupied. Furthermore, the final C3 category building denotes needing minor repairs (MCGM 2005).

Dilapidated buildings and their redevelopment-related operational problems are one part of the story in Mumbai. The densely populated city where quality and affordable housing is still a distant dream is the other side. Realtors often cut corners to make what is called substandard buildings and putting the buyers at risk. These circumstances create the situation of structural failure and ultimately falling the quality of housing (Penttala 2009). Even one can wonder how some structures in India are standing long, and some are collapsing within a few years. Typically building failures are categorized into two broad groups: (i) Physical (structural) failures which result in the loss of specific characteristics, e.g., strength; and (2) performance failures which means a reduction in the function below an established acceptable limit (Douglas and Ramson 2007). In India, most of the buildings collapse due to structural failures, such as the use of poor-quality raw material, bad design, faulty construction, foundation failure, and extraordinary loads (Mahadik and Jaiswal 2014). Further, haphazard growth, increasing concrete city, heavy rainfall during monsoon, failure of early warning system before monsoon, and efflorescence on inner building elements, as well as water influx through the building shell increases the possibility of building failure (Bhagat et al. 2006; Gautam et al. 2016; Mansur Hamma-adama 2017).

However, the irony of the city is that despite declaring the buildings as dangerous and not fit for possession, yet lakhs of families residing in these crumbling houses and are not willing to vacate the premises.

> People tend to feel fear that they will not be allowed to go back to their homes once if they shift into transit camps (Resident of chawl, Bandra, male, 72 years). This probably shows the trust deficit of residents in the authorities. Recently, the Bombay High Court (HC) ordered that tenants of dilapidated buildings must shift out and cannot demand transit camp accommodation. Though the tenants are promised alternative houses or rents, but once the inhabitants vacate, the developers give them a hard time. The tenants are meant to be housed in an alternate building by the developers. And the developers and builders are expected to pay rent to the tenants. In many cases, they fail to pay rent (Resident of Bandra, Chawl, Female, 46 years).

However, in several cases, if residents shifted to transit camps then they might have to forced to live there for decades (Resident of Bandra, Chawl, Female, 46 years) as redevelopment projects got stuck due to reasons such as the delay in redevelopment projects are identified as bureaucratic delays in getting clearance for the projects, fight between landlords and tenants, or builders and tenants over related terms and conditions of redevelopment. Moreover, it has also been seen that housing profoundly affects a broader aspect of family and community life and wellbeing (Jiboye 2014).

It connects people from each other and forms a broad network of social capital. So, despite their dissatisfaction with the dilapidated house, people staying in the same place for long duration are satisfied with their life confounding within their neighbourhood (Galster 1987). Likewise, residents are willing to risk their lives rather than to shift to MHADA or transit houses. The study is examining the issues related to the dilapidation and leakage of buildings in Greater Mumbai.

13.2 Data and Methods

The present study is carried out in Greater Mumbai. Household-based primary data was collected from three different zones, namely, Bandra, Dadar, and Chembur (Fig. 13.1). In each area, a landmark was fixed to create a buffer zone of 5 km aerial distance to collect the desired sample. The sample collection was done from five different localities (housing typologies), which differ in-terms of built forms and socio-economic attributes of the dwellers. Thus, to capture the variation across localities, High-rises building households (HRBH), Old Colonies, Chawls, Slums, and Slum Rehabilitated Houses (SRH) were selected.

This study used cross-sectional self-reported data gathered using *quota sampling* method from 450 households of three selected areas (150 households from each). In each area further from five different localities, the samples were collected by fixing the quota of 90 household samples. These 90 households were identified with the help of a sampling frame (Fig. 13.2) of house lists. It was prepared during the initial field visits and pilot survey. The fixing of sample quota helps to avoid high non-responses (particularly from a high-rise building and old colonies, which appears during pretesting of a questionnaire) to retain the desired number of samples, besides keeping in mind the time and money constraints. On the second level, buildings that were 5 years or older were mapped. Third, random sampling was used to select households for conducting the survey. Respondents were the head of the household, and in their absence, any adult aged 18 years and above were interviewed.

A walk-through observation of the structure and the condition reported by the respondent was recorded to assess the condition of the building. The closed-ended questions were translated from English to Hindi. It was pre-tested on 30 households to check the suitability of the instrument. Later, necessary modifications were completed before conducting the main survey. The period of data collection was six months between January and June 2016. After the fieldwork was completed, the data was entered using CSPro 6.3 software. Bivariate and multivariate statistical techniques were performed using STATA 14.1 to assess the objectives of the study. For collecting qualitative information on different issues related to building seepage, 5 Key informants interviews (KIs), 5 In-Depth interviews (IDIs) in each locality, and 4 Focus Group Discussions (FGDs) were performed. All the guidelines and protocols were followed. Later, the qualitative data were analysed using ATLAS.Ti 5.0 software.

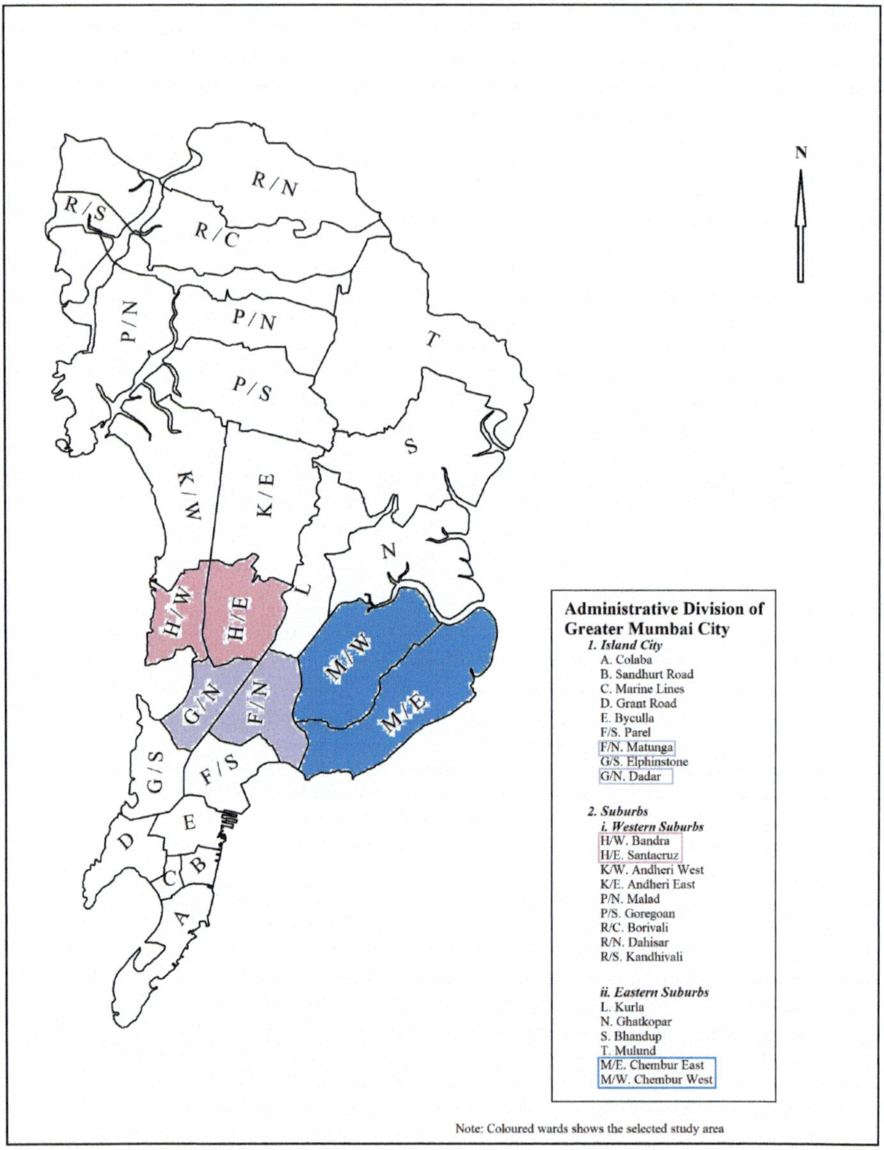

Fig. 13.1 Administrative divisions in Greater Mumbai (*Source* Brihanmumbai Municipal Corporation BMC)

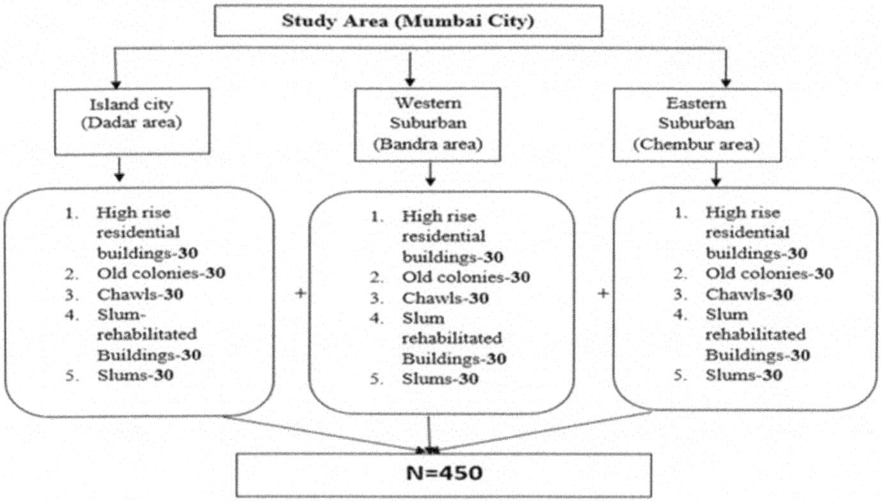

Fig. 13.2 Procedure for sample collection (*Source* Authors' perception)

The study was conducted with the approval of the Students Research Ethics Committee (SREC) of the International Institute for Population Sciences (IIPS), Mumbai, India. The ethics committee examined the methodological, technical, and ethical aspects of the study. In addition, before conducting the interviews, informed consent was also taken from all the participants.

13.3 Operational Definitions

Slums: Slum refers to the residential areas where dwellings are unfit for human habitation by reasons of dilapidation, overcrowding, faulty arrangements and design of such buildings, narrowness or faulty arrangement of the street, lack of ventilation, light, or sanitation facilities or any combination of these factors which are detrimental to the safety and health (Census 2011) (Fig. 13.3).

High-rise residential buildings houses (HRBH): High-rise residential buildings refer to tall buildings specially structured for residential purposes. In India, a building greater than 75 feet (23 m), generally 7–10 stories, is considered as high-rise. The *Municipal Corporation of Greater Mumbai (MCGM)* proposed that any building with a height of 30 m (nine floors) be categorized as high-rise. It is basically a multi-dwelling unit. It may be owned or rented by tenants. We considered those buildings having a minimum of nine floors as HRBH. The dwelling unit here is considered as flats or apartments, which is part of the tall building structure. The dwelling units here can access water, bathroom, and toilet facilities of their own within the households (Fig. 13.4).

Fig. 13.3 Picture of a typical slum area in Bandra zone (*Source* Authors)

Fig. 13.4 A snapshot of high-rise building in Chembur zone (*Source* Authors)

Chawls: A chawl is a type of building found in Mumbai. They are often 1–2 or sometimes 3-storied with about 10–20 tenements, referred to as *kholis*, which literally mean room on each floor. A usual tenement in a chawl consists of one all-purpose room that functions both as a living and sleeping room. Families on each floor have or may not have to share a common block of toilets (Fig. 13.5).

Old colonies: Here, old colonies refer to the colonies which established long back. This mainly refers to those colonies which are created by the people who have

Fig. 13.5 Picture of a chawl in Matunga zone (*Source* Authors)

migrated long ago. For our study purpose we will select one among the following colonies—Parsis/Sindhi/Gujarati's/Telgu's as they are considered among one of the oldest settlers of Mumbai (Fig. 13.6).

Slum Rehabilitated Authority Houses (SRH): The Government of Maharashtra has launched a comprehensive slum rehabilitation scheme by introducing an innovative concept of using land as a resource and allowing incentive floor space index (FSI) in the form of tenements for sale in the open market, for cross-subsidization of the slum rehabilitation tenements which are to be provided free to the slum-dwellers (Fig. 13.7).

Personal relationship: This particular category includes question-related to relation with relatives, friends, and satisfaction with family. It captures how closely a household connected with its relatives and friends. It also considers the frequency of meeting with them.

Social network support: This is the category most commonly associated with social capital. The questions here consider the nature and extent of a household member's feel supported and seek advice and suggestion from relatives, neighbours, and friends. Financial assistance received or given to relatives, neighbours, and friends have also been captured here.

Civic engagement: This category includes question-related to the participation of households in various types of social organizations and the range of contributions that one gives and receives from them.

Trust and cooperative norms: In addition to the canonical trust question asked in a remarkable number of cross-national surveys, this category seeks to procure data on trust towards neighbours/friends. It also asks questions related to the sense of community household feels in the locality and whether females feel safe at the locality.

Fig. 13.6 An old colony at Dadar zone (*Source* Authors)

Fig. 13.7 A slum rehabilitated house in Chembur zone (*Source* Authors)

13.4 Results

13.4.1 The City of Ancient Buildings

The city has over thirty thousand residential buildings, which are older than at least 30 years. Of these, more than ten thousand buildings are about 70 years old or more (MHADA 2017). Figure 13.8 shows the average age of building in different localities. The localities consist of Chawl has the highest concentration of old buildings followed by old colonies and slums buildings. It has also been found that the buildings of the chawls, slums, and old colonies have already crossed the benchmark set by MHADA for redevelopment (refer to Table 13.1).

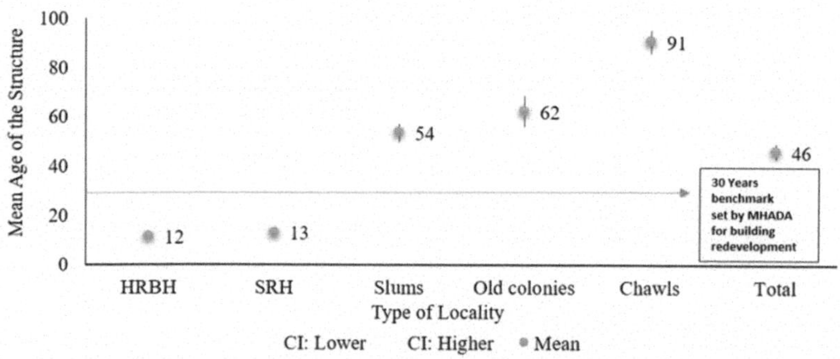

Fig. 13.8 Mean age of the house by locality, Mumbai 2016 (*Source* Primary data: Jan–Jun, 2016, Mumbai)

Table 13.1 Percentage of housing categories by the benchmark of redevelopment age, Mumbai 2016 (N = 450)

Locality	Fit structure (below 30 years)	Aged structure (30 years and above)
SRH	100.0	Nil
HRBH	97.8	2.2
Old colonies	34.4	65.6
Slums	16.7	83.3
Chawls	Nil	100
Total	49.8	50.2
N	**224**	**226**

χ^2 310.8, Pr = 0.000

Note Fit structure refers to those buildings which are less than 30 years old, and Aged structures are those building which are more than 80 years old

Source Primary Data: Jan–Jun, 2016

Table 13.2 Percentage of houses affected with seepage with rain, Mumbai 2016 (N = 176)

Locality	House not affected with seepages	House affected with seepage
Old colonies	93.3	6.7
HRBH	72.2	27.8
Chawls	65.6	34.4
Slums	41.1	58.9
SRH	32.2	67.8
Total	**60.9**	**39.1**
χ^2 91.3, Pr = 0.000		

Source Primary Data: Jan–Jun, 2016

13.4.2 Seepage Affected Households

Rain seepage is the most widespread and less understood source of moisture. Apart from water seepage/leakage from walls and roofs in Mumbai during monsoon. The problem of seepage is further compounded by Mumbai's high humidity and abundant rainfall. Water seepage through external walls is mainly due to joint and cracks in the walls, penetrating dampness, rising dampness due to capillary action, poor construction practices, separation gaps, honeycombing, etc. In this context, data pertaining to whether households face seepage problems or not has been collected. The study result indicates that around 176 households among 450 households in the study area reported having a problem of seepage (Table 13.2).

Irrespective of any weather term or housing typologies, seepage is a common problem everywhere. But the severity of seepage is high among monsoon time. In spite of newly constructed slum rehabilitated buildings, seepage is high. Seepage, cracks, and fungus are also found to be high in slums due to the precarious condition of the houses. High-rise houses, chawls, and old colonies too reported about seepage issue.

> There are many chemical solutions available in the market which can deal with the problem, if it is done properly during the construction stage, leakage and seepage problem can definitely be avoided (Resident of Bandra, HRH, Male, 34 years).

13.4.3 Association Between House Age and Seepage

Building decay is very closely associated with rain seepage. Table 13.3 tries to show the relationship between dwelling age and the effect of seepage. Our study suggests that seepage is not necessarily concerned with old buildings, rather also prevalent among new buildings. The study highlights that 53% of residential structure which is affected with seepage were below 30 years. At the same time, 47% of the residential structure are 30 years or more. Interestingly it was found that HRH and SRH were

Table 13.3 Percentage of houses affected by seepage by the benchmark of redevelopment age and, Mumbai 2016 (N = 176)

Locality	Below 30 Years	30 years and above age structure
HRBH	100	Nil
SRH	100	Nil
Slums	15.1	84.9
Chawls	Nil	100
Old colonies	Nil	100
Total	**53.4**	**46.6**
N	**94**	**82**
χ^2 148.7, Pr = 0.000		

Source Primary Data: Jan–Jun, 2016

also reported about seepage problem, even though they do not cross the benchmark age of redevelopment. Qualitative results also show the concerned issue.

> For someone who is paid-up hard-earned money to buy a home, "Why should the walls be damp, or the ceiling leak" (Residents of Matunga, Slum Rehabilitated Houses, Female, 45 years).

The correlation between house age and seepage problem has been assessed. Generally, it is underestimated that the old buildings are more prone to seepage. But our study depicts that seepage was strongly correlated with newly constructed buildings also. The SRH, normally which are 5–7 years old, do even face seepage. However, other localities were also positively associated with seepage except in high-rise households (Table 13.4). Irrespective of housing age, seepage remains one of the devastating problems.

> Actually no builder wants to build something shoddy or shabby, simply because it affects his reputation, and that has a direct effect on future business. But the problem of seepage is one of the persistent issue (Old colony, General Secretary, Male, 62 years)

Table 13.4 Correlation coefficient of house age and seepage by residential categories, Mumbai 2016

Type of localities	Correlation coefficient (r)
HRBH	−0.01
Slum	0.05
SRH	0.62
Chawls	0.17
Old colonies	0.04
G. Mumbai City	−0.09

Source Primary Data: Jan–Jun, 2016

13.4.4 Background Characteristics of Dwellers Residing in Seepage Affected Household

The ownership of the house and the respondent's migration status don't have a significant role in the households' seepage condition. Table 13.5 shows the ownership and migration status of the locality type by seepage-affected household. In general, it is the notion that inflow of migrants to Mumbai city and persistence of high standard of living usually forced people to stay in sub-standard housing condition and adjust to that situation (Bhagat 2017). However, the findings here are contradicting to the notion. The fact is that the locals are more adjusted in the seepage-affected house in each locality compared to the migrant household. Similarly, the owner is usually settled in seepage-affected houses compared to the renter, except in the chawls locality. Here, 80% seepage affected households are occupied by renters.

In the context of above Table 13.5, which shows the ownership of household and migration status, few qualitative insights from the study were cited. It clearly depicts that residents especially staying in chawls lack homeownership, and they don't even know who their owner is, as most of the chawl-owners have shifted abroad. Few chawl-dwellers are even staying there free of cost as now there is no owner to pay the rent. So in this situation, residents got accustomed to receiving notices from municipal corporations for the redevelopment. Currently, in spite of staying in the dilapidated buildings, they don't want to shift (refer to Fig. 13.9).

> We don't even know who is our owner (landlord), as this house is our ancestral house. The majority of residents here even do not have ownership of their rooms as it was passed by their relatives, most of them, not even blood relations, but all are staying there for many years now. Most of the owners are difficult to trace as they have moved to other countries, and few are no more. In this situation, authorities are bound to send or issue notices to occupants. Not surprisingly, the residents are accustomed to receiving this type of notices (46 years, Male, Chawl residents).

> My uncle (maternal) was bachelor, was residing alone in the small single room in pagadi system in chawl which was purchased by my nanaji (Grandfather) in 1947. My uncle passed away a few years back. He had a ration card and light bill and all other related documents

Table 13.5 Percent of the house affected by seepage by selected characteristics, Mumbai 2016 (N = 176)

	HRBH	Slums	SRH	Chawls	Old colonies	Total
Migration status						
Local	52.0	41.5	78.7	64.5	100.0	61.9
Migrant	48.0	58.5	21.3	35.5	0.0	38.1
Ownership of house						
Owner	100.0	60.4	82.0	12.9	0.0	63.1
Renter	0.0	39.6	18.0	87.1	100.0	36.9
N	**25**	**53**	**61**	**31**	**6**	**176**

Source Primary Data: Jan–Jun, 2016

Fig. 13.9 Notice by MCGM in a dilapidated building alerting the unsafety to stay (*Source* Authors)

> in his name. We have all original receipts of the room purchased from the owner (Not the landlord), which never got transferred to nanaji or uncle. The current landlord of the chawl is the son of the second purchaser owner landlord of the chawl. He uncle chawls second landlord had stopped taking rents some 25 years back. My mother is the only sister, and my uncle has nominated my mother for the above room in chawl (25 years, male, Chawl resident).
>
> Till 2012, we have paid the rent by cheque, but the landlord is not giving us the receipt for the rent paid. Post-2012, the landlord is not ready to accept the rent nor giving the rent receipt. Now, how important is the rent receipt to prove the ownership of the house that we also don't know (35 years, female, chawl residents).

On the other hand, many social issues stem from the history of poor quality of housing. Earlier research has also tried to show that the connection between housing and education and the existing causal relationship between these variables. For instance, good housing conditions facilitate better scholar results or good education results give better housing opportunities? It has also been proved that good quality of housing may be considered a result of educational success (Alexiu et al. 2010). The review of scientific literature also explored the relationship between housing conditions, education, and social exclusion for vulnerable groups (Somerville 1998) and found that a low level of education relates to disadvantaged housing conditions (Alexiu et al. 2010).

Similarly, our study also found this result, respondents having seepage houses and having a background of no schooling belong mainly from slums and Slum Rehabilitated Houses (poor housing condition) (Table 13.6). Similarly, in the context of seepage-affected households with employment type result shows that most of the seepage-affected household heads are employed in private and self-employed work (70%, N = 123) (Table 13.7).

Despite households having high seepage and dilapidated housing conditions, residents still want to continue and risk their lives in a hazardous situation. Pertaining to this, the study examines the contribution of the building type to the lifestyle and

Table 13.6 Percentage of seepage affected household by education, Mumbai 2016 (N = 176)

Locality	No school	1–9 year	10–12 year	more 12 year
HRBH	0.0	0.0	48.0	52.0
Slums	15.1	54.7	26.4	3.8
SRH	13.1	26.2	52.5	8.2
Chawls	0.0	19.4	48.4	32.3
Old colonies	0.0	0.0	0.0	100.0
Total	**9.1**	**29.0**	**41.5**	**20.5**
N	**16**	**51**	**73**	**36**

Source Primary Data: Jan–Jun, 2016

Table 13.7 Percentage of seepage affected household by employment type, Mumbai 2016 (N = 176)

Locality	No employment	Private and self-employed	Public/Government
HRBH	76.0	4.0	20.0
Slums	5.7	86.8	7.6
SRH	13.1	78.7	8.2
Chawls	9.7	83.9	6.5
Old colonies	66.7	33.3	0.0
Total	**21.0**	**69.9**	**9.1**
N	**37**	**123**	**16**
$\chi^2 = 76.2$, Pr = 0.00			

Source Primary Data: Jan–Jun, 2016

culture of the people living in seepage-affected clustered houses (Slums and Chawls). The interview with residents reveals that social support networks and mixed uses within the neighbourhood are critical in the lives of residents. On the other hand, the building form of chawls and slums contributes substantially to the close-knit social life of the residents. The qualitative result also depicts that social connectivity is the foremost reason for the dwellers to live in hazardous residential buildings.

13.4.5 Building Live Loads and Seepage Problem

The loads on a building, maybe a residential or any type of structure, are mainly classified into vertical and horizontal loads. The vertical loads are the dead and live loads. The horizontal loads are the earthquake and the wind loads. This is the general classification of loads. Here we are not dealing with horizontal loads but will only concentrate on the vertical aspect.

Live loads are temporary loads that are either movable or moving loads without any acceleration or impact. These are also called as imposed loads and are changeable and dynamic. These include loads such as vehicle traffic, the occupants of the building, including weights of movable partitions or furniture, etc. The intensity of these loads may vary depending on the time of the day.

For instance, a residential building may experience increased live loads during the morning or night and weekends but much smaller loads during the afternoon and working hours as most of the members may move outside for work. For example, in gym building, usually, there are increased live loads at morning and evening time, but smaller loads in the afternoon time as people will be at work. These loads keep on changing from time to time. These loads are assumed accordingly by the one who designs the building. It is considered to be one of the major loads in the design.

Table 13.8 indicates the vertical live loads through the mean of residential density in terms of population. Study result indicates that households with a high mean (2.22) of vertical loads are more affected with seepage problem, compared to a households with less vertical loads that have less or no seepage problem (1.20). However, this result is not consistent at the locality level. Slum households show higher seepage households with more vertical loads followed by chawls, slum rehabilitated houses, old colonies, and high-rise households.

Table 13.9 shows the differential in the high level of social capital utilization in the seepage of affected houses. It found that there was no locality that consistently utilizes a high level of social capital parameters. However, dwellers from high-rise building houses have a comparatively high level of social capital utilization.

Table 13.8 Mean vertical live loads by seepage affected household, Mumbai 2016 (N = 450)

Locality	No seepage		Seepage	
	Mean	SD	Mean	SD
HRBH	0.35	0.06	0.37	0.07
Old colonies	0.30	0.07	0.42	0.04
SRH	0.73	0.07	0.43	0.15
Chawls	0.41	0.10	0.47	0.15
Slums	1.18	0.60	1.07	0.37
Total (Overall)	**1.20**	**1.41**	**2.22**	**1.80**

Note **Live loads** are temporary loads that are either movable or moving loads without any acceleration or impact. Including the weight of movable partitions, person or furniture etc.
Source Primary Data, Jan–Jun, 2016

Table 13.9 Percent of seepage affected household by social capital, Mumbai 2016 (N = 176)

Locality	Personal relation	Family support	Civic engagement	Trust and corporation
HRBH	76.0	80.0	80.0	72.0
Slums	43.4	47.2	86.8	52.8
SRH	19.7	8.2	27.9	9.8
Chawls	19.4	35.5	77.4	45.2
Old colonies	50.0	66.7	16.7	83.3
Total	**35.8**	**36.9**	**61.4**	**40.3**
N	**63**	**65**	**108**	**71**
χ^2	29.9, Pr = 0.000	46.2, Pr = 0.000	55.4, Pr = 0.000	42.3, Pr = 0.000

Source Primary Data: Jan–Jun, 2016

Table 13.10 Ranking of factors by respondents for residing in seepage affected hazardous building

Rank	Reasons
1	Social connectivity
2	Location at city centre
3	No/cheap rents
4	Similar caste group
5	Common seepage problem everywhere

Note Based on self-rated response gathered in qualitative information
Source Primary Data, Jan–Jun, 2016

13.5 Main Ascertained Factors to Reside in Seepage Affected Houses Based on Qualitative Findings

Factors were determined which forced dwellers to reside in seepage-affected houses. Social connectivity is cited as the foremost important factor, followed by location of household in the city centre, and no/cheap rents (Table 13.10).

> I am a landlord of a 50-year-old building that has Pagdi[1] Tenants. There is no rule that favours us, and so at the moment, we get some 100 to 150 rent for a 250 carpet area room from our tenants. We have to chase for property tax. We have to request that nobody parks cars inside (very small compound), the building is in dilapidated condition. We did a BMC authorized audit, and they gave us a letter that the building is endangered. But nobody willing to vacate. Half of the tenants keep the room locked and stay elsewhere, but when we ask them to vacate, they don't.

> We stay in a Pagdi system for more than 25years. 7-8years earlier, our landlord has changed. 3 years ago, my dad (under who's the room belongs) expired. So landlord has stopped taking

[1] Pagdi system: The pagdi system ensures the tenant that the rent for the property remains nominal despite the rise in inflation or any other market fluctuations. Many areas in South Mumbai have Pagdi rented flats, wherein tenants pay rents as low as Rs. 500 for a property with market rates going as high as Rs. 60,000 or more.

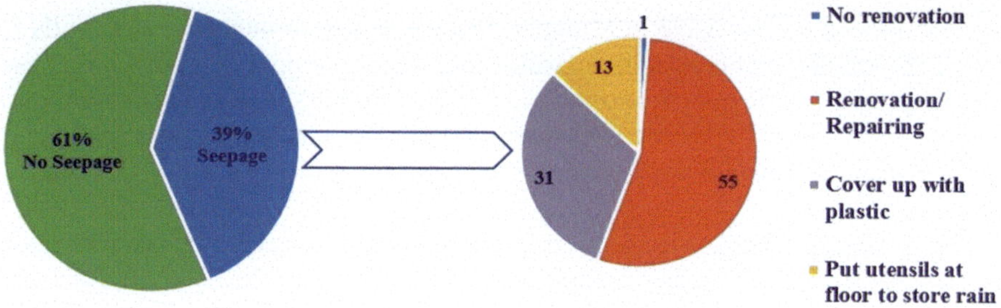

Fig. 13.10 A coping mechanism for seepage issue (*Note* Self-reported by respondent) (*Source* Primary Data: Jan–Jun, 2016)

> rents. Our room had leakages from upper floors, and all paint was coming out. It was causing some health issues to my mom as plaster Crust (papdi) from the wall caused breathing issues. So we decided to do plaster on that area and fix it by putting tiles over it so that there would be no more leakages. The work started in the morning, and in the afternoon, it was stopped as people from the landlord said that the landlord's permission is not taken. We were not doing any structural changes and were just doing a plaster of the wall. By that time, the outer portion of the old plater was taken out. Landlord's people have taken photographs of half work done. Now that the work has stopped, the condition of the house is worse (32 years, male, chawl residents, Bandra).

With the arrival of the monsoon season, Mumbai receives more than 300 mm rainfall in a day which causes leaking roofs, wet patches, and cracked walls. It has been found that 176 (39%) of the household in the study area are facing seepage issues (Fig. 13.10). Most of the households dealing the leakage through the painting of house, renovation, repairing, and few put utensils on the floor to store rainwater, use plastic to cover the top to restrict water to percolate inside the house.

> There are a lot of temporary applications available in the market, or we can do with ourselves, but this all application can address seepage only for 6 months to 2–3 years (Slum and Chawls, resident).

The Maharashtra Housing and Development Authority (MHADA) had declared more than 14,000 buildings in dilapidated conditions. In this situation, shifting occupants to transit or other buildings is a complicated and sensitive issue. Most of the respondents added that they leave those transit tenements due to lack of facilities (FGD in Chawls, Dadar Zone).

13.6 Discussion

This study shows the interrelationship, factors, and correlates those force dwellers to risking their lives by residing in seepage-affected houses. Chawl localities are one of the highest locations for the concentration of old buildings in Mumbai. Rain seepage is a common problem in Mumbai. Irrespective of any housing typologies or weather, the term seepage is pervasive in Mumbai. Water seepage through external walls is mainly due to joint and cracks in the walls, penetrating dampness, rising dampness due to capillary action, poor construction practices, separation gaps, honeycombing, etc. High wind speeds experienced by high-rise buildings also increase the likelihood of water seepage. Despite newly constructed slum rehabilitated buildings, seepage is high.

The ownership of the house and the migration status of the respondent doesn't have a significant role in the seepage condition of the household. Mostly slums, chawls, and slum rehabilitated houses are in seepage-affected situations. In this context, it is essential to address the situation of chawls, where the tenants are running in pagdi system. Chawl residents even do not have ownership of their rooms as it was passed by their relatives, most of them, not even blood relations, but all are staying there for many years now. Most of the owners are difficult to trace as they have moved to other countries, and few are no more. In this situation, authorities are bound to send or issue notices to occupants. Not surprisingly, the residents are accustomed to receiving this type of notices.

Secondly, meager rents received by the chawl owners vary between Rs. 500–1500/- in a month. Some of the households even don't pay as owners are difficult to trace. They have shifted to some other country, which is also one of the most contributing important reasons for poor maintenance of residential structures. The rents received by the chawls landlords located in the city centre is less than expected. Due to the poor condition of chawls, many were converted to slums as well by the government. Further, people residing in the seepage-affected household are with lower educational levels face a greater risk of exclusion from housing policies. Other factors such as lack of knowledge of financial issues related to housing are also vital. Despite households having high seepage and in dilapidated condition, residents still want to continue and risk their lives in hazardous situations. The building form of chawls and slums contributes substantially to the close-knit social life of the residents. The qualitative result also depicts that social connectivity is the foremost reason for the dwellers to live in hazardous residential buildings. In many cases, it is also noticed that old buildings in the city area (especially Chawls) which were earlier used for residential purposes but have now been changed to commercial, storage room, or business use. Those structurally old bearing type beams and flooring now cannot bear loads only because the purpose of their utility has changed.

13.7 Conclusion

The study has identified the reasons for vulnerability in people's lives living in dilapidated buildings across different housing typologies in Greater Mumbai. Due to the long rainy season, many structures across the city become life-threatening for its residents. Seepage is a common problem irrespective of housing typologies. Apart from old buildings, the seepage problem is widespread in newly constructed buildings as well. Residents of slums and slum rehabilitated houses were found to be more susceptible to life and standard of living. Unplanned settlements, low quality construction materials and high live loads also hamper emergency services in the region. People also do not want to relocate it due to excessive linkage and dependency with their existing areas. Thus relocating the people in transit tenements is a complicated and sensitive issue. It is desired to bring an effective arrangement for people living in these localities to make their lives safer and healthier. These arrangements should not only focus on the physical aspects of the living arrangements (i.e., structure planning, constructing material) but also considering the social aspects of the dwellers.

References

Ahmad S (2003) Reinforcement corrosion in concrete structures, its monitoring and service life prediction: a review. Cement Concr Compos 25(4–5):459–471

Alexiu TM, Ungureanu D, Dorobantu A (2010) Impact of education in terms of housing opportunities. Procedia Soc Behav Sci 2(2):1321–1325. https://doi.org/10.1016/j.sbspro.2010.03.194

Bhagat R (2017) Migration, gender and right to the city. Econ Polit Wkly LI I(32):35–40

Bhagat RB, Guha M, Chattopadhyay A (2006) Mumbai after 26/7 deluge: issues and concerns in urban planning. Popul Environ 27(4):337–349. https://doi.org/10.1007/s11111-006-0028-z

Bowen WM, Gleeson RE (2018) The evolution of human settlements: from pleistocene origins to anthropocene prospects. Springer, USA

Dias WPS (1994) Structural appraisal of reinforced concrete buildings from in-situ material properties- some issues and insights. Transactions, Institution of Engineers, Sri Lanka, pp 129–145

Dias WPS (2003) Useful life of buildings. University of Moratuwa, Moratuwa, June. http://www.slaasmb.org/USEFUL-LIFE-OF-BUILDINGS.doc. Accessed March 2012

Dias WPS (2013) Factors influencing the service life of buildings. Eng J Inst Eng Sri Lanka 46(4):1. https://doi.org/10.4038/engineer.v46i4.6801

Dias WPS, Jayanandana ADC (2003) Condition assessment of a deteriorated cement works. ASCE J Perform Constr Facil 17(4):188–195

Dias WPS, Sivasubramaniyam S (1989) Assessment of floor slabs in the Bandaranaike Wing of the Colombo General Hospital. Engineer (Sri Lanka). September, pp 27–36

Douglas J, Ramson WH (2007) Understanding building failures. Taylor & Francis

Firdaus G, Ahmad A (2013) Relationship between housing and health: a cross-sectional study of an urban centre of India. Indoor Built Environ 22(3):498–507

Galster GC (1987) Duke University Press. Homeowners and neighborhood reinvestment

Garner G (2006) The ecology and inter-relationship between housing and health outcomes. In: International conference on infrastructure development and the environment, pp 1–21

Gautam D, Rodrigues H, Bhetwal KK, Neupane P, Sanada Y (2016) Common structural and construction deficiencies of Nepalese buildings. Innov Infrastruct Solut 1(1):1–18. https://doi.org/10.1007/s41062-016-0001-3

Jiboye AD (2014) Significance of house-type as a determinant of residential quality in Osogbo, Southwest Nigeria. Front Archit Res 3(1):20–27. https://doi.org/10.1016/j.foar.2013.11.006

Krieger J, Higgins DL (2002) Housing and health: time again for public health action. Am J Public Health 92(5):758–768

Mahadik AB, Jaiswal MH (2014) Structural audit of buildings. Int J Civil Eng Res. ISSN 2278-3652

Mansur Hamma-adama TK (2017) Causes of building failure and collapse in Nigeria: professionals' view. Am J Eng Res 6(12):289–300

MCGM (2005) Brihanmumbai Mahanagarpalika Section 4 Manuals as per provision of RTI Act 2005 of 'N' Ward

Penttala V (2009) Causes and mechanisms of deterioration in reinforced concrete. In: Failure, distress and repair of concrete structures. Woodhead Publishing, pp 3–31

Services L (1888) Section 353 - B. In: Mumbai Munic. Corp. Act. https://www.lawyerservices.in/Mumbai-Municipal-Corporation-Act-1888-SECTION-353-B-Structures-Stability-Certificate-1. Accessed 11 March 2021

Shinde GS, Arwikar SJ (2018) A review on 'Structural audit of residential building'

Somerville P (1998) Explanations of social exclusion: where does housing fit in? Hous Stud 13(6):761–780. https://doi.org/10.1080/02673039883056

Suryawanshi SS, Vishe VV, Sah DP, Sharan R (2018) Structural audit of RCC building. Int J Adv Res Ideas Innov Technol (IJARIIT), ISSN 1370–1374

Sweansen A, Chang, Pao C (2020) Construction building. Encyclopædia Britannica, inc.

Thompson M (2017) Contesting 'dilapidated dwelling'. In: Cairns G, Artopoulos G, Day K (eds) From conflict to inclusion in housing: interaction of communities, residents and activists. UCL Press, London, pp 199–226. https://doi.org/10.2307/j.ctt1xhr55k.19

Tiwari P, Rao J (2016) Housing markets and housing policies in India

Van Doorn L, Arnold A, Rapoport E (2019) In the age of cities: the impact of urbanisation on house prices and affordability. In: Nijskens R, Lohuis M, Hilbers P, Heeringa W (eds) Hot property. Springer, Cham

Yadav V, Bhagat RB, Yadav AK (2018) Neighbourhood characteristics and residential satisfaction: a study of greater Mumbai City. socialspectrum.in 4(4):191–204

Dr. Vidya Yadav is working as an Assistant Professor with the P.G. Department of Geography in College of Commerce, Arts & Science, Patna (Patliputra University), Bihar. She obtained her Doctoral degree in Population Studies from the International Institute for Population Sciences (IIPS), Mumbai. She holds an M.A. degree in Applied Geography from Ravenshaw University, Cuttack, and a Masters in Population Studies from IIPS. Her research interest primarily focuses on urban issues, which include urban planning, migration, living conditions of slum dwellers, and environmental concerns. Apart, she has been publishing and presenting many of her research works on various national and international platforms.

Dr. Arun Kumar Yadav is currently working as a State Programme Coordinator (M&E) EAII advisor, Evidence Action's technical partner in India. Before this, he was working as an Assistant professor at the Central University of Karnataka, Kalaburagi. He obtained his doctoral degree in Population Studies from the International Institute for Population Sciences, Mumbai. He was also associated with the Institute for Economic Growth, Delhi for a short period, where he actively contributed to various research activities. His research area is mainly inclined towards livelihood, migration, urbanization, and public health issues. He jointly published five research papers.

Dr. Rahul Mishra obtained a doctoral degree in population studies from the International Institute for Population Sciences, India. Afterwards, he worked as a senior project officer in the fifth round of the National Family Health Survey. Later, he joined the Institute for Economic Growth in Delhi, where he worked as a junior consultant in the Comprehensive National Nutrition Survey. The project was led by the Ministry of Health and Family Welfare in collaboration with the United Nations Children Fund, and it was the first national representative nutrition survey of children and adolescents in India. In both these national-level projects, he carried out various types of responsibilities. He has jointly published six research papers and is presently inclined towards researching the health and living conditions of the vulnerable population.

Dr. Ram Babu Bhagat is a Professor and Head, Department of Migration and Urban Studies, International Institute for Population Sciences (IIPS), Mumbai, India. His research interest includes migration and urbanization; demography, ethnicity, and politics analysed through the lens of space and place as a critical framework of analysis. He has served as Consultant to UNESCO-UNICEF (India Initiative on Migration), International Organization of Migration (IOM), and Population Division, United Nations. Prof. Bhagat was a member of the IUSSP panel on Demography of Armed Conflict, 2006–09 and the Co-ordinator of ENVIS Centre on Population and Environment funded by the Ministry of Environment and Forests, Government of India from 2007–12. He was a coordinating lead author of the migration chapter of Hindukush Himalayan Monitoring and Assessment Programme (HIMAP) at the International Centre for Integrated Mountain Development (ICIMOD), Kathmandu. He has also been a project team member of the 'IMAGE Project: Comparing Internal Migration Around the Globe' located at Shanghai University. His recent books incudes 'Climate change, vulnerability and migration', published by Routledge 2018 and 'Migration and urban transition: A development perspective', is also published by Routledge in 2020. He is currently the Chief Editor of Demography India- An Official Journal of IASP.

Part V
Challenges and Options of Sustainable Livelihood

Chapter 14
Livelihood of the Displaced: A Study on Selected Areas of Bangladeshi Immigrants in West Bengal, India

Utpal Roy and Subham Pramanick

Abstract Migration is one of the significant socio-cultural, economic as well as political issues determining the demographic pros and cons of a region. From the time of independence of India (1947) to the period of termination of the Bangladesh liberation war, i.e., in 1971, the flow of cross-border Bangladeshi migrants was very high and a continuous process. Even after the independence of Bangladesh, this infiltration was not stopped, and it has been a critical discourse even in the recent past. In this study, different investigative insights are incorporated in order to analyse the ex-post-facto livelihood strategy of mainly and solely the Bangladeshi immigrants in India under unfamiliar and distinct social, economic, and political context. It also reveals the comparison of adopted strategies of living in respect to before and after migration. A significant change in education, societal status, political participation, cultural assimilation is prominent. The problems experienced during living by the immigrants in both places are different in nature, i.e., religious discrimination at place of origin and socio-economic and political issues in place of destination. The present chapter attempts to evaluate the situation in which they have been living in an unfamiliar social and cultural set-up at the present place and notice the temporal dynamicity of those uprooted people.

Keywords Discrimination · Secularism · Citizenship rights · Cultural assimilation · Dynamicity

14.1 Introduction

Migration is the attempt of maneuvering from one location (i.e., source) to other location (i.e., destination) with the purpose of staying at the destination permanently or living for a longer or shorter period. In the widest sense, human migration can be of two kinds—Internal Migration and International Migration. The former one means the human movement exactly within the political boundary of a country, whereas

U. Roy (✉) · S. Pramanick
Department of Geography, University of Calcutta, Kolkata, India

M. Mishra et al. (eds.), *Regional Development Planning and Practice*,
Advances in Geographical and Environmental Sciences,
https://doi.org/10.1007/978-981-16-5681-1_14

the latter one signifies the trans-political boundary migration stream, i.e., from one nation to other one. In regard to India, the instance of international migration, which is most concerning for a prolonged time in the politico-economic and social fabric, is the cross-border migration of Bangladeshi migrants in the nearby Indian state of West Bengal, which has extensive boundaries with Bangladesh of more than 2216 km. The root of the issue could be traced back to India's memoir of Independence in 1947 when the Dominion of Pakistan was born by dividing India. Pakistan had two parts; East and West. The occurrence of partition has imprinted a deep-seated wound on the political trajectory in India. According to the United Nations High Commissioner of Refugee (UNHCR 2000) calculation, 14 Million Hindus, Sikhs, and Muslims were uprooted at the time of the partition, and it was taken into account as the biggest human migration in the saga of humankind. Dutta (2005) opined that the partition of Bengal in 1947 heralded forced irregular migration from erstwhile East Pakistan. It is also estimated by her that, about 15 million Bangladeshi nationals live in India without having a proper citizenry identity. According to Boyd (1989), international migration stemmed from political conflicts; frequently exists minimally illustrated through economic modeling. Family connections and other social networks perform a significant pivotal role in migration decisions, including the choice of place of destination. Although, Entorf (2002) presents substantial proof of unlawful networks, deployment of irregular migrants in proscribed executions such as smuggling functions. There is a lot beyond economics and 'push–pull', which has been well implacable, as suggested by Radu and Straubhaar (2012) in explaining international migration flow, especially the undocumented one.

In 2010, the Bangladesh Supreme Court restored secularism (which was once abolished in 1977) as one of the constitution's basic tenets, but Islam remained the state religion (Kumar 2011). Most of the citizens of Bangladesh observe various secular festivals at different times throughout the year. The ethos of secularism in South Asia is in many ways different from that of Western versions that assert complete separation of church and state. Rather, it is the freedom of individuals to practice the faith he or she desires without being subject to any form of state or communal discrimination (Zakaria 2015). This declining trend supports the 2nd postulation under the head of volume of migration as put forward by Lee (1966), i.e.,*'the volume of migration varies with the diversity of people'*. As most of the existing Hindus faced a feeling of minority, their rate of exile from the native land is increased (Kumar 2011).

Broader push and pull factors are held responsible for the cross-border movement of Bangladeshi immigrants in the state of West Bengal (Datta 2005). During the field study in 2019, the authors empirically observed that various factors among the remarkable economic push factors motivate people to leave Bangladesh. The major are *instability and economic depression, poverty, lack of employment opportunity, struggle for livelihood, forced grabbing of landed property from the minority group, and lack of industrialization* in Bangladesh.

Among the demographic factors, population explosion in Bangladesh and the poor Human Development Index (HDI), according to Sharat and Madan (1995), can be the most vital reason for irregular immigration from Bangladesh to West

Bengal in India. Besides, Sengupta (2015) rendered Educational curricula liable, which were structured at par with Islamic preaching and reduction of provisions enjoyed by Hindu minority communities, for the irregular migration of the Hindu minority population. Social insecurity is another cause here. Besides, many other factors have been identified by Roy (2003) as well as by PTI (2008), like the political imbalance, trepidation of riots and terrorism in Bangladesh, inhuman attitude and functions of the political leaders, non-existence of democratic rights, Muslim religions domination, religious instigation by political leaders, insecurity feeling of Hindus are the vital issues that operate as political push factors. Ever since the partition of the subcontinent and the formation of East Pakistan and later Bangladesh, their nationals came to India with or without valid documents. Kumar (2011) asserted that over 4.7 million Hindus had sought refuge in India, mostly in West Bengal, up to 1971. The volume of the opposite flow of Muslims to East Pakistan is, however, unknown. Numerous people returned back after the Liberation War in 1971, but a substantial undocumented portion, as traced by Guha Roy and Sen (2002), stayed and assimilated with the mainstream of India's life. This massive undocumented migration imparts a grave danger, according to Nath (2003), to our national security, social harmony, and economic well-being. Undocumented migration from Bangladesh has been a great concern in the border areas of districts in West Bengal, a state in India. The antisocial activities in the border area have been disharmonizing everyday life and posing a security threat to the nation. Banerjee (2003) speculated that the Hindu families in this border area of Nadia district are displaced twice, once due to the partition of Bengal and secondly due to the conflict issues between Muslim and Hindu irregular migrants. In this context, researches have been done by Das and Ansary (2018) to interrogate the status quo of both the inter-state and Bangladeshi immigrants by means of occupational engagement and availability of amenities, along with the factors and reasons for migration. Thus it is a very robust fact to state that the districts which act as the buffer between India and Bangladesh face most of the menaces of the international migration from Bangladesh to India.

In the world context, Global North (Economically developed) and Global South (Economically under-developed/ developing) are the two transitional global economic divisions (although this division closely contradicts the geographic northern and southern hemisphere) separated by *Brandt Line*. Global South to Global South migration possesses a vital place. However, this field has not been investigated and analysed under the umbrella of policymaking and planning policies. Rather a traditional view that the mainstream international migration tends to be from Global South to Global North is till now persistent. Joseph and Narendran (2013) asserted that increased migration streams among developed and developing nations in the last few decades have modified static boundaries globally. When it comes to cross border migration, Hujo and Piper (2007) demonstrated that migration to the North/relatively wealthy countries continues to be the focus of attention of policymakers, academics, and media, although about one-third of the global migrant stock both originated and was living in the Global South. This number was slightly smaller than the number of international migrants originating in the South and living in the North. At par with the estimate by Ratha and Shaw (2007), merely about 40 percent of all migrants from

Table 14.1 Trend of change of Hindu population to total population in Bangladesh (1901–2011)

Year	Percentage	Year	Percentage
1901	33.00	1961	18.50
1911	31.50	1971	13.50
1921	30.60	1981	12.13
1931	29.40	1991	10.51
1941	28.00	2001	9.02
1951	22.05	2011	8.96

(*Source* Census of Bangladesh 2011)

developing countries dwell in wealthy Organization for Economic Co-operation and Development (OECD) nations, 13 percent dwell in wealthy non-OECD countries, like Saudi Arabia, the United Arab Emirates, Hong Kong Special Administrative Region (SAR) and Singapore, whereas nearly 47 percent dwell in other developing nations. The share of South-South migrants in total emigrant counts stands typically large in the poorest nations. What is particularly interesting is that 80 percent of those flows take place between neighboring nations. This shows great potential in investigating this topic.

Here in this chapter, the overall livelihood strategies of the immigrant families will be investigated in a retrospective and prospective manner, highlighting the present situation of them interrogating and thereby associating the cause-effect satire that is significant to this arena. The present analysis has shown that a considerable Hindu population, which had generated a significant shortfall observed in Bangladesh, has migrated to India due to religious issues followed by unemployment and poverty at the source. Ironically, their present status of livelihood is not much better in the destination as they have added a surplus population in an already heavily populous country. Their socio-cultural match with the aborigines of destinations is roughly balanced, which sometimes gets disturbed by biased political interference. The respondents feel better in India regarding inferiority comparison, and interestingly, they do not even dare to think to compare their life in a superiority comparison.

The percentage of the Hindu population has continuously decreased in Bangladesh for the last 110 years (Table 14.1 and 14.2). On the contrary, the emergence of a high proportion of the Muslim population has been observed between 1975 and 2010. This indicates a havoc decline in the Hindu population in Bangladesh. A substantial count

Table 14.2 Population share by major religious groups in Bangladesh (1975 to 2010)

Religious Groups (In Percentage) / Year	1975	1990	2010
Muslim	84	87	89.5
Hindu	15.6	12.4	9.5
Christian	0.1	0.1	0.3
Buddhist	0.3	0.5	0.7

(*Source* Census of Bangladesh 2011)

of these uprooted people has taken shelter in the Indian buffering districts adjacent to Bangladesh–India international boundary, i.e., Canning-I and II, Basanti-I and II, and other adjacent Community Development Blocks (CD Blocks) of South 24 Parganas and Bashirhat-I, Hasnabad, Bongaon, Habra, and some other neighbouring CD Blocks in the North 24 Parganas district. A counterpart justification can be given with the help of population growth data adopted from the Census. For instance, West Bengal exhibits a decadal growth rate (2001–2011) of 13.93 percent as per the Census of India 2011. However, the growth rate of Canning-I and Canning-II is 19.72 percent and 22.40 percent, respectively, as per Census of India 2011. Moreover, the Basirhat-I and Hasnabad CD Block exhibit growth rates of 16.16 percent and 14.50 percent, respectively (Census of India 2011). Astonishingly, the average population growth rate in Bangladesh is 2.4 percent as of 2010. Simultaneously, the neighbouring Bangladeshi district Satkhira exhibits a growth rate of 6.50 percent (Census of Bangladesh, 2011). All these statistical figures validate the fact that indeed there are some issues of immigration in India, especially in the buffering districts of West Bengal, which is principally sourced by the emigration from Bangladesh. Irregular migration puts a threat to the cultural identity and internal security of the country.

As part of its one of the salient security measures, it is significantly mentioned by Chung (2010) that Japan made fingerprint registration compulsory for the foreigner and the Koreans settled there for three generations or more in succession have been exempted from that. This was considered as recognition where security is ensured through citizenship. In the case of international labour migration, it was investigated by Martin (2004) that non-recognition of civic rights frequently results in agitations and partitions. However, Ghosh (2012) conducted a study in Indian context and found that, after accommodating the refugees of the 1971-war, India's approach of managing Bangladeshi migration was not initiated for bestowing citizenship. They obtained elector's identity without being lawfully naturalized.

Several journals, articles, books, and internet sources have been consulted before the detailed investigation of the Bangladeshi immigrant families' adaptation strategies in some villages of South 24 Pargana District. This facilitates a wide array of mutually related issues and facts about the topic. Mashihur Rahaman Khan (1972) in his paper *'Pattern of External Migration to and from Bangladesh, 1901–1961'* opined that there was a significant change over in the flow of external migration from Bangladesh during 1901–1961. Haakon Lein (2000), in her paper *'Hazards and 'forced' migration in Bangladesh'*, presented the fact that poverty and landlessness in rural Bangladesh led to migration to and settlement in disaster-prone areas along the main rivers or low-lying areas in the Bay of Bengal. According to her, people living in these areas experience loss of land, property, and income opportunities due to natural disasters; they are forced to seek their livelihood and housing in urban areas. To explain the role of any migrant stream in the regional conflicts, the historical and ethnographic accounts of such migration and settlements as well as interviews with recently repatriated migrants, are most relevant.

Moreover, the current theories and opinions of frontier or state relations are also very significant. Ammarell (2002) found many instances where the migrants have frequently acted as the factor of assimilation in connection with existing economically strong communities. Migration has contributed to the richness in the diversity of cultures, ethnicities, and races in developed countries. Individuals who migrate experience multiple stresses that can impact their mental well-being, including the loss of cultural norms, religious customs, social support systems, adjustment to a new culture, and changes in identity and self-concept. Indeed, the rates of mental illness are increased in some migrant groups. Bhugra and Backer (2005) emphasized the need of the awareness of Psychologists to the unique stresses and cultural aspects that impact immigrants and refugees to well meet the requirements of this rising and vulnerable population. The large influx of Hindu migration from Bangladesh to India is still a burning issue in contemporary politics and economics, and other disciplines. The present trend is showing an increasing Hindu migration from Bangladesh to India because the situation of Bangladesh is responsible for the displacement of Hindus from Bangladesh. A projection estimated by Barkat (2008), exhibits that no Hindus will be left in Bangladesh after 30 years from present day if the current rate of 'exodus' continues as a mean value of 632 people from the minority community abandon the Muslim-majority nation every day. Livelihood strategies act as one of the prime factors of motivation in migration. Besides, aspirations and norms also help to dictate the choice of migration in preference to other strategies.

White (2009) analysed that the boom phase of educational attainment promulgates the younger cohort of the society to maneuver for better professional engagement scope and consequently well livelihood facility. The influence of physical factors is also not negligible. For example, climate change-induced migration cannot be attributed as a failure to adapt but as a strategy to increase household resilience. But in this instance, at par with Kartiki (2011), the migration is usually planned and supported and not due to distress. Environment and Political Economies exert a profound influence on the livelihood strategies and thus the pattern of migration. Between the national and household factors, community or 'meso-level' changes in political economy and livelihoods in southwestern Bangladesh illustrate that to understand the impacts on people and nations of climate change-related environmental changes—changes that are expected to include rising sea level, saline inundation, and increased likelihood and intensity of cyclones in Bangladesh—we need to understand the dynamics of the built and natural environment and the political economies this sustain. Ackerly et al. (2015) analysed the impact of Micro-level political economies upon the sources of income and livelihood available in distressed environmental situations, and thus effect how better the people in them can adapt to altering environmental situations.

14.2 Study Area

For the study and analysis of Bangladeshi immigrant families' adaptation strategy in selected areas of the South 24 Parganas district, villages of Canning-I and Canning-II

Community Development Blocks are chosen. The reasons for the selection of two CD Blocks are that,

- *Firstly,* both are located at the extreme south of the district with proximity to the India–Bangladesh international border.
- *Secondly,* the authors' pilot survey and primary investigation revealed that these two blocks have the highest concentration of residing Bangladeshi immigrant families in the district. Moreover, here some families are found who had settled earlier and some others have migrated in recent times.

14.2.1 About Canning I and II Blocks

Since 1781, the substantial clearance of forest in this area started (Kumar 2011), and subsequently, a huge amount of indigenous as well as people from Bangladesh (erstwhile East Pakistan) moved and migrated, respectively (Kumar 2011). Moreover, some recent stream of migration has been observed here due to some socio-economic or political pitfalls.

These places are chosen due to many-fold reasons. As of the Census of India (1991 2001, and 2011), these two CD Blocks show the highest population growth rate in 1991–2001 and 2001–2011 decades. Between 1991 and 2001, the population growth rate of Canning-I and Canning-II was 19.76 percent and 22.62 percent, respectively. In 2001–2011, these figures were 19.72 percent and 22.40 percent, respectively. The district's average population growth rate was 15.81 percent in 1991–2001 and 14.86 percent in 2001–2011 (Census of India 2011). Satkhira, which is the bordering district of Canning-I and Canning-II at the Bangladesh part, exhibited a growth rate of 6.50 percent (Census of Bangladesh 2011). Indeed, Bangladesh has never introduced a strict government policy that can reduce its population growth by controlling the fertility rate. This evidence hits a question of the certain one (i.e., migration) of the three vital population attributes (i.e., Fertility, Mortality, and Migration). It is also true to mention that the selected study area will represent the overall mosaic comprehensively. Table 14.3 shows the overall trend of immigration from Bangladesh to the State of West Bengal, especially in North and South 24 Parganas districts.

Although it is a temporally declining trend, still the immigration rate is by no means negligible. For example, the difference in the percentage of immigrants in the context of West Bengal and North and South 24 Parganas districts for 1991–2001 and 2001–2011 is 7.37 percent and 9.42 percent, respectively. It indicates a significant increase in the differential percentage in the two decades.

14.3 Database and Methods

This study is based on the primary data which has been collected from *Canning-I* and *Canning-II* CD Blocks (Fig. 14.1). At first, as a pre-field study, a pilot survey

Table 14.3 Percentage of Immigrants from Bangladesh to West Bengal and North & South 24 Parganas

Year	West Bengal scenario	N & S 24 Parganas Scenario
	Percentage of Bangladeshi Immigrants	Percentage of Bangladeshi Immigrants
1961–1971	21.51	40.29
1971–1981	24.8	33.44
1981–1991	24.66	25.42
1991–2001	23.5	30.87
2001–2011	21.54	30.96

Source Adopted from Chandrakantan (2014)

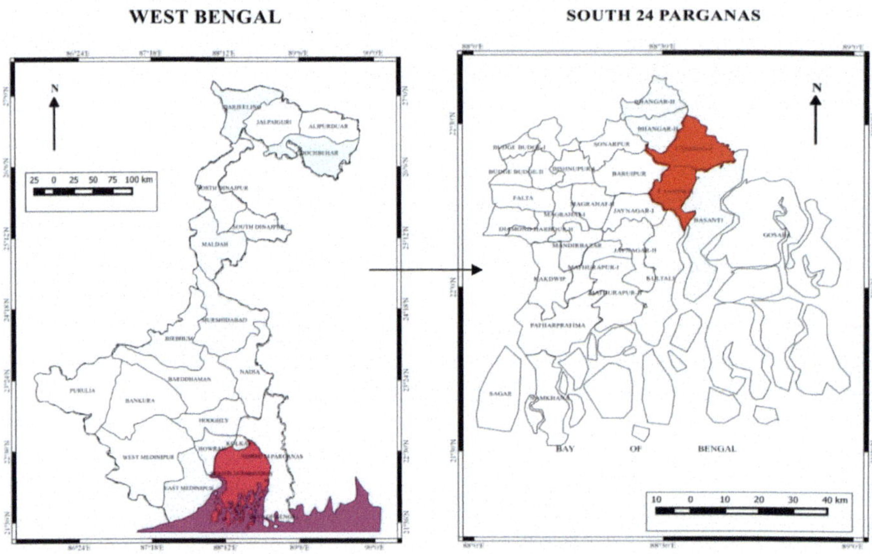

Fig. 14.1 Location map of the study area (*Source* Census of India 2011)

was done to make a survey plan and target the required samples and take necessary authorization. After that, the study was conducted based on a purposive sample study. Such 291 samples (families) were selected through purposive sampling and surveyed to fulfill the targeted goals. After the data collection was over, as a post-field action, they were first tabulated in their respective format in the SPSS data editor. The appropriate analysis has been done, including necessary diagrams both in SPSS and MS-Excel to place the findings and conclusion (Fig. 14.2). The location map has been prepared in Q-GIS 2.18 (Las Palmas) software.

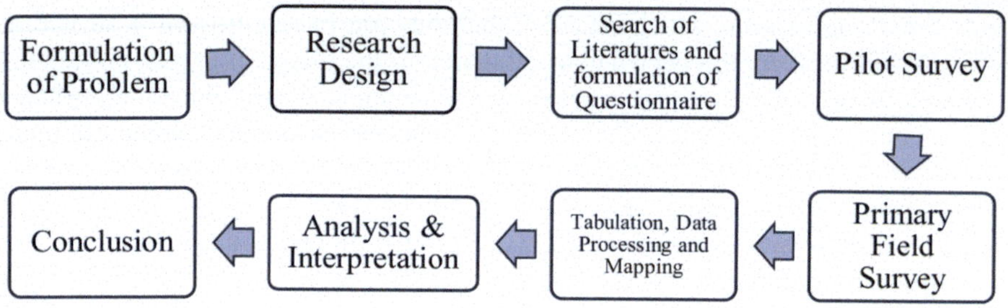

Fig. 14.2 Flowchart showing the research framework (*Source* Compiled by authors)

14.4 Objectives of the Study

The present study attempts to envisage a detailed investigation of the Bangladeshi immigrant families presently residing in India to understand the principal causes of the immigration process. The study has extended to examine the forces of attraction toward destinations and detraction at sources. Besides, the authors record the problems at the present place of living and the adopted strategies in detail. The study assesses the migration processes and adaptation strategies and cross-validated it with other established models to conclude whether this migration process agrees with any established pattern or present a unique pattern itself.

14.5 Conceptual Framework

The people who ring alarm bells about Bangladeshis reckon that there might even be 20 million of them working in India, though that calculation seems to be on the high side. Nevertheless, Tumbe (2018) opines that the 20-million figure has no methodological foundation pinning it up. Based on the Census, he estimated that the count of people born in Bangladesh and dwelling in India declined from 3.7 to 2.7 million between 2001 and 2011 (though, this is a partial figure as many people who came across in 1947 and 1971 had died). Moreover, Sarkar (2001) explored that declining trend was of pan-India and Bangladeshi immigration is hugely kinship-oriented. Thus, it can be inferred that the immigrants always have a socio-cultural and economic retrospective base and presently holds a new socio-cultural and economic base based on the former one.

In the beginning, the poor Hindu migrants had tried to take shelter in their relative's house or nearby. During their field survey, the authors have noticed that the economically poor migrants had to struggle a lot for their livelihood, citizenship, and permanent shelters. The key reason for which the political parties protect the irregular migrants is, according to Nandy (2019), that the illegal migrants are considered potential vote banks that need to be nurtured and sustained. Primarily they facilitate the stay of the irregular migrants in India by providing them with Voter Identity

Cards and Ration Cards, which establish that they are Indian citizens with voting rights. Whenever there has been a demand to deport the illegal migrants, politicians of almost all affiliations protest against it. Above all, it is very uncommon that a 'poor' country like India has set herself as the receivers of undocumented migrants from other poorer countries. Instead, in the apt words of Ramachandran (2005), commonly they should be the senders of 'Asylum-seekers', 'irregular' migrants, and professional emigrants destined for more prosperous countries in the West.

In the present paper, three fundamental and inevitable dimensions have been considered to grasp the overall scenario of livelihood strategies, i.e., *social, economic, and political* backbone. In the case of the uprooted people, who have migrated from their native country (i.e., Bangladesh, a country in the Global South) to the immediately neighboring country (i.e., India and other countries in the Global South), the analysis from such a viewpoint deserves a massive significance to understand the overall cause–effect relationship between these two countries. Moreover, these three arrays of human life also help to compare the pros and cons of those people's livelihood patterns easily in their previous habitat along with the present one. This aspect will break the stereotype viewpoint in international migration research, i.e., migration streaming from Global South (i.e., Less Developed Countries) to Global North (i.e., More Developed Countries). For making the analysis understandable, these three broad heads along with another head named 'other' are further sub-divided into related attributes. The logic has been applied here for considering the selected variables that are generally chosen to measure the living status and challenges of migrants at the place of arrival. Although other variables could have been effective in this discussion, principal determinants are selected for assessing the overall condition of living of those newcomers at the present place of living. It could have been better if other secondary variables were chosen in understanding the current living standard and related issues. For avoiding the debatable and controversial aspects, some selected variables have been considered here. These are as follows:

- **Social Variables**: Reason for Migration, Relationship with the place of origin, Average family size, Social coherence, and conflict with neighbourhood and Problems at the present living place.
- **Economic Variables**: Remittance and Financial support, Right to landholding and other assets' account, Professional engagement.
- **Basic Right Related Variable**: Citizenship rights.
- **Others**: Age-Sex Composition and educational profile.

These variables are sorted in orderly manner so that they could explain and illustrate the objectives in an arranged as well as composite manner. Social, Economic, Basic rights-related, and other variables are selected to express the overall view of the factors of immigration processes and interrogate the respondents' adaptation strategies.

As stated, the interrogation of these facts will be stated from the aspects of popular migration theories and models, i.e., Ravenstein (1885), Lee (1966), and Norris (1972).

14.6 Social Variables

'*Social variables*' indicate those that are utilized for the experiment of an overall societal condition endowed specifically to analyse a certain social issue from a broader perspective.

14.6.1 Reason for Migration

The psychological trauma of fleeing their homeland and leaving family members behind can also complicate refugees' adjustment to their new environment. Very after the Independence of the Indian Union in 1947, the two other countries took birth from erstwhile undivided India, namely, East Pakistan and West Pakistan. Until 1971, the tug of war remained unresolved for both the Pakistans because there was a problem of long-distance and biased attitude of the nodal administration, i.e., West Pakistan towards East Pakistan. This was due to the religious contradiction between the two distinctive religion-based communities residing in East and West Pakistan. Ultimately this problem remained the prime factor to generate other factors of disturbance and led towards mass emigration. Sharma (2015) found in her study that most of the migrants who came to Assam have come from mostly the closely proximal located parts of Bangladesh, i.e., Rangpur and Dhaka. The present study finds that most of the study area's immigrants belong to Khulna and Barishal in respect to their native origin (31.80 percent) followed by Satkhira (15.80 percent). Other places include Bagherhat, Rajshahi, Comilla, Faridpur, and Dhaka. This trend more or less closely fits with some of the propositions devised by Ravenstein (1885):

> We have already proved that the great body of migrants only proceeds a short distance and that there takes place consequently a universal shifting or displacement of the population, which produces 'currents of migration' setting in the direction of the great centers of commerce and industry which absorb the migrants…. The inhabitants of the country immediately surround a town of rapid growth flock into it. (Ravenstein 1885, p. 198).

Here, most migrants are from Khulna, Barishal, and Satkhira of Bangladesh, which lies respectively at a distance of approximately 150, 240, and 80 km by road. The other places that contribute small figures are at an average distance of more than 300 km from the study area.

14.6.2 Nature of Migration

The nature of migration depends on several factors such as the flow pattern of migrants, reasons for the move, the number of people moving, and time spent for migration. The present study observes the occurrence of different forms of migration—individual, couple, single-family migration, and joint family migration (Table

Table 14.4 Family type of the migrants

Family type (In Percentage)			
Individual	Couple	Single family	Joint family
7	9	58	26

(*Source* Field Survey by the authors in 2019)

14.4). The single-family migration is observed with the highest priority, involving 58.60 percent to total migrants, and the individual migration counts for the least—only 4.50 percent of the total. The joint family and couple migration has occurred respectively for 23.1 percent and 7.90 percent of the total migration cases in the study. The leading cause underlying this trend of priority to single-family migration is to accept the higher chance to mobile themselves in the small unit rather than in a larger extent of big families. There is a visa problem in the case of many families migrating together where it is easy to get a visa for a single-family. Mobilizing in a single-family is easier than families with many members. This trend aptly supports Ravenstein's (1885) statement, who stated that most of the migrants usually retain their large family in native land or take his atomic family (i.e., Father-Mother–Offspring/s) in maximum. (Atomic family).

14.6.3 Reasons for Migration from Source

This section is formulated keeping in mind the basic tenets of the model of Lee (1966). Lee (1966) emphasized 4 (Four) specific factors based on which a migration stream is generated—*firstly*, 'factors associated with the area of origin'; *secondly*, 'factors associated with the area of destination'; *thirdly*, 'intervening obstacles' and *fourthly*, 'personal factors'. Many identified factors regulate the flow of migration (Datta 2005). The present study reveals that the majors are having lack of jobs, the homo-ethnic climate in destination, political turmoil, etc. In this section, the first factor is analysed in a comprehensive aspect. Reasons or factors for the immigration in India from Bangladesh or, say, the emigration from Bangladesh holds various factors ranging from natural hazards to human-induced violence (Tumbe 2018).

About 50.70 percent of the surveyed Bangladeshi immigrants (Table 14.5) in India came due to religious issues both during pre and post-independence. 18.50 percent of the surveyed immigrants opined unemployment as the prime cause for their migration. 11.60 percent, 5.50 percent, and 4.10 percent of migrants came due to insecurity, lack of job opportunity, and due to flood in Bangladesh's coastal region, respectively. Minority status in the source and better job prospect in the destination stands as the second and third reason for their migration. In a juxtaposed manner, this result sharply coincides with some of 'the disutilities that act as push factors on the potential migrant', as elaborated by Norris (1972, pp. 296–298). These disutilities are 'Religious Persecutions, Political Persecutions, Racial or Social Persecutions, Economic Crises, and Temporal Changes'. The reasons which show the prime dominance, i.e.,

Table 14.5 Reasons for migration

Reasons	Percentage of immigrants		
	First reason	Second reason	Third reason
Storm Surge	1	1.7	3.4
Flood	4.1	5.8	3.8
Soil Salinity	1.4	2.4	1.7
Crop Failure	1.4	3.1	1.4
Insecurity due to Minority Status	11.6	19.9	8.6
Feeling of Outside the Mainstream	1.7	4.8	4.8
Religious Issues	50.7	12	5.1
Unemployment	18.5	18.8	15.1
Lack of Job Opportunity	5.5	18.5	14.4
Better Future Prospect in Destination	2.7	6.2	21.6
Better Wages at Destination	0.3	3.1	11.6
River Bank Erosion	0	0	0.3
Others	1	0	1

(*Source* Field Survey by the authors in 2019)

high percentage in first reason (Insecurity due to minority status, Religious Issues, and Unemployment) comes under the broad head of 'Religious Persecutions', 'Political Persecutions', and 'Economic Crises'.

14.6.4 Year of Migration

The Bangladeshi cross-border immigration in India is a long-time phenomenon since the Independence of India and is still ongoing. Although the intensity of different reasons changed over time, the gross outcomes remained the same, i.e., mass-scale immigration in India. Samir Guha Roy (2003) of the Indian Statistical Institute called these estimates 'motivated exaggerated'. After examining the population growth and demographic statistics, Roy (2003) states that while a vast majority are irregular immigrants, significant amounts of internal migration are sometimes falsely thought to be immigrants. An analysis of the numbers by Roy (2003) revealed that around 91,000 Bangladeshis irregularly crossed over to India every year during 1981–1991.

Among the samples studied, the fact that came out of the analysis is that most migrations have taken place after the Independence of Bangladesh. Before this event, the average migration was less. A steady rise in migration has been observed since 1980 onwards, with the highest value observed in between the year 1987 to 1990 (Fig. 14.3). However, after this period, there was a drastic fall in this event, although not fully nil. This is probably due to the stringent administrative control of the Border Security Force (BSF) as well as the Indian Army in the border as well buffer

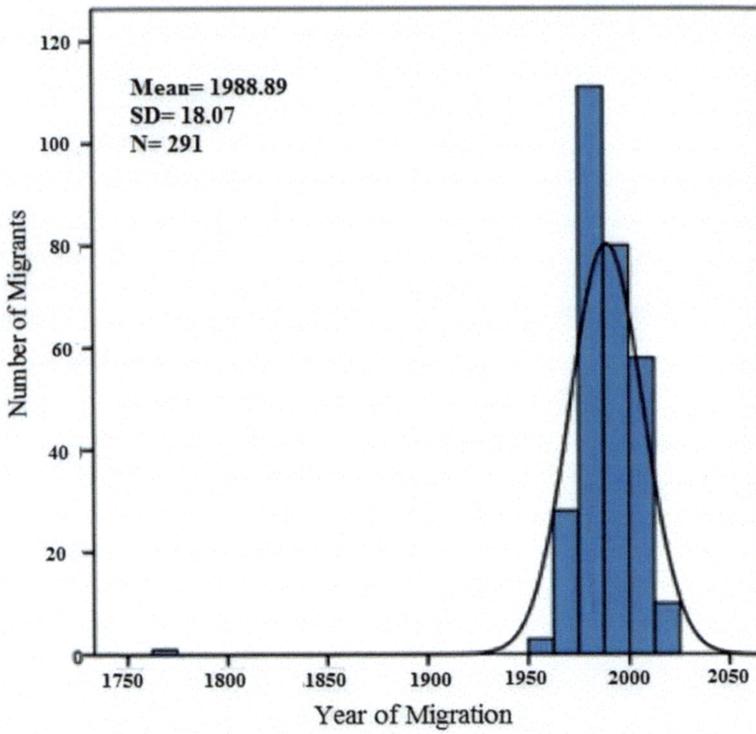

Fig. 14.3 Year-wise frequency of migrants (*Source* Field Survey, 2019)

regions, according to the respondents. This result closely exemplifies the proposition of Lee that states:

> …unless severe checks are imposed, both volume and rate of migration tend to increase with time (Lee 1966)

A rising trend of migration is observed between the periods of 1980 and 2000 due to lack of surveillance in the international border, presence of numerous penetrable tracts, etc. But after 2000, there was a steady decrease in the trend due to the imposition of stringent administrative control on the border. With this situation, another concept of Lee (1966) matches very well, i.e., 'Intervening Obstacles'. The stringent administrative control regarding the border crossing acted as a large *intervening obstacle* for the later stage migration aspirants.

14.6.5 Reasons for Selection of Destination

Out of Lee's (1966) four selected factors of migration as given in Sect. 14.5.1.3, the factors associated with the area of destination are the backbone of the formulation of this section.

In the event of human migration, the most important influencing factors are *Push* and *Pull factors.* In the case of the Bangladeshi migrants in India, *Pull factors* are the principal determinants because they attract migrants to immigrate. According to the respondents, domiants push factors are broadly related to the Socio-economic condition along with the political situation, which stands comparatively better than the source region. The largest share of the respondents has opined for Job Opportunity (27.4 percent) followed by the presence of native people (14.7 percent) and peaceful life (13.4 percent) (Fig. 14.4). This clear-cut percentage-wise specification of pull factors indicates the 1st characteristics of Lee's theory (1966, p. 56) i.e., *'Migration is selective'*. This portion also largely resembles one of the destination's pull factors, as stated by Norris (1972, p. 299). Out of five pull factors mentioned by him, one, i.e., *'Economic opportunities including jobs'*, is similar to the highest percentage group, i.e., 'Job opportunity'. One of the sources of knowledge about the destination for a migrant, according to Norris (1972), is *'personal communication with friends or relatives already living at the destination'*, which perfectly matches with the second-highest percentage class, i.e., 'Relatives'. Pull factors are exactly similar phenomena, like the North Pole's attraction towards the South Pole of a magnet. This is exactly the opposite of the same poles' repelling property to each other (Datta, 2005).

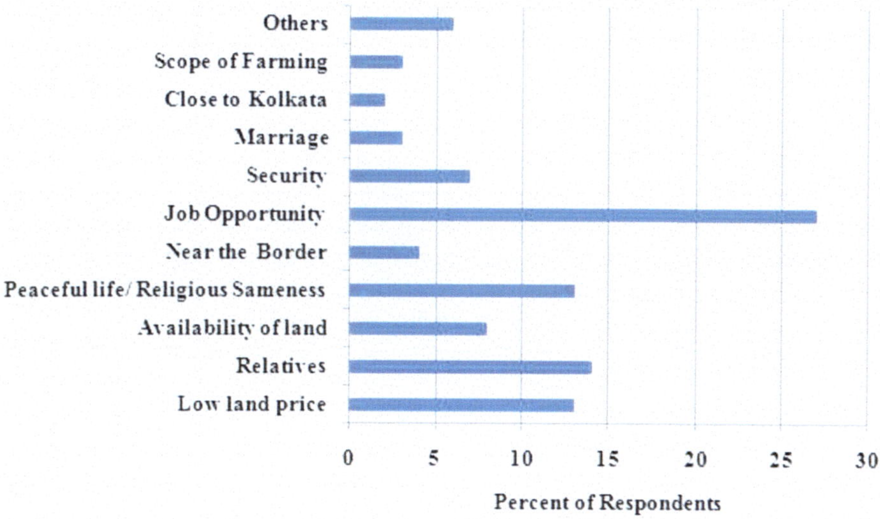

Fig. 14.4 Reasons for selecting the destination *Source* Field Survey by the authors in 2019

14.7 Connection with the Neighbours or Relatives at the Place of Origin

The relationship tie-up status with the native has been analysed by investigating the respondents from two conditional sub-aspects. While analysing *the respondents' visiting frequency, it is astonishing* that mostly all of the respondents (90.4 percent) have not visited their native land and are not willing to visit anymore. Other too has not interested to visit frequently. Mainly the matter of peaceful life and job security in India and the vice-versa condition in Bangladesh is the major backset behind these respondents' attitude. *The number of family members present in Bangladesh* has also been found to be in a varied numerical count. Nearly 50 percent of the sample shows that there is nobody present in Bangladesh presently. 22 and 13 percent respectively answer that they have 1–3 persons and 3–6 persons living in their native land, and 18 percent confess that more than 6 persons are living in Bangladesh. It has been found that there is a very little tendency to visit their native country (it happens in case the number of relatives living in Bangladesh is 6 or more than that). Thus the counter migration stream effect is not so much impactful. This trend shows a very significant deviation from the mainstream theoretical consensus put forward in the name of '*Migration Counter stream*' by Lee (1966) and '*Return Migration*' by Norris (1972). Thus it stands as an exception.

14.8 Average Family Size

The average family size of Bangladeshi immigrants stands more or less 5 (five) among the surveyed respondents (Lin and Paul, 1995). The *average family size* in *Bangladesh* decreased to 4.4 persons from 4.8 persons reported in 2001 and 5.5 reported in 1991 (Sharma 2015). While the average family size is concerned, most people belong to single-family and nuclear families (i.e., 3 to 4 members), accounting for 95 percent of the total sample. The remaining 5 percent belong to joint families comprising of 6 to 11 members. Thus an inverse relationship can be seen between the family size and the number of family members. Moreover, most of them are daily wage laborers, and thus their income is not sufficient to sustain a joint family. Immigrants report about their efforts to give their children a better future and become aware of family planning activities. It accounts for the reduction in the family size after migration.

14.9 Social Coherence and Conflict With Neighbourhood

14.9.1 Family Tie-up With Neighbourhood

Cultural assimilation or cultural integration follows the occurrence of migration through which a newcomer immigrant adjusts him/herself to the new ambiance in which he or she has recently entered. In the case of Bangladeshi immigration, the immigrants have established a socio-cultural relationship with their neighbourhood.

> Ever since the partition of India in 1947, successive waves of people facing hostile conditions, persecution, intolerance, and adverse economic situations in what constitutes present-day Bangladesh have found sanctuary in India (Tripathi, 2016).

While some of them later returned to their homes in Bangladesh, the majority chose to assimilate within India. In the study, it is found that most of the immigrants (72 percent) have established a healthy neighbourhood comprising both co-immigrants and existing Indians. Nearly 18 percent of the respondents believe that cultural assimilation will ruin their home culture and lose their native identity and maintain their socio-cultural tie-up only with their native people. Such kinds of psychological issues are the major hindrances in cultural development through the cultural mix. 10 percent of the respondents opined that they do not like the neighbours and will try to move to a better locality if possible. So, 28 percent of the overall respondents are not assimilated or integrated with the natives present in the destination. It partially supports one of the characteristics of migration as put forward by Lee:

> The characteristics of migrants tend to be intermediate between the characteristics of the population at origin and the population at the destination (Lee 1966, p. 57)

14.9.2 Socio-Cultural Participation

Cultural programs are the most effective way to make socio-cultural bonding among families more rigid and strong. Sharma (2015), in her study in Assam, significantly noted that.

> …Son or daughter of 40 percent of respondents are married to local people. This seems to suggest that their efforts to assimilate with the local population can't simply be attributed to the compulsions or expediency to get them legally accepted; rather, in many cases, some assimilations have indeed taken place driven by their "heart". When asked about the future of their children in Assam, most of them (60 percent) are generally optimistic (Sharma 2015)

Nearly 14 percent of the respondents are not interested in participating in any kind of cultural program and 76 percent of the people presently involve themselves in various cultural programs but not as active members. A very few migrants involve themselves as the organizer or any active part of a cultural program. The socio-cultural relationship is very dynamic there and presently in a juvenile stage that may evolve with time. In the case of children, the percentage of participation in the cultural program is nearly 61 percent. The rest of the children cannot participate due to language barriers, religious hindrances, and poverty. It also partially supports one of the characteristics of migration as put forward by Lee:

> The characteristics of migrants tend to be intermediate between the characteristics of the population at origin and the population at destination Lee (1966, p. 57)

14.9.3 Nature of Support Obtained from the Present Living Place

The sense of *community* grows and evolves with the dynamic nature of the support and mutual understanding among the families living within the particular spatial extent. It is found that in many ways, the immigrants have obtained help, i.e., job, house, land, and other aspects. Ninety percent of the respondents have opined that they have obtained help from their neighbourhood in their needs. Although such bits of help is very much formal, i.e., monetary or family level help is very less, which signifies their failure to become culturally assimilated or integrated with the existing or the other migrants also partially supports one of the characteristics of migration by Lee (1966, p. 57) as discussed in the previous section (i.e., Sect. 14.9.2).

14.9.4 Socio-Cultural Conflict

Lin and Paul (1995) did such an investigation and found most of the respondents asserting that most of them are well settled in the present locality of the destination at Delhi. According to them:

> As the more and more relatives are buried near the banks of the Yamuna River, and the emotional attachment to Delhi becomes more deeply rooted—Lin and Paul (1995)

Because there is a co-existence of both Bangladeshi immigrants and native Indians, social and cultural conflict and dichotomy are substantially low. Only 3 percent of the total respondents have faced any conflict-related challenges. The rest of them hardly suffers from any kind of socio-cultural conflict. Even 77 percent of the total respondents agree to give marriage of their children with the existing Indians. It may be inferred that the matter of socio-cultural standpoint of the immigrants matches with the existing native Indians to a considerable extent.

14.10 Economic Variables

'Economic Parameters' indicate those variables which are utilized for overall economic condition endowed specifically to analyse a certain economic issue from a broader perspective.

14.10.1 Remittance/Financial Support

'Remittance/ Financial Support' means the economic backbone necessary to stimulate the dynamicity of daily life both in the short run and long run (Sharma 2015). During the 1970s and 1980s, a large pool of migrant labourers from Bangladesh engaged in Kolkata's construction sector who had no options in revealing their identity (Datta 2005). Yet recent World Bank data (Bilateral Remittance Matrix 2014) show, out of the $7.6 billion of outward remittances from India, 54 percent or $4.16 billion was to Bangladesh alone in 2014. Almost every year, 50 to 55 percent of India's total outward remittances are to Bangladesh (Acharya 2016). In the present study, it is found that nearly all of the respondents are dependent on their occupations they are engaged in India (99 percent). This is because most of them have migrated to India to find a secure job and consequently a better lifestyle than in Bangladesh previously. Moreover, the reason for non-remittance from the Bangladeshi counterparts of the Bangladeshi migrants is due to the lack of minimum employment and consequently impoverished economic conditions.

14.10.2 Land and Other Asset Holdings

14.10.2.1 Variability of Land Ownership in Place of Sending and Receiving

Sharma (2015), in her study, found that there was a shortfall in the land holding of the Bangladeshi immigrants in Assam.

About 83.5 percent of the migrants owned land in Bangladesh compared to 72.2 percent in Assam. It appears that, more migrants had land ownership when they were in Bangladesh, and it took them a while to build up enough capital in Assam. However, some migrants among them who would permanently own the land that they are living in after paying their share of the land price to the principal landowner in installments. Here the understanding among them is that one person (principal owner) would own a plot of land and five other families would settle on it by paying the principal owner in installments to claim the land title in their names in near future" (Sharma 2015)

Table 14.6 Change in land ownership for homestead

Quantity of Land	Percentage for Bangladesh	Percentage for India	Percentage Difference
No Land	0.70	1.00	0.30
Less than 3 Kattah	23.60	26.00	2.40
3 to 5 Kattah	26.70	41.10	14.40
5 to 10 Kattah	15.80	21.60	5.80
10 to 20 Kattah	16.40	7.20	– 9.20
Greater than 20 Kattah	16.10	2.40	– 13.70

(*Source* Field Survey by the authors in 2019)

However, the quantitative class-wise measurement has not been done in that study. In this present study, it is observed that most of the residents in the study area are associated with the primary economic sector. Thus they exhibit a variable nature of land holdings, both at the source (Bangladesh) and the destination (India) (Table 14.6).

Table 14.6 shows the difference in the quantity of landholdings in source and destination. The critical finding is that upto a land holding of 10 Kattah, there is a general rise with the highest increase in the category of 3–5 Kattah. However, in the case of holdings larger than that, there is a sharp and sudden fall. One of the main reasons for this is the decreased size of the family and the deficiency of the bulk amount of monetary savings that could be spent on buying fixed assets. In Bangladesh, people used to live in a joint family, but the family size has decreased after shifting to India in search of work, food, security, etc. Moreover, they are engaged here in such works, which gives very marginal monetary returns. Due to this, it is impossible for the immigrants to spend a lot of money on holding a large quantity of land.

The scenario of holding agricultural land is relatively less in India compared to Bangladesh, as reported by the respondents. Table 14.7 exhibits that the percentage

Table 14.7 Change in ownership of agricultural land

Quantity of agricultural land	Percentage in Bangladesh	Percentage in India	Percentage Difference
No Land	23.30	74.30	51
Less than 20 Kattah	25.00	8.20	–16.8
20 to 40 Kattah	16.40	5.50	–10.9
41 to 60 Kattah	9.90	1.40	–8.5
61 to 80 Kattah	8.20	0.70	–7.5
81 to 100 Kattah	6.50	0.00	–6.5
Greater than 100 Kattah	10.60	0.70	–9.9

(N.B. 1 Kattah = 66.89 Square Metre) (*Source* Field Survey by the authors in 611 2019)

of respondents having no agricultural land has increased largely (51 percent). In all the other quantity categories, a significant decrease in agricultural land holdings has been observed. This is because, in the destination place, the price of land selling value is nearly unaffordable for those immigrants having no such monetary support. The little bit number of them having quite large amount of land, although their are exceptions.

14.10.2.2 Ownership of the Houses

House is the essential asset for a person specifically who has migrated on an international scale. 'It is worth mentioning that migrants' houses are not always well-built, permanent structures of wood or concrete. In most cases, they have mere shelters with no other indoor amenities' (Sharma 2015). In the present study, the thing has been studied giving substantial importance. 92.4 percent of the total respondents possess houses. Most of the immigrant families owned their houses (80 percent), and some others (18 percent) reside in rented houses.

A matter worth mentioning is that while purchasing the land in India, nearly 38 percent of them have opined that they managed to purchase the land without taking any external help, i.e. brokers or promoters. 32 percent of the respondents have taken help only from their relatives in India. However, the matter of extreme interest is that 30 percent of the respondents have not unraveled their monetary source of land purchasing. Most of the Houses (56.6 percent) are typical hut types whose roofs are built up of earthen tiles, and the walls are made up of either mud or mud-covered with polyester sheets. It indicates the impoverished standard of living of most of the families.

14.10.2.3 Assets Other than Land and Houses

For the proper functioning of daily life, many other fixed and liquid assets are required than land and houses. Sharma (2015) has found in her study in Assam that the percentage of people possessing assets is diversified. For instance, most of the respondents hold their own houses (92.4 percent), followed by bicycles (80.4 percent), land (72.2 percent), etc. But the spatial comparison is not made in the study mentioned above. In the case of analysing the Bangladeshi immigrants in this regard, the present study finds that most of the households availed the LPG connection and cable TV connection. A few respondents are found purchased Motor Cycles, Bi-Cycles, Refrigerators, and Pumps. The asset-possession of the immigrants gets better during their living in India than that of Bangladesh (Table 14.8).

Table 14.8 Comparison of Assets in Pre and Post Migration Scenario (Percentage of Household)

Motor Cycle		Refrigerator		Pump	
Past	Present	Past	Present	Past	Present
2	6	1	3	1	2
Gas		TV		Others	
Past	Present	Past	Present	Past	Present
5	65	6	70	8	38

(*Source* Field Survey by the authors in 2019)

14.10.3 Professional Engagement

14.10.3.1 Present Job Status

While analysing the job status of the immigrants and also tracing their pre- and post-scenario-based occupational change, a varied picture has been observed. There is substantial change over in the sector-wise distribution of the respondents in the study of Sharma (2015). According to her, there was a percentage increase in the Self-employed category (37.3 percent to 59.5 percent). But the Business (15.2 percent to 10.1 percent) and Unemployed category (34.8 percent to 15.2 percent) has been reduced. The count of daily labourers and housewives has been increased in the post-migration condition. On the other hand, the count of cultivators has been decreased sharply. The increase of daily labourers and consequent decrease in cultivators is due to the fact that most of them have lost their fertile land in Bangladesh and have failed to purchase new agricultural land to that amount. A few samples got married after migration. This has further increased the number of housewives.

During the post-migration phase, the population's count as daily labour has increased (Table 14.9). Also, interestingly, the increasing number of females have been counted as 'housewives'. The former is because it is relatively easy to employ self as daily labour than any other, i.e., cultivator, Micro enterprise.

14.10.3.2 Occupation Related Problems

The migrants of Bangladesh have faced so many problems which are not very unique rather, these are very common. In a study conducted by Singh (2002), it has been found that he has introduced the *Place-Work-Folk* concept as put forward by *Geddes* in Urban Geography and also compared the spatial extent with that of *Lebensraum,* the concept of living space as put forward by *Ratzel.* A living space with its aboriginal natives stands out to be somewhat odd for a new stranger or, say, immigrants in all aspects. Sharma (2015) has also found some sort of adjustment problems regarding occupation. In this paper, it has been observed that mainly the basic tenets of the problems are Residential mobility, Political Pressure, Unskilled laborers, etc. As a

Table 14.9 Comparison of occupational structure

Occupation Type	In Bangladesh (In Percentage)	In India (In Percentage)
Cultivator	10	2.1
Agricultural Labour	1.7	2.2
Daily Labour	3.9	20.9
Business	0.4	3.6
Micro Enterprise	0	0.4
Fishery	0.7	0
Domestic Service	2.4	1.5
Housewife	9.8	26.8
Service	0.4	2.9
Others	3.3	11
NA	67.4	28.6

(*Source* Field Survey by the authors in 2019)

result, they lack proper remuneration for their work and also failed to secure any lucrative job involvement.

14.11 Basic Right Related Variables

'*Basic Right Related Variables*' indicate those variables which are utilized for the experiment of an overall condition about fundamental rights explicitly endowed to analyse a specific political issue from a broader perspective.

14.11.1 Citizenship's Rights

Citizenship is a person's status recognized under the custom or law as being a legal member of a sovereign state or part of a nation. A person may have multiple citizenships, and a person who does not have citizenship of any state is said to be stateless. *Citizenship right* is very important to newcomer and immigrants. It is important that citizenship is not just a benefit to immigrants but also boosts the nation's economy. 'Irregular migrants are excluded from the acquisition of citizenship through birth, registration, or naturalization in India. There is no program to grant irregular migrants or their children legal status or citizenship' (Ahmad 2017). But in September 2015 and July 2016, the central government exempted certain groups of irregular migrants from being imprisoned or deported. These are irregular migrants who came into India from Afghanistan, Bangladesh, or Pakistan on or before 31 December 2014, and belong to the Hindu, Sikh, Buddhist, Jain, Parsi, or Christian religious communities. Irregular migrants from this group cannot be imprisoned or deported for not having valid travel documents' (Chaturvedi 2016).

14.11.1.1 Availability of Cards and Certificates

The status of these Bangladeshi immigrants' citizenship identity is determined based on the availability of different identity cards (i.e., Ration card, Voter card, Aadhaar card, etc.) and certificates (i.e., Caste Certificate) in them. Tripathi (2016) states:

> Detecting irregular immigrants from Bangladesh is a daunting task. The subtle differences in the accents, dialect, and features between an Indian Bengali and a Bangladeshi are not easily discernible. The fact that most Bangladeshis already hold ration cards, voter identity cards, or even the unique-identity Aadhaar cards further compounds the difficulty. Ironically, an irregular Bangladeshi immigrant is more likely to be equipped with an Indian identity document than an Indian Bengali who may take his or her Indian citizenship for granted.

In the present study, it is clear that almost all of the respondents possess a Ration card, Voter card, and Aadhaar card. The number of BPL cardholders and Caste certificate holders is also significant in number, although much less than the possession of Voter Card and Aadhaar Card. The percentage of respondents having club membership has also been analysed, and the number of such respondents is minimal. The reason for such an outcome is that most of them cannot carry out such things which they think as 'luxurious' and 'extravagant' in terms of their economic condition.

14.11.1.2 Broker's Involvement

In case of obtaining a Ration card, Voter Card, Bank Account, and Aadhaar card, most of the respondents have obtained those in their own interest. However, in the case of Job card, BPL card, Caste Certificate, and Club membership, others' influence has been observed. Several respondents have opined that obtaining a Job card, BPL card, Caste Certificate involves some sort of corruption and biases. The Political context also plays a significant role in this regard. All these matters are also mentioned in a newspaper report by Lobo (2019). It depicts that.

> At least 20 citizens of Bangladesh were recently arrested by Tamil Nadu police. Interestingly, these irregular immigrants allegedly had not only acquired ration cards and driving licenses but had also managed to get authentic Aadhaar cards and, through Aadhaar, passports.

A report by the Press Trust of India (PTI, 2008) quoted the then Parliamentary Standing Committee on Home Affairs stating that the illegal migrants have been able to secure ration cards, driving licenses, voter cards, and even the Permanent Account Number (PAN) cards.

Table 14.10 reveals that most of the respondents possess the basic requirements for self-identity, i.e., Ration Card, Voter Card, Aadhaar Card, etc. Other conditional identities, i.e., BPL Card, Job Card, and Caste Certificate, are possessed by 22.6 percent, 30.8 percent, and 21.2 percent of the respondents respectively. Thus it is evident

Table 14.10 Possession of various means of physical and social identity (In Percentage)

Possession	Ration Card	Voter Card	Aadhaar Card	BPL Card	Job Card	Club Member	Caste Certificate
Yes	97.6	97.9	96.6	22.6	30.8	3.4	21.2
No	2.1	2.1	3.4	7.2	63	87	70.9
N.A	0.3	0	0	7.2	6.2	9.6	9.8

(*Source* Field Survey by the authors in 2019)

Table 14.11 Percentage types of Helping Persons for getting Citizenship Right-related documents

Helping Persons	Ration Card	Voter Card	Aadhaar Card	Job Card	BPL Card	Caste Certificate	Club Membership	Bank Account
Self	47.30	48.30	56.60	11.6	6.5	7.2	2.4	69.9
Friends	0.70	0.70	0.30	0.3	0	0	0.7	0.70
Relatives	2.40	2.40	1.70	0.7	0.3	1	0	1.70
Neighbour	0.70	0.70	0.70	0.3	0	0.3	0.3	1.40
Political Party	16.40	15.80	13.70	7.5	7.2	6.8	7.5	1.0
Broker	3.40	2.40	0.70	1	0.7	0.3	1	0.3
Others	29.1	29.80	24.3	78.4	85.3	83	78.4	25

(*Source* Field Survey by the authors in 2019)

that most of them have been successful in securing their minimum identity to lead a normal life in place of destination. However, the standard of living is not satisfactory, which is reflected in the percentage holding of Club membership.

The medium/s through which the means of physical and social identity have been obtained are shown in table 14.11. The basic three means, i.e., Ration Card, Voter Card, Bank Account and Aadhaar Card has been obtained mostly by self-help followed by 'OTHERS'. In case of Job Card, BPL Card, Caste Certificate, and Club membership, this 'OTHERS' is highest. This OTHERS' explanation was not obtained lucidly but it has been heard from some of the respondents that this is actually illegal or unfair means.

14.12 Other Indicators

14.12.1 Age-Sex Composition

Migration exerts a strong influence over a destination place's age-sex structure and thus changes the existing structure (Pelletier and Spoorenberg 2015). A random search of literature proves that there is probably no such separate age-sex structure

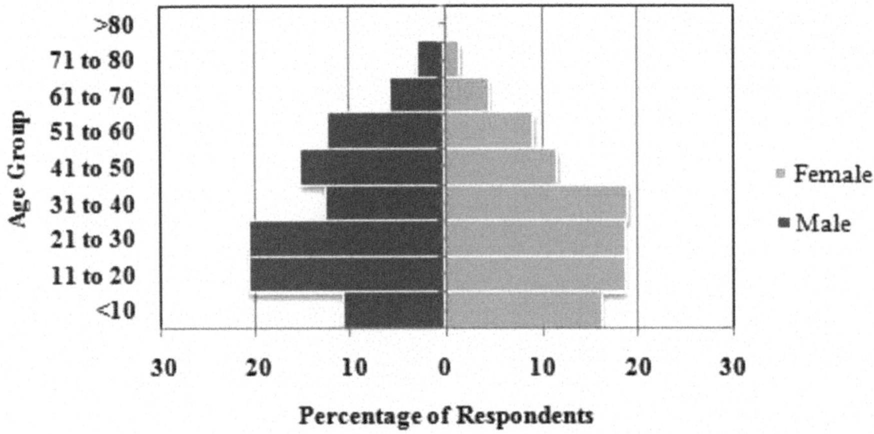

Fig. 14.5 Age-Sex Structure of respondents (*Source* Field Survey by the authors in 2019)

of Bangladeshi migrants (Sharma 2015). From the age-sex pyramid (Fig. 14.5), it can be observed that most of the population is in the youthful to late youth stage. The whole sample's average age is within 21 to 40 years, which signifies avast potential of demographic dividend for utilization.

This pattern of age sex strongly conforms to one of the characteristics of the law of migration as put forward by Lee (1966), stating:

> …the heightened propensity to migrate at certain stages of the life cycle is important in the selection of migrants'. Another one is 'Migration is Selective

It also matches well with one of the principles of migration given by Ravenstein (1885), i.e., 'most of the migrant people are maturely aged'.

14.12.2 Pre and post-migration Education

The 'before' and 'after' scenario is determined by taking into account the highest educational category immediately before the migration and the highest/ pursuing educational category of the family after settling down in the destination respectively. During the survey. 59.55 percent of the respondents turn out to be illiterate followed by 12.53 percent of the respondents who have obtained education within class V-VIII. Educated more than XII in Bangladesh is the lowest (2.48 percent). Before migration, most of the people were engaged in agricultural activities, and it was their main source of livelihood. These outcomes can be attributed to the fact that some of them were poverty-stricken, had a low standard of living, faced Political turmoil, etc., in the place of source. However, after migration, a difference can be noticed in terms of educational status. Though the majority of the people were still illiterate (46.57 percent). About 13.07 percent of people studied/ studying within class IX to X after migration, and 6.54 percent of the migrants had higher education. After migrating to India, immigrant families have got comparatively better-living facilities and they have comparatively better employment opportunities than before, and so they are taking the initiative of educating their children to give them a better future. People are engaging themselves in various secondary activities in the cities and now understand the need for education through various awareness programs. This supports much of the available pieces of literature, such as of Nandy (2019) where he has mentioned that availability of land, better economic opportunities, education, and health facilities, and a similar cultural landscape have attracted these migrants to settle in India. Through Sarba Siksha Avijan and the establishment of new schools in remote villages, the government is trying to spark the education of first-generation learners. The amount of illiteracy has decreased by 12.98 percent, and the amount of higher educated is increased by 4.06 percent, which is a bit positive sign.

14.12.3 Problems at the Present Place of Living

The problems faced by the migrant families according to the intensity ranked as reported by them mentioned in Table 14.12. To most of them, Water logging as their principal problem, followed by Electricity Shortage. In the case of second-order ranking, most of the opinion indiates Soil Salinity, followed by Flood. Mostly shared, Low Wages as the third-ranked problem, followed by Political Bias and Criminal Activities. As a whole, this region may be marked as vulnerable to natural calamities and sensitive to socio-political discourses. The problem of Crop Failure, Electricity cutoff, Lack of Drinking Water, and other subsequent problems also act to disturb the normal rhythm of the life of the migrants.

Most of them confess that they want to move towards the proximity of the major city (i.e., Kolkata) to avoid these obstacles. This scenario acts as the 'Counterpart concept' of *Intervening Opportunities,* i.e., *Intervening Obstacles.* Norris (1972) mentioned that migrants always search for betterment, which, in a dynamic sense, is called the *intervening opportunities.* According to Lee (1966), *intervening obstacles* are always avoided by the migrants. Thus, it may be synthesized from these two concepts that intervening opportunities stand out to be a viable alternative to avoid the intervening obstacles.

14.13 A Quantified Approach Towards the Level of Adaptability

Adaptation is one of the foremost process tools to understand the pros and cons of the human adjustment strategies under the context of human migration. The wholesome

Table 14.12 Ranking of problems (Intensity)

Reasons	Actual figure		
	1st Rank	2nd rank	3rd Rank
Cyclone	31	31	6
Embankment Breaching	0	1	2
Flood	153	35	19
Soil Salinity	2	40	6
Crop failure	0	0	6
Political Bias	13	33	39
Criminal	12	27	38
Lack of Job opportunity	12	30	34
Competition for Job	0	8	14
Low wage	6	19	41
Water logging	35	17	23
Lack of drinking water	4	14	17
Electricity cut-off	17	28	31
Inaccessible condition of road	4	5	8
Deprivation from social amenities	2	3	0
Others	0	0	7
Total Sample	**291**	**291**	**291**

(*Source* Field Survey by the authors in 2019)

process of adaptation can be broadly divided into three sub-processes, i.e., *Integration, Assimilation,* and *Accommodation.* They have been quantified to generate a numerical term to express the surveyed families' level of adaptability. For this, the four most significant variables are selected under each of the subdivisions (i.e., *Integration, Assimilation,* and *Accommodation*), and a 5 point score scale is applied to each of them. Each subdivision holds 20 point score.

Thus the total score (i.e., Adaptation Index) stands as 60 points. Figure 14.6 depicts that the moderate stage is prevalent in all the indices except the Integration Index, where nearly 44 percent of the respondents are in the advanced stage, followed by the moderate one (43 percent). In the case of the Assimilation Index, the advanced stage is in the second position (24 percent), whereas, in the case of the Accommodation Index, the advanced stage is in the last position (5 percent).

Figure 14.7 shows the level of adaptability on a quantitative scale out of a total 60-point scale. Most of the families are moderately adapted (85 percent) followed by the infancy stage (12 percent). The mean value is approximately 37. Thus it can be inferred that although they have adapted significantly, there are still substantial gaps and vacuums to be adapted or modified by them as per their norms and standards.

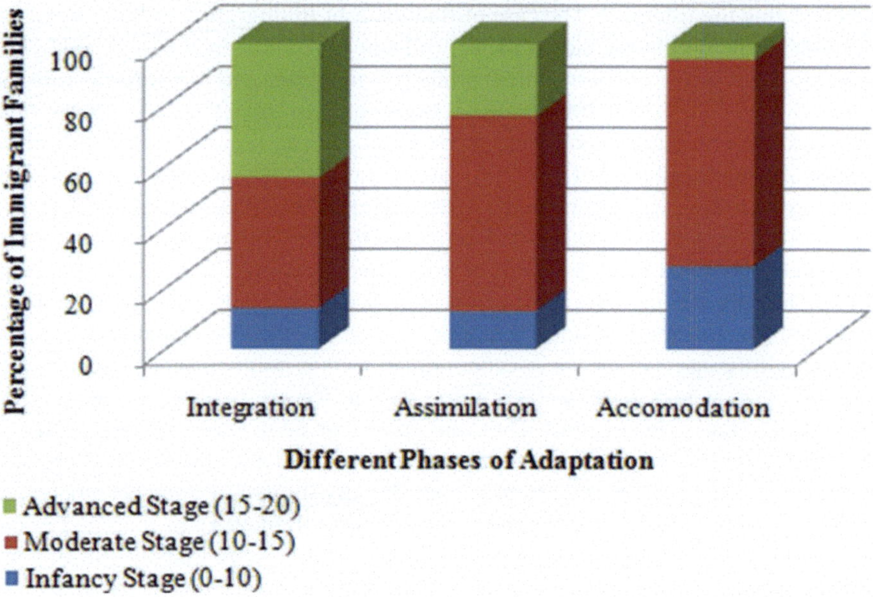

Fig. 14.6 Adaptation status of immigrant families during integration, assimilation, and accommodation at the present living place (*Source* Compiled by authors)

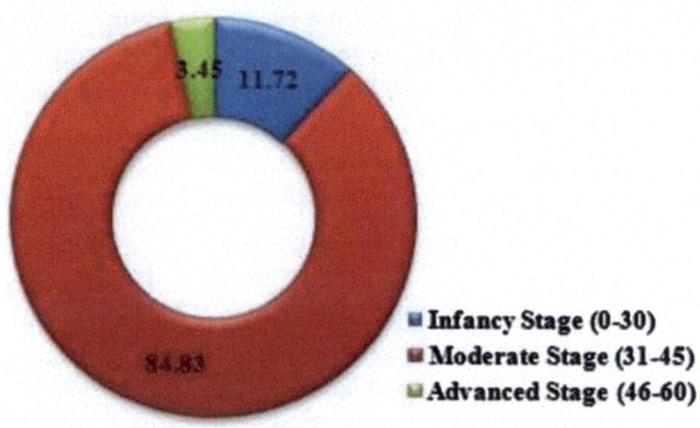

Fig. 14.7 Stage wise situation of Adaptation Index (*Source* Compiled by authors)

14.14 Conclusion

The study shows that there is a consistent shortfall of the Hindu population in Bangladesh from 1901 to 2011. Moreover, there is a constant rise in the Muslim population and an oppositely decrease in the Hindu population even after the Independence of Bangladesh.

The natural increase of population, relocation of administrative boundaries, and net migration are the major contributors of population increase, with the first and last one being the foremost. The case of Bangladeshi Immigration in India stands as a prolonged socio-cultural and politically influenced phenomenon stimulated by the religious persecution. In spite of civic protection measures and measures to curb the rapid influx of Bangladeshi Immigrants, it is very hard to arrest this event unless there would like any measure to solve the problem. Moreover, they are opting this place because it is such a place that provides a small sort of peace, harmony, and scope of livelihood that they even lack in their native land due to the different reasons. A Dhaka University Professor, Dr. Abul Barkat, projected that the Hindu population would be completely abolished from Bangladesh within the upcoming 30 years (Barkat et al. 2014). The only medicine for this deep-rooted disease is inculcating the value sense of humanity as the best individual's identity among people. Although this study thus highlights the context of the socio-economic, as well as political supra-structure and deep-structure at present as well as a duo-spatial perspective (i.e., Bangladesh as source and India as Destination). Thus, this study can aid in unraveling the degree of present-day applicability of such models and the observed deviations and exceptions. Moreover, as the study area's extent is considerably limited, it unravels a potential aspect to conduct studies underpinned with a similar framework or modified as per requirement. It may take a form of project in any dual or more trans-countries level where issues of similar international as well as internal migration use to persist.

Moreover, the parameters utilized here and the selected models' evaluation can also be judiciously inculcated in such a venture, preferably to a more critical and analytical extent. Indeed, the study performed here can construct an overall scenario and identify push and pull factors deployed in such migration. In the decision-making process, this may help assessing the level of cultural mix. The institutional bodies that try to reduce the flow of migrants inside can understand root causes and subsequently take steps to restrict the flow. Moreover, this endeavor in itself is depicting a matter that such a micro-level approach. However, the limitation of parameters and even the selection is study area-specific, can feasibly portray the impact of migration in that concerned place, which can ultimately help the planning initiatives be specific in formulating remediation for the place.

Acknowledgements Both the authors are very much indebted to Ms. Riyanka Chatterjee for her immense help and support during the pre-field and during field operation of this research work. Without her cooperation and support, it would have been a very strenuous and tuff challenge to conduct the field survey of Bangladeshi immigrants. She helped us a lot by devoting herself as the smooth medium between the respondents and us for communication and convinced them many times to participate in our survey. Therefore, we express our sincere gratitude to her and expect this kind of technical help in our future endeavors.

References

Acharya N (2016) Why Bangladesh leads in remittance outflow from India. Business Standard, Kolkata. Accessed 27th May 2019.

Ackerly BA, Anam M, Gilligan J (2015) Environment, political economies and livelihood change. AHDPH, Dhaka

Ahmad T (2017) Laws concerning children of undocumented migrants: India. Library of Congress. https://www.loc.gov/law/help/undocumented-migrants/india.php#_ftn8

Ammarell G (2002) Bugis migration and modes of adaptation to local situations. University of Pittsburgh Press, pp 51–67

Banerjee B (2003) Infiltration and border problem. Paper presented in the workshop at the Indian Statistical Institute on the Undocumented Migration From Bangladesh to West Bengal organised by Population Studies Unit. Kolkata, India

Barkat A, Osman A, Sengupta S (2014) Reinforcing ties: enhancing contributions from Bangladeshi Diaspora Members. Int Labour Organ (ILO) Country Office of Bangladesh. Dhaka. Bangladesh

Barkat A (2008) Political economy of unpeopling of indigenous peoples: the case of Bangladesh. Department of Economics, University of Dhaka, Dhaka. Bangladesh

Bhugra D, Backer MA (2005) Migration, cultural bereavement and cultural identity. Special Publication, World Psychiatry, USA

Boyd M (1989) Family and personal networks in international migration: Recent developments and new agendas. Int Migr Rev 23(3):638–670

Census of India (2011) District census handbook—South 24 parganas. Ministry of Home Affairs. Kolkata: Directorate of Census Operations. Retrieved January 20, 2020, from https://censusindia.gov.in/2011census/dchb/DCHB_A/19/1917_PART_A_DCHB_SOUTH%20TWENTY%20FOUR%20PARGANAS.pdf

Census of India (2001) District census handbook—south 24 parganas. Ministry of Home Affairs. Directorate of Census Operations, Kolkata Retrieved January 10, 2020, from https://lsi.gov.in:8081/jspui/bitstream/123456789/5377/1/38703_2001_SOU.pdf

Chaturvedi A (2016) Simply put: understanding issues of illegal migration in new citizenship bill. Indian Express. http://indianexpress.com/article/explained/understanding-issues-of-illegal-migration-in-new-citizenship-bill-3101213/, Archived at https://perma.cc/3ZR6-JZXJ

Chung EA (2010) Immigration and citizenship in Japan. Cambridge University Press, Cambridge

Das B, Ansary R (2018) Bangladeshi and inter-state migrants: differential adaptation and acceptance by the locals in West Bengal, India. Spatial Demogr 6:159–178. https://doi.org/10.1007/s40980-017-0040-1

Das P (2016) Illegal migration from Bangladesh–deportation, border fences and work permits. IDSA Monograph Series No. 56, New Delhi; Institute for Defence Studies and Analyses (ISBN-978-93-82169-69-7)

Datta P (2005) 'Push-pull factors of undocumented migration from Bangladesh to West Bengal: a perception study. Qual Rep 9(2):335–358 Retrieved from https://nsuworks.nova.edu/tqr/vol9/iss2/9

Entorf H (2002) Rational migration policy should tolerate non-zero illegal migration flows. Int Migr 40(1):27–43

Ghosh P (2012) India's 'Mexican' problem: Illegal immigration from Bangladesh. Int Bus Times. Retrieved from http://www.ibtimes.com/indias-mexicanproblem-illegal-immigration-bangladesh-213993.

Guha Roy S, Sen K (2002) 'Patterns of health care utilization and expenditure: a household survey in an Indian state: West Bengal 1999. ISI, Kolkata.

Hujo K, Piper N (2007) South-South Migration: challenges for development and social policy. Development 50(4):1–7

Joseph J, Narendran V (2013) Neither Here nor There: An Overview of South-South Migration from both sides of the Bangladesh-India Migration Corridor. Int Instit Soc Stud Erasmus Univ, Rotterdam:1–39

Kartiki K (2011) Climate change and migration: a case study from rural Bangladesh. Gender Develop 19(1):23–38. https://doi.org/10.1080/13552074.2011.554017

Khan MR (1972) Pattern of external migration to and from Bangladesh (1901–1961). Bangladesh Econ Rev 2:599–632

Kumar A (2011) Bangladeshi migration in South 24 Parganas District of West Bengal", ISBN-978–3846530382. Lambert Academic Publishing, Germany

Lee E (1966) A theory of migration. Demography 3(1):47–57. Retrieved July 30, 2019, from www.jstor.org/stable/2060063

Lein H (2000) Hazards and 'forced' migration in Bangladesh. Norwegian J Geogr Dept Geogr Norwegian Univ Sci Technol Norway 54: 122–127. ISSN: 0029-1951

Lin SG, Paul MC (1995) Bangladeshi migrants in Delhi: Social insecurity, State power, and captive vote banks. Bull Concerned Asian Scholars 27(1):3–20. https://doi.org/10.1080/14672715.1995.10413070

Lobo SN (2019) Getting Aadhaar not-so-difficult task for illegal immigrants? The New Indian Express. Retrieved June 5, 2020, from. https://www.newindianexpress.com/cities/chennai/2019/jun/02/getting-aadhaar-not-so-difficult-task-for-illegal-immigrants-1984797.html

Martin P (2004) New NAFTA and Mexico-U.S. Migration: The 2004 Policy Options. Agric Resour Econ 8(2):1–4

Nandy D (2019) Socio-political and security perspectives of Illegal Bangladeshi migrants in West Bengal: the impact on Indo-Bangladesh relations. J His Arch Anthropol Sci 4(3):123–129. https://doi.org/10.15406/jhaas.2019.04.00190

Nath DC (2003) Undocumented migration in India with special reference to West Bengal: A big security threat. Paper presented in the workshop at the Indian Statistical Institute on the Undocumented Migration From Bangladesh to West Bengal organized by Population Studies Unit. Kolkata, India

Norris Robert E (1972) Migration as spatial interaction. J Geogr 71(5):294–301. https://doi.org/10.1080/00221347208981672

Pelletier F, Spoorenberg T (2015) 'Evaluation and analysis of age and sex structure'. Presented in Regional Workshop on the Production of Population Estimates and Demographic Indicators (Population Division) UN Department of Economic, and Social Affairs, Addis Ababa, Equador

Press Trust of India (2008) Parliamentary Standing Committee Report on Bangladeshi Immigration'. Times of India. Accessed 20 July 2019

Ramachandran S (2005) Global migration perspectives-indifference, impotence and intolerance: transnational Bangladeshis in India, No 42. Global Commission on International Migration, Switzerland

Ratha D, Shaw W (2007) South-South Migration and Remittances, World Bank Working Paper No. 2. The World Bank, Washington DC

Roy G (2003) Some observation on Bangladeshi migration to India with special reference to West Bengal. Paper presented in the Workshop on Undocumented MigrationFrom Bangladesh to West Bengal organized by Population Studies unit held in Indian Statistical Institute, Kolkata

Sarkar JP (2001) Bangladeshi Migration to West Bengal: a cause of concern. Centre for Studies in Regional Development, JNU, New Delhi, pp 1–34

Sharma M (2015) A study of migration from Bangladesh to Assam, India and its impact, Ph.D. Thesis. Department of Geography, Environment and Population, Faculty of Arts, The University of Adelaide, Australia.

Sengupta D (2015) The partition of Bengal: fragile borders and new identities. Cambridge University Press, UK

Singh P (2002) Demographic movements: the threat to india's economy and security. Low Intensity Conflict Law Enforcement 11(1):94–115. https://doi.org/10.1080/0966284042000268328

Tripathi S (2016) Illegal immigration from Bangladesh to India: toward a comprehensive solution. Retrieved September 5, 2020, from https://carnegieindia.org/: https://carnegieindia.org/2016/06/29/illegal-immigration-from-bangladesh-to-india-toward-comprehensive-solution-pub-63931

Tumbe C (2018) India moving: a history of migration. Penguin Viking, New York, USA

UNHCR (2000) Rapture in South Asia. In The state of the world's refugees 2000: Fifty years of humanitarian action

White A (2009) Internal migration, identity and livelihood strategies in contemporary Russia. J Ethnic Migr Stud 35(4):555–573. https://doi.org/10.1080/13691830902765095

Zakaria A (2015) The footprints of partition. Harper Collins, New York, USA.
Ravenstein EG (1885) The laws of migration. J Stat Soc London 48(2) 167-10.2307/2979181

Utpal Roy is an Associate Professor of Geography at the University of Calcutta, India. His research interest covers population dynamics and geography of health and healthcare. He specializes in field-based research on migration and other issues related to population dynamics and has published several papers on population and urban environment. He has been awarded a Ph.D. from Vidyasagar University.

Subham Pramanick has obtained his undergraduate (B.Sc. Honours) and postgraduate (M.Sc.) degree achieving 1st class from the University of Calcutta with National Scholarship from MHRD, (Presently Ministry of Education) Govt. of India, New Delhi. He has teaching experience in undergraduate and postgraduate levels for more than one year. He has published several research papers in notable journals, the research interest of whose are specifically emphasized on Urban Transport, Urban Population Dynamics, Social Hazards, quantitative analysis of urbanization etc. Presently he is pursuing a Ph.D. as a UGC-JRF in the Geography Department, University of Calcutta.

Chapter 15
Household Vulnerability of Tribal People to Climate Change in the Part of Dooars Region, West Bengal, India

Pamela Deb and Rameswar Mukherjee

Abstract This study applied the Vulnerability Index at the household level (HVI) to assess the vulnerability to climate variability among the major Scheduled Tribes (STs) in the Dooars region of West Bengal, India. These tribal communities are Oraon, Munda, and Santal, who mainly work as tea plantation labourers. Few of them are crop rearers. A total of 650 households were selected based on purposive sampling, and the sample size was chosen by the proportional allocation method. The HVI was calculated by the data collected from the households, and then a composite indicator index was prepared by assigning a relative weight to each indicator. The final outcome is a weighted average index which is further described with the IPCC dimension of adaptability, sensitivity, and exposure. The study area is located in the foothill zone of Shiwalik Himalaya of West Bengal, which made it a high rainfall and flood-prone zone. Besides, very high rainfall in the hills causes flash floods. This rainfall and floods make the tribal population vulnerable who are predominantly living in rural areas with a poor resource base. This study found that the Maynaguri block is the most vulnerable block of the study area, particularly concerning adaptability and exposure. On the other hand, Falakata is the least vulnerable block due to its high adaptability.

Keywords Vulnerability Index · Adaptability · Sensitivity · Exposure

15.1 Introduction

Studies on household-level vulnerability among the tribal population in response to climate change are rare. Addressing this gap can enhance the knowledge on the drivers of climate change vulnerability, effective adaptation, and resilient techniques to minimize the risk of the tribal communities who are considered marginalized

P. Deb
Department of Geography, Visva-Bharati, Santiniketan, West Bengal, India

R. Mukherjee (✉)
Department of Geography, Samsi College, Malda, West Bengal, India

M. Mishra et al. (eds.), *Regional Development Planning and Practice*,
Advances in Geographical and Environmental Sciences,
https://doi.org/10.1007/978-981-16-5681-1_15

sections of the society. According to the Census of India (2011), around 89.97% of Scheduled Tribes (STs) still live in rural areas that are mostly associated with primary economic activities, e.g., agriculture, forest-produces collection, lumbering, herding, plantation work, etc. These economic activities are directly or indirectly dependent on the climate. Hence, the variability of climate significantly affects their lives. The problems of slower economic growth, lack of education, poverty, and malnutrition are acute among the tribal communities of India. Hence it is observed that they have a dearth of knowledge or capabilities on preparedness, adaptation, and mitigation techniques which are necessary to cope up with climate-induced risk factors. It makes them highly vulnerable to climate change.

Vulnerability denotes the ability or inability of individuals and social groups to respond, in the sense of cope with, recover from, or adapt to any external stress placed on their livelihoods and well-being (Kelly and Adgar 2000). The Intergovernmental Panel on Climate Change (IPCC), however, conceptualized vulnerability from the perspective of climate change and extreme events and denotes it as the degree to which a system is susceptible to and unable to cope with the adverse effects of those extremes such as droughts, floods, and cyclones (IPCC 2007).

Various approaches had been adopted so far by researchers around the world to study vulnerability. For instance, in the single stressor approach, the negative impact of natural calamities on human society was determined (Alexander 1991; Lindell et al.1997; Eakin and Luers 2006). This approach lacked the expanding view on the impact of the man-environment relationship and the socio-cultural impact of the natural hazard (Torry 1979; Palm and Carroll 1998; Cutter et al. 2000). Hence, a broader view was adopted in this domain where multiple aspects were added, including physical, social, cultural, and political. This led to the development of the index-based approach, which considers several sub-components or indicators aggregating the contributing variables of vulnerability (Pandey and Jha 2012; Adu et al. 2018; Hahn et al. 2009; Ghosh and Ghosal 2020). The approach developed by IPCC (2001) has become the most popular among researchers. It set three contributing factors to vulnerability: exposure, sensitivity, and adaptive capacity (IPCC 2001). Exposure relates to the extent to which a particular system may be exposed to climatic stresses or variations (IPCC 2007). On the other hand, sensitivity determines the response of a given system to climate change and may be shaped by socioeconomic and ecological conditions of the system (IPCC 2007). Adaptive capacity in the context of climate change has been defined by the IPCC (2007, p. 869) as:

> the ability of a system to adjust to climate change (including climate variability and extremes) to moderate potential damages, to take advantage of opportunities, or to cope with the consequences.

Researches on vulnerability based on IPCC approach were carried out at both national and regional levels. The studies which took place at national levels were based on census data (Eakin and Bojorquez-Tapia 2008). In these studies, local-level information was missing. Apart from that, many studies also took into consideration the field data collected through household surveys. But in these studies, it was observed that the researchers preferred to collect information only from few sample villages (Sam et al. 2017; Ghosh and Ghosal 2020; Ahsan and Warner 2014). In the previous researches, the study on the vulnerability index of households did not integrate the most vulnerable segment of the society, i.e., the tribal population.

With this backdrop, the present study attempted to assess the vulnerability of the tribal people in parts of the Dooars region at the household level who have limited means of livelihood opportunities. The concerned tribal population of this study includes Oraon, Munda, and Santal, who comprise the major Scheduled Tribes of the study area. The majority of them are plantation labourers in the tea gardens. Several studies were carried out specifically on the tribal communities of this region from where the information about the poor economic condition of the tribes came out (Nandi 2015; Bhowmik 2016). But these studies did not unearth the vulnerability of these tribal populations from climatic variability. The study area, which forms a major part of the Dooars region of West Bengal, is located in the foothill zones of the Outer Himalaya and intricated by a dense mesh of drainage networks. Due to this location, it receives heavy and prolonged rainfall hence become extremely flood-prone.

In the present study, the household vulnerability of the major Scheduled Tribes (STs) of the Jalpaiguri and Alipurduar district, part of Dooars, has been assessed. This paper aims to find out the degree of vulnerability by exploring the household characteristics of the tribal population and their exposure to climate change.

15.2 Study Area

The Dooars region in West Bengal denotes the area between the river Teesta and the river Sankosh. Administratively it includes the Jalpaiguri, Alipurduar, and Kochbehar districts of West Bengal (Fig. 15.1). The word Dooars means door. During the pre-colonial period (before the arrival of the British), these passageways served as routes for business and communication between Bhutan and the plains of Bengal (Bhawal 2016). However, the study area includes the present Jalpaiguri district and Alipurduar district comprising thirteen community development (CD) blocks. The Alipurduar is a newly formed district that was separated from the Jalpaiguri district in 2014, but for obtaining individual ST data from the latest census (Census of India 2011), we have considered undivided Jalpaiguri district.

A major part of the Dooars region comes under the piedmont alluvial plain except for the north and north-eastern part where dissected hills are found. These hills are parts of Shiwalik Himalaya. The Jayanti-Sinchula range, which is located along the Bhutan border, is the highest part of the area. In the south and south-eastern part, the younger alluvial plain is observed, which is composed of fine-grained deposits. Being located in the plain region, the rivers form floodplains on both sides of the rivers.

In this area, two river systems are noticed: the Brahmaputra and the Ganga system. The major rivers are Mahananda (Ganga basin), Teesta, Jaldhaka, Torsha, Raidak,

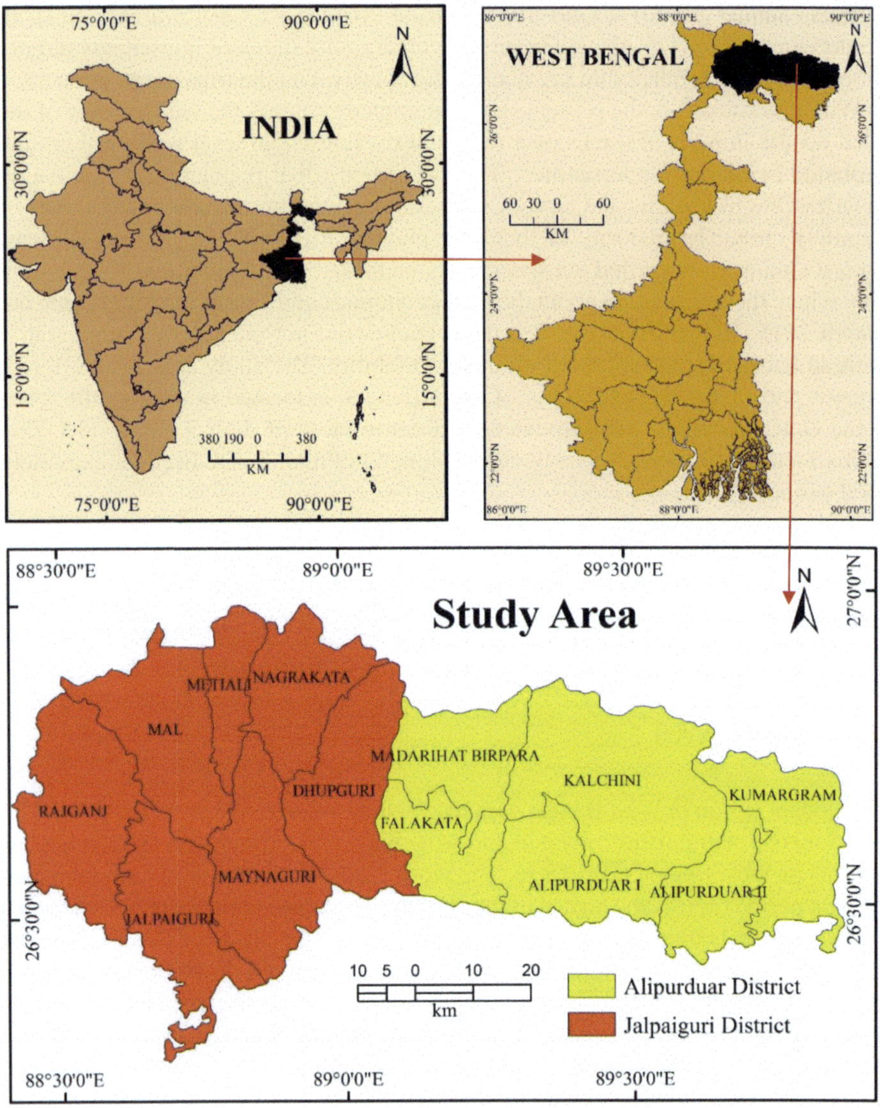

Fig. 15.1 Location map of the study area (*Source* Prepared by the authors)

and Sankosh (Brahmaputra basin). Mainly in the lower reach of the area, rivers are known to shift their courses frequently due to the unconfined valley setting.

The physical environment of this region involves a humid subtropical climate. The highest mean maximum temperature is 34 °C registered in May, and the lowest mean minimum temperature is experienced in January (11 °C). As the region is located in the foothill region of the Himalayas, high rainfall is the obvious characteristic here.

The mean annual rainfall is 3466 mm (Deb, 2015). Most of this rainfall is received from May to September. As a result of that, sudden and high voluminous floods are widely experienced during this period (Gruning 1911; Chakraborty 1987; Rakshit 2003).

As per the Census of India (2011), the study area has the highest number of ST population in the state of West Bengal, i.e., 731,704. The diverse tribal communities inherit their origin from the Dravidian, Proto-Australoid, and Indo-Mongoloid stocks. Overall, 40 tribal groups have been identified in this area. The major STs of this area are Oraon, Munda, and Santal, who altogether share 67.34% of the total ST population. These three tribes came during the colonial era to mainly work in the tea gardens as plantation labourers. Other than this occupation, in the reserved forest as forest labourers, in the tribal colonies (mainly for Santals) as cultivators, and as labourers for constructing railway tracks, they were engaged. Other important STs are Mech, Rabha, Lodha/Kheria, Garo, Lepcha, and Toto. The three major STs are traced in the tea garden villages of the district. Their other locations are marked in the forest and agricultural villages.

The Oraon, Munda, and Santal tribes are part of the *Modesia* community. *Modesia* is a term used to denote the tribes from central India, including Oraon, Munda, Santal, Mahali, Chik Bariak, Lohars, Asur, etc. Although they are heterogeneous in terms of social segregation, their common origin and their occupation brought them under one umbrella. The society of these tribes who live in the tea gardens is different from the ones who live outside the tea gardens. The former group is less rigid in terms of social segregation, following their custom and belief compared to the latter (Deb 2015). The Oraon and Munda are Dravidian tribes, and the Santals are of Proto-Australoid origin.

15.3 Methodology

The present study was carried out based on a primary field survey by using purposive sampling method. The survey was done on the 45 tribal-dominated villages from all the 13 CD blocks of this area. The preliminary information about the concerned tribes was collected from the census data as well as from the voter list issued by Chief Electoral Officer, West Bengal. A total of 650 households from all the 13 CD blocks of the Jalpaiguri (and Alipurduar) district of the Dooars region were surveyed. The proportional allocation method was applied for selecting the sample size.

A comprehensive schedule was prepared for the collection of primary data at the household level of the concerned ST population. Apart from that, the focused group discussions and informal interviews were also conducted to find out the root causes of household vulnerability of the concerned STs and their level of suffering.

15.3.1 Analytical Framework

The Household Vulnerability Index (HVI) has incorporated the household characteristics, livelihood scenario, health scenario, and climatic exposure. Through this approach, the household adaptation techniques and risks to climatic exposure can be evaluated (Hahn et al. 2009; Sam et al. 2017). The index comprises of six major domains: socio-demographic profile, livelihood strategies, health condition, water, food, natural hazards (Table 15.1). Each major domain has further sub-domains. Thus, these six major domains consist of a total of seventeen subdomains (Fig. 15.2). These subdomains were developed based on the existing literature on household vulnerabilities (e.g., Hahn et al. 2009; Pandey and Jha 2012; Ahsan and Warner 2014; Sam et al. 2016; Sam et al. 2017; Ghosh and Ghosal 2020). Apart from that, the focus group discussions and informal interviews conducted in each block gave us the insight to choose these. Hence, the sub-domains have been modified in accordance to relevance to the household vulnerabilities of the rural tribal communities to the climate change.

The household vulnerability index (HVI) of the study area was prepared on the basis of the composite indicator framework method. It consisted of three major dimensions: adaptability, sensitivity, and exposure. These were further subdivided into six domains. As the indicators used under each domain were measured at different scales, therefore it was essential to standardize all of them. For getting the index score for each indicator, the following formula provided by UNDP (2007) was used:

$$\text{Indicator index score} = \frac{X_i - X_{min}}{X_{max} - X_{min}} \tag{15.1}$$

Where, X_i = Actual value.

X_{min} =Minimum value.

X_{max} = Maximum value.

For getting the domain score, the mean indicator index score of each domain was calculated.

Table.15.1 IPCC dimensions of vulnerability

IPCC dimensions of vulnerability	Domains
Adaptability	Socio-demographic profile
	Livelihood strategy
Sensitivity	Health condition
	Access to water
	Provision of food
Exposure	Climate induced natural hazard

Source Adopted from IPCC (2001)

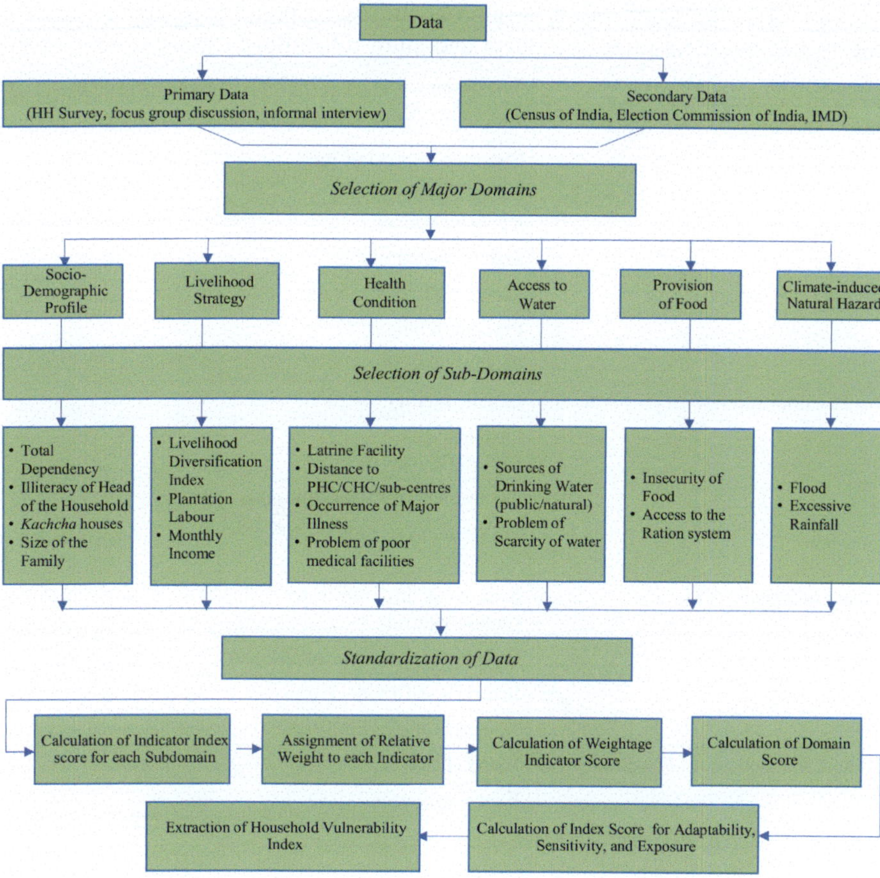

Fig. 15.2 Methodological chart of the present study (*Source* Authors' perception)

For preparing the HVI it was necessary to give weight to all the selected criteria (Antwi-Agyei et al. 2013). Therefore, relative weight for each criterion was provided (Table 15.2), and the weightage indicator score was obtained from the following formula:

$$\text{Weightage indicator score (WIS)} = \text{Indicator index score} \times \text{Relative weight} \tag{15.2}$$

Once WIS of each selected criterion was obtained, the domain score (DS) was calculated by averaging the weightage value of each domain:

$$\text{Domain Score(DS)} = \frac{\sum WIS}{n} \tag{15.3}$$

Table.15.2 Major domains and sub-domains designed for the preparing household vulnerability index

Major domain	Sub-domain	Relative weight
Socio-Demographic Profile(SDP)	Total dependency	6
	Illiteracy of head of the household	4
	Kachcha houses	5
	Size of the family	3
Livelihood strategy(LS)	Livelihood diversification index	6
	Plantation labour	3
	Monthly income	5
Health Condition (HC)	Latrine facility	5
	Distance to PHC/CHC/sub-centres	4
	Occurrence of major illness	6
	Problem of poor medical facilities	6
Access of Water (WT)	Sources of drinking water (public/natural)	4
	Problem of the scarcity of water	3
Provision of Food (FO)	Insecurity of food	5
	Access to the ration system	4
Climate induced natural hazards (CLNH)	Flood (FL)	6
	Excessive rainfall (RF)	6

Source Computed by the authors

The index scores for adaptability, sensitivity, and exposure were calculated as follows:

$$\mathrm{DM}_{Adaptability} = \frac{\sum(DSofSDP + DSofLS)}{n} \tag{15.4}$$

$$DM_{Sensitivity}\frac{\sum(DSofHC + DSofWT + DSofFO)}{n} \tag{15.5}$$

$$DM_{Exposure} = \frac{\sum(DSofFL + DSofRF)}{n} \tag{15.6}$$

After that, the household vulnerability index was calculated by combining the scores for adaptability, sensitivity, and exposure:

$$\text{Household vulnerability Index} = \frac{\sum(DM_{Adaptive\ capacity} + DM_{Sensitivity} + DM_{Exposure})}{n} \tag{15.7}$$

The livelihood diversification index was prepared by using the formula of Simpson diversification index (SDI):

$$\text{SDI} = 1 - \sum_{i=1}^{n} P_i^2 \tag{15.8}$$

Where n is the total number of source of income; and

P_i is the income proportion of the i-th income source

After calculating the index, the value was subtracted by 1 to bring uniformity to the results of all the selected indicators.

For assessing the variability of the rainfall, the monthly average rainfall data was collected from the India Water Portal[1] database for the period of 1901 to 2002. Z score was prepared to show the rainfall variability. The information on the occurrence of floods in the study area was collected from reports published by the Irrigation and Waterways Department, Govt of West Bengal (2004, 2014, 2020), and research articles (Sarkar 2008; Ghosh and Ghosal 2020).

15.4 Result and Discussion

15.4.1 Variability of Rainfall

Over the assessed period (1901–2002), greater variability of the rainfall has been observed. The annual average rainfall ranges from 266.02 mm (1984) to 130.87 mm (1957). Around 80% of the total annual rainfall occurs in the monsoon season (June to September). More than 300 mm rainfall has been recorded in the monsoon season over the assed period. The highest amount of rainfall has been recorded in July (1105.93 mm). In 102 years of study period, the frequency of more than 500 mm rainfall is 72 times. In June, August, and September maximum occurrence of the average monthly rainfall ranges from 300 to 500 mm. The Z score of the monthly rainfall in the monsoon season reveals that 45 times it registers positive and 57 times negative. The extremely higher variability of rainfall has been recorded in 1955 (2.91), followed by 1984 (2.83). Since 1901–2002, a higher magnitude of rainfall has been reported 25 times. From 1917 to 1922, 1949 to 1955, and 1984 to 1989 for 3 times as well as in 1954 and 1955 consecutive years, an excessive amount of rainfall has been noted.

---

[1] See for details: https://www.indiawaterportal.org/.

15.4.2 Occurrence of Flood

The study region is notorious for the high flood situation. The river systems originate from Darjiling, Sikkim, Bhutan, Tibetan Himalaya, and submontane country creates havoc floods in the Dooars. The rivers like Teesta, Torsa, Raidak, Sankosh are largely responsible for flood hazards here. Among all these rivers, the Teesta river is considered as the most devastating. There are several reasons for the occurrence of high magnitude floods here: sudden and torrential rainfall caused by cyclonic activities; the incidence of heavy rainfall for the longer period; abrupt reduction of slope in the Himalaya-Gangetic plain interface; nearly flat riparian tract, unconfined valley setting in plain, etc. The Dooars remain heavily moist for most of the time of the year, usually because of the frequently occurring high-intensity rainfall. The rivers coming from the Himalayan region bring a huge volume of water discharge. In addition to that, prolonged rainfall in the headward area of the catchment aggravates the amount of discharge which eventually causes floods in the Dooars. This region experienced devastating floods in 1886, 1890, 1898, 1899, 1906, 1910, 1915, 1968,1993, 2003, 2004, 2007, 2009, 2010, 2015, and 2017 (Mahalanobis 1927; Lieten and Datta 1995; Irrigation and Waterways Department, Govt of West Bengal 2004, 2014, 2020; Sarkar 2008; Das 2009; Roy 2011; Ghosh and Ghosal 2020).

15.4.3 The Analysis of Household Vulnerability

The analysis of household vulnerability has taken place in two ways: domain-wise and IPCC-dimension-wise.

15.4.3.1 Socio-Demographic Profile

Social fabric plays the most important role related to how a community will respond to any natural hazard. In this social fabric, the socio-demographic profile is an essential element. A good socio-demographic profile can enhance the adaptive capacity of the households (Sam et al. 2017). In this study, among all the blocks, Maynaguri is the most vulnerable (0.750). It is followed by Alipurduar I (0.642), Nagrakata (0.584), and Alipurduar II (0.516) (Fig. 15.3.a).

For understanding the economic dependency of any population, the dependency ratio is calculated. It shows the number of dependents per 100 workers. A high dependency ratio increases the vulnerability of households. It advocates that economically active individuals have many others to support. Hence, resources for coping with natural disasters would be more limited. In this study, the dependency ratio is 63.52%. The highest index value is observed in the Maynaguri block (1.000). Apart from that, Rajganj also has a higher dependency ratio (0.957). The least dependency ratio is observed in Falakata (0.000) (Table 15.3).

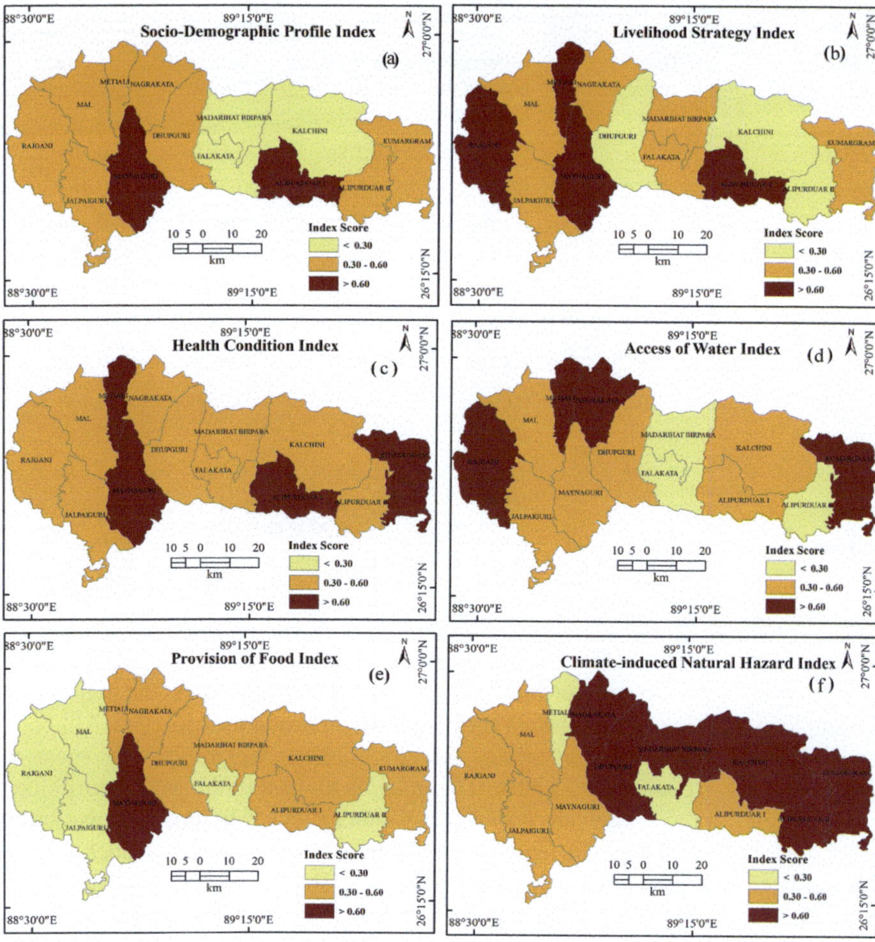

Fig. 15.3 Major domain wise index map: **a** Socio-demographic profile index, **b** Livelihood strategy index, **c** Health condition index, **d** Access of water index **e** Provision of food index, **f** climate-induced natural hazard

Literacy has a key role in coping up with difficult situations. If the head of the household is literate, he/she has a better ability to communicate with the government authority regarding climate-related risks. The householders also can equip themselves with new adaptative technologies to make them resilient. Here, most of

the household heads are illiterate (68%). The highest index value for illiterate households is observed in Alipurduar I block (1.000). It is followed by Mal (0.882), Rajganj (0.850), and Matiali (0.829). The index value is the lowest (0.000) for the Maynaguri block (Table 15.3).

Table.15.3 Socio-demographic profile

Blocks	Total dependency	Illiteracy of head of the household	Kacchcha houses	Size of the family	Average
Kalchini	0.252	0.202	0.137	0.467	0.265
Mal	0.187	0.882	0.242	0.533	0.461
Dhupguri	0.201	0.390	0.029	0.633	0.313
Madarihat-Birpara	0.164	0.047	0.000	0.467	0.169
Nagrakata	0.566	0.756	0.216	0.800	0.584
Kumargram	0.130	0.455	0.328	0.733	0.412
Matiali	0.338	0.829	0.046	0.767	0.495
Falakata	0.000	0.099	0.179	0.283	0.140
Alipurduar II	0.081	0.763	0.487	0.733	0.516
Alipurduar I	0.279	1.000	0.689	0.600	0.642
Jalpaiguri	0.347	0.567	0.176	0.333	0.356
Rajganj	0.957	0.850	0.059	0.000	0.467
Maynaguri	1.000	0.000	1.000	1.000	0.750

Source Computed by the authors

In the study area, most of the houses are *semi-pucca*[2]in nature. As the majority of the sample population lives in the tea gardens, they get residential quarters. Although they are living in semi-pucca houses, the condition is deplorable. The share of *kachcha* houses accounts for 32.62%. The index is highest in the Maynaguri block (1.000) where all the houses are *kachcha.*[3] Another block Alipurduar I, also has a higher index value (0.689). The least value is for the Madarihat-Birpara block (0.000) where only 15% of households have *kachcha* houses. The walls of the *kachcha* houses are made of mud and bamboo and have tin or thatch or tile roofs. Some temporary shacks have been identified which are constructed by plastic or polythene sheet, tin, mud, and waste building materials. The *kachcha* houses are extremely vulnerable for any households, especially those who are living in the Dooars area, which is characterized by heavy rainfall and recurrent flood.

Family size is also an important indicator. It is generally accepted that if a household has a large number of family members, it is more vulnerable. The higher number of dependents, difficulty to access adequate houses, and basic infrastructure for house, education, health care facilities can make them vulnerable. Here, the Maynaguri is the most vulnerable (1.000), which is followed by Nagrakata, Matiali, Kumargram, and Alipurduar II. In this respect, the least vulnerable block is Rajganj (0.000) followed by Falakata and Jalpaiguri.

---

[2] A house that has its walls made up of concrete but roof is made up of tin/asbestos/tile/bamboo/timber etc.

[3] A house which have not have concrete roof/wall/floor and made of mud,grass,thatch,reeds,bamboos etc.

15.4.3.2 Livelihood Strategy

People engaged with diverse livelihoods have a better chance of survival during a natural disaster. It would help them to find out more defensive strategies (Van Den Berg 2010). The households whose all income comes from only one source are more vulnerable compared to those who diversify their assets, incomes, and activities to reduce risk in case of the failure of any one of them (Barrett et al. 2001). The score for the livelihood diversification index is highest in Matiali (1.000), Rajganj (1.000), Maynaguri (1.000) (Fig. 15.3b). It means in these blocks, the tribal households are dependent on only one source of income. The diversification index is lowest in Alipurduar II block (0.000) (Table 15.4). In this block, the major livelihood options observed are agriculture, tea plantation labour, and labour other than the tea plantation sector.

In the study area, more than 85% of households are dependent on the wages of tea garden authority. The index value of the households who are solely dependent on the plantation is highest in Rajganj (1.000) and Maynaguri (1.000) where all households are tea plantation labour. The value is also higher in Matiali (0.939), Nagrakata (0.838), Kumargram (0.826), Alipurduar I (0.809) and Falakata (0.698). The least value is observed in Alipurduar II (0.000) where around 23% of households are engaged with agriculture. But here, agriculture is practiced at the subsistence level. Another block where the value is low is Kalchini (0.288). In this block, household members, apart from being plantation labourers, are also engaged as labourers in other sectors, businesses, and services. However, the nature of business includes small grocery shops, rice beer shops (*handia*), tea shops, and service professions,

Table.15.4 Livelihood strategy

Blocks	Livelihood diversification index	Plantation labour	Monthly income	Average
Kalchini	0.000	0.288	0.406	0.231
Mal	0.167	0.639	0.099	0.302
Dhupguri	0.063	0.431	0.181	0.225
Madarihat-Birpara	0.063	0.567	0.480	0.370
Nagrakata	0.375	0.838	0.273	0.495
Kumargram	0.167	0.826	0.054	0.349
Matiali	1.000	0.939	0.000	0.646
Falakata	0.375	0.698	0.493	0.522
Alipurduar II	0.167	0.000	0.280	0.149
Alipurduar I	1.000	0.809	0.358	0.722
Jalpaiguri	0.375	0.350	0.688	0.471
Rajganj	1.000	1.000	0.688	0.896
Maynaguri	1.000	1.000	1.000	1.000

Source Computed by the authors

including sub-staffs (like security guards) in the tea garden offices. Due to poor educational background, they are not able to get high-graded jobs. It is noteworthy that in most of the blocks, the index value is higher (near to 1.000), which means that the majority of the household working members are plantation labourers. They work as tea plantation labourers in the field as well as in the factory. The tea plantation economy is highly dependent on climate. Besides, sickness in the tea industry of the Dooars region, lockouts, withdrawal of various facilities from the tea garden labourers by the tea garden authorities, irregular wages has made this livelihood option vulnerable. Again, in the Maynaguri block, the index value is the highest (Table 15.4).

Overall, the monthly income among these tribal households is extremely low. Around 76.92 percent of household income is below Rs. 4000/month. Based on the monthly income scenario, Maynaguri (1.000) is the most vulnerable block. Higher index values are also observed in Jalpaiguri (0.688) and Rajganj (0.688). The least value is found in the Matiali block (0.000). It is followed by Mal (0.099), Dhupguri (0.181), Nagrakata (0.273), and Alipurduar II (0.280).

15.4.3.3 Health Condition

Matiali block has the highest health vulnerability index (0.750). In this block, the major drawbacks are greater distance to PHC/CHC/Sub-centers (1.000) and occurrence of major illness (0.881) (Fig. 15.3c). Besides, there is a shortage of latrine facilities which signifies poor sanitation of this block. The problem of not having latrines is severe in the whole study area (Table 15.5). Around 75.69% of households do not possess latrines in their houses. Hence, open-air defecation is very common among them. No community toilets have also been found during the survey. Among the blocks, Rajganj and Maynaguri have the highest score for latrine facilities (1.000). The least score is observed in the Jalpaiguri block (0.000).

The next variable is the distance from the Primary Health Center (PHC)/Community Health Center (CHC)/Sub-center. The higher distance indicates higher vulnerability. Overall, 40.15% of households are located beyond 1 km radius of the health centers. In this respect, Dhupguri, Matiali, and Jalpaiguri are highly vulnerable (1.000) (Table 15.5). The least vulnerable blocks are Mal, Falakata, Rajganj, and Maynaguri (0.000).

In the study area, around 73.08% of households raised their grievances for poor medical facilities. Many blocks score one (1.000) in this section, including Kumargram, Alipurduar II, Alipurduar I, Rajganj, Jalpaiguri, and Maynaguri. Apart from these, the Kalchini block also scores high (0.758). The least vulnerable blocks are Dhupguri and Nagrakata. In Dhupguri, the sample villages are near to Dhupguri Municipality, which makes them less vulnerable than others. Because, for any emergency, they can go to the urban area. On the other hand, in Nagrakata, the sample villages are near the government hospital. It is to be mentioned that, according to the Plantation Labour Act, 1951, in each tea garden, there should be a hospital with doctors and basic medical facilities (Bhowmik 1996). But in the field, the noticeable

Table.15.5 Health condition

Blocks	Latrine facility	Distance to PHC/CHC/Sub-centre	Occurrence of major illness	Problem of poor medical facility	Average
Kalchini	0.476	0.248	0.794	0.758	0.569
Mal	0.704	0.000	0.704	0.515	0.481
Dhupguri	0.417	1.000	0.500	0.000	0.479
Madarihat-Birpara	0.611	0.200	0.611	0.030	0.363
Nagrakata	0.695	0.467	0.778	0.000	0.485
Kumargram	0.643	0.357	0.655	1.000	0.664
Matiali	0.623	1.000	0.881	0.520	0.756
Falakata	0.651	0.000	0.620	0.408	0.420
Alipurduar II	0.445	0.282	0.598	1.000	0.581
Alipurduar I	0.804	0.618	0.000	1.000	0.605
Jalpaiguri	0.000	1.000	0.000	1.000	0.500
Rajganj	1.000	0.000	0.333	1.000	0.583
Maynaguri	1.000	0.000	1.000	1.000	0.750

Source Computed by the authors

hindrances are - the absence of doctors, ambulance, lack of essential medicines. The problem of insufficient medical facilities also persists in the non-tea garden villages (Table 15.5).

The score for the occurrence of major illness is highest in the Maynaguri block (1.000). In this block, from all the houses, the occurrence of major illness has been recorded. The most prevalent diseases in the study area are tuberculosis, diarrhoea, and typhoid, which are specifically caused by malnutrition, open-air defecation, and recurrence of the flood.

15.4.3.4 Access to Water

Water is one of the essential elements of any household. Apart from drinking, water has the most important role in the domestic works of a house. In the study area, most of the households don't have their own source of drinking water. The plantation labourers use community tube well/well/tap. The panchayats have constructed taps in many places. In the tea gardens, both the village panchayat and the tea garden authorities facilitate the construction of taps and wells in the labour lines. The possession of private wells or tube wells is not very frequent here. Hence, they are dependent on the public or natural source of water (Table 15.6). In Jalpaiguri, Rajganj, and Maynaguri blocks, all the households are dependent on the natural or public source of water (1.000) (Fig. 15.3d). The scarcity of drinking water is also a major concern

Table.15.6 Access to water

Blocks	Source of drinking water (public/natural)	Problem of scarcity of drinking water	Average
Kalchini	0.585	0.410	0.497
Mal	0.771	0.367	0.569
Dhupguri	0.369	0.300	0.335
Madarihat-Birpara	0.159	0.333	0.246
Nagrakata	0.503	0.700	0.602
Kumargram	1.000	0.339	0.670
Matiali	0.957	0.472	0.714
Falakata	0.040	0.302	0.171
Alipurduar II	0.000	0.051	0.026
Alipurduar I	0.663	0.118	0.390
Jalpaiguri	1.000	0.200	0.600
Rajganj	1.000	1.000	1.000
Maynaguri	1.000	0.000	0.500

Source Computed by the authors

in some blocks, including Rajganj (1.000), Nagrakata (0.700). The lack of a sufficient number of tap/ well/ tube well is the cause of such scarcity. Overall, the blocks vulnerable to water are Rajganj, Kumargram, Matiali, and Nagrakata.

15.4.3.5 Provision of Food

Due to low income, food habits are not healthy among these households. They take rice, lentils, and vegetables. The proper intake of protein and fat is absent among them. Many of them are unable to eat three times a day. Hence tuberculosis is a common disease among them. All the households in the Maynaguri block raised their problem of food insecurity (1.000) (Fig. 15.3e) (Table 15.7). The reasons for food insecurity are their low average income, delay in getting a wage, not getting ration properly. It is to be mentioned that around 80.6% of households are dependent on the ration[4] for food.

15.4.3.6 Climate-Induced Natural Hazard

Jalpaiguri is a flood-prone district located in the foothills of the Himalayas, and it gets copious rainfall. The major rivers of this area are Teesta, Torsha, Jaldhaka,

[4] A system where poor people get food and non-food items at subsidized prices. which is procured and maintained by the Food Corporation of India, a Government-owned corporation and in some instances Tea Garden authority.

Table.15.7 Provision of food

Block	Insecurity of food	Access to the ration system	Average
Kalchini	0.392	0.320	0.356
Mal	0.119	0.459	0.289
Dhupguri	0.143	1.000	0.572
Madarihat-Birpara	0.333	0.774	0.554
Nagrakata	0.286	0.387	0.337
Kumargram	0.235	0.369	0.302
Matiali	0.012	0.682	0.347
Falakata	0.000	0.480	0.240
Alipurduar II	0.452	0.066	0.259
Alipurduar I	0.72	0.455	0.588
Jalpaiguri	0.048	0.258	0.153
Rajganj	0.524	0.000	0.262
Maynaguri	1.000	0.258	0.629

Source Computed by the authors

Raidak, and Gadadhar. The blocks Jalpaiguri, Rajganj, and Maynaguri are extremely vulnerable to flood as these blocks are located by those mighty rivers, and the terrains are almost flat. Apart from that, the sample villages of these blocks are situated at the lower reaches of those rivers. The rivers at their mature stage overtop the banks and create floods. Alipurduar I (0.900), Kumargram (0.800), and Alipurduar II (0.733) blocks are also in the list of highly vulnerable blocks (Table 15.8). In terms of extreme

Table.15.8 Climate induced natural hazard

Block	Flood	Excessive rainfall	Average
Kalchini	0.644	1.000	0.822
Mal	0.689	0.258	0.474
Dhupguri	0.667	0.677	0.672
Madarihat-Birpara	0.644	0.871	0.757
Nagrakata	0.611	0.742	0.676
Kumargram	0.800	0.710	0.755
Matiali	0.000	0.581	0.290
Falakata	0.511	0.032	0.272
Alipurduar II	0.733	0.516	0.625
Alipurduar I	0.900	0.290	0.595
Jalpaiguri	1.000	0.000	0.500
Rajganj	1.000	0.000	0.500
Maynaguri	1.000	0.000	0.500

Source Calculated by authors

rainfall, Kalchini block is the most vulnerable block (1.000) (Fig. 15.3f). Madarihat-Birpara (0.871), Nagrakata(0.742), Kumargram (0.710), Dhupaguri (0.671) are some other blocks, where high rainfall is received. These blocks, due to the nearness of the Bhutan Himalaya, experience a heavy amount of rainfall. Overall, in terms of experiencing natural hazard, Kalchini is the most vulnerable block (0.822), which is followed by Madarihat-Birpara (0.757), Kumargram (0.755), Nagrakata (0.676), and Dhupguri (0.672). The least vulnerable blocks are Falakata (0.272) and Matiali (0.290).

15.5 Household Vulnerability Index

The HVI was used to identify the vulnerability of communities by considering the IPCC dimensions of climate change, i.e., adaptivity, sensitivity, and exposure. Here the same data was used and then grouped under IPCC dimensions to estimate the Vulnerability Index. The various dimensions of vulnerability have been given in the table (Tables 15.9, 15.10 and 15.11). Maynaguri has the highest vulnerability in terms of adaptability which is followed by Rajganj (3.138) and Alipurduar I (3.008). The least vulnerable block is Madarihat-Birpara (0.986) (Fig. 15.4a). The index value for sensitivity is highest for Maynaguri (Fig. 15.4b). A high index value of latrine facilities, family size, dependence on natural or public or natural sources for water, food insecurity, and the risk of flood make Maynaguri block highly sensitive to household vulnerability.

Table.15.9 Adaptability to climate-induced natural hazard

Block	Socio-demographic profile	Livelihood strategy	Average
Kalchini	1.102	0.830	0.966
Mal	1.864	1.104	1.484
Dhupguri	1.203	0.798	1.001
Madarihat-Birpara	0.643	1.332	0.988
Nagrakata	2.474	1.951	2.212
Kumargram	1.610	1.231	1.420
Matiali	1.968	2.939	2.453
Falakata	0.535	2.105	1.320
Alipurduar II	2.043	0.707	1.375
Alipurduar I	2.730	3.286	3.008
Jalpaiguri	1.558	2.018	1.788
Rajganj	2.359	3.918	3.138
Maynaguri	3.500	4.333	3.917

Source Computed by the authors

Table.15.10 Sensitivity to climate-induced natural hazard

Block	Health	Water	Food	Total	Average
Kalchini	3.170	1.445	1.619	6.233	2.078
Mal	2.708	1.009	1.215	4.932	1.644
Dhupguri	2.271	1.712	2.358	6.340	2.113
Madarihat-Birpara	1.926	2.182	2.381	6.489	2.163
Nagrakata	2.501	2.044	1.489	6.035	2.012
Kumargram	3.643	0.509	1.325	5.476	1.825
Matiali	3.879	0.794	1.393	6.067	2.022
Falakata	2.356	2.374	0.960	5.690	1.897
Alipurduar II	3.235	2.077	1.262	6.574	2.191
Alipurduar I	3.123	0.851	2.711	6.685	2.228
Jalpaiguri	2.500	0.300	0.636	3.436	1.145
Rajganj	3.250	1.500	1.310	6.060	2.020
Maynaguri	4.250	0.000	3.016	7.266	2.422

Source Computed by the authors

Table.15.11 Exposure to climate-induced natural hazard

Block	Flood	Excessive rainfall	Weightage total	Weighted average
Kalchini	3.864	6.000	9.864	4.932
Mal	4.134	1.548	5.682	2.841
Dhupguri	4.002	4.065	8.067	4.033
Madarihat-Birpara	3.864	5.226	9.090	4.545
Nagrakata	3.666	4.452	8.118	4.059
Kumargram	4.800	4.258	9.058	4.529
Matiali	0.000	3.484	3.484	1.742
Falakata	3.066	0.194	3.260	1.630
Alipurduar II	4.398	3.097	7.495	3.747
Alipurduar I	5.400	1.742	7.142	3.571
Jalpaiguri	6.000	0.000	6.000	3.000
Rajganj	6.000	0.000	6.000	3.000
Maynaguri	6.000	0.000	6.000	3.000

Source Computed by the authors

However, in terms of exposure, the Kalchini block has the highest vulnerability(Fig. 15.4c). This block has both high rainfall and the risk of flood. Due to its locational virtue, this block receives high-intensity rainfall, which makes this block highly vulnerable.

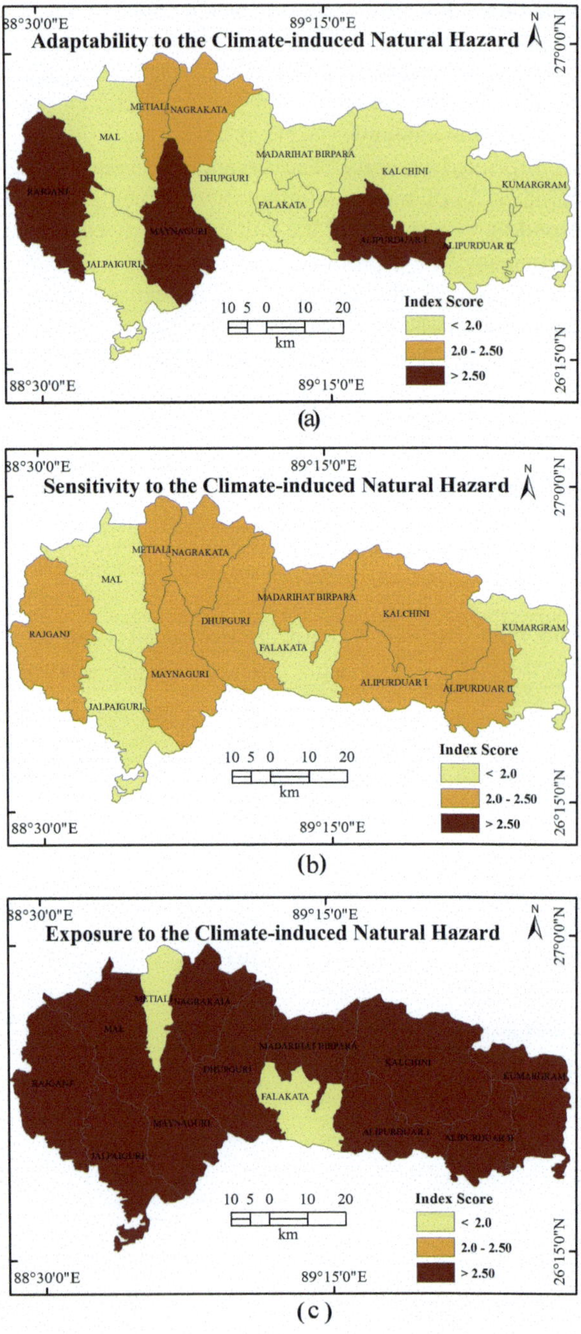

Fig. 15.4 Adaptibility, Sensitivity, and Exposure to climate-induced natural hazard: **a** adaptability to climate-induced natural hazard, **b** sensitivity to climate-induced natural hazard, **c** exposure to climate-induced natural hazard (*Source* Prepared of authors)

The household vulnerability index which was calculated after adding the relative weight to each domain, shows that Maynaguri is the most vulnerable block (3.113) (Fig. 15.5). It is followed by Alipurduar I (2.936), Rajganj (2.791), Nagrakata (2.761), Kalchini (2.659), Kumargram (2.592), and Madarihat-Birpara (2.565). It is seen that the Maynaguri block, in each of the domains, shows either the highest or higher score. In this block, the sample village is a tea garden village where the authority has not given the minimum provision. The absence of semi-pucca residential quarters, latrines, unavailability of ration, inadequate medical facilities are some of the infrastructural lacunae in this block. The next most vulnerable block is Alipurduar II which acquires the highest score in illiterate household heads, poor medical facilities, and livelihood diversification index. This block also scores high in size of the family, *kachcha* houses, latrine facility, the occurrence of flood, food insecurity, and source of natural drinking water. Several houses reported about the major flood that took place in 1993 in this block, which in some incidents, compelled them to change their residence. The Rajganj is the third most vulnerable block where the vulnerability exists in terms of higher dependency ratio, illiterate household head, livelihood diversification index, dependence only on the plantation, low monthly income, lack of latrine facilities, dependence on the natural or public source of drinking water, scarcity of water and occurrence of a flood. The least vulnerable block is Falakata because the household members have higher adaptability.

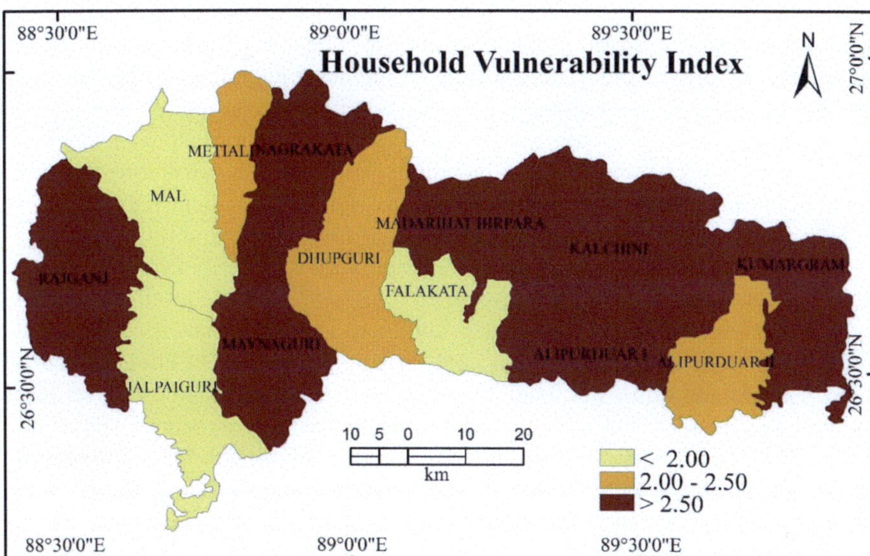

Fig. 15.5 Household Vulnerability Index (Mapping bythe authors based on their own calculation)

15.6 Conclusion

This study endeavours for the vulnerability status of the major tribal communities at the household level. The study attempts to construct an index, based on their socio-demographic profile, livelihood strategy, water and food provision, health condition, exposure to natural hazards. This study is chiefly based on primary data, which is free from the pitfalls of secondary data-based work. Further, the provision of relative weightage to the indicators makes it more transparent.

The impact of natural hazards or disasters on any community largely depends on their resource structure (economic and human). If they have a poor economic base, poor resource base, low level of literacy, they are more prone to be vulnerable. The study illustrates that most of these tribal communities have a very poor socio-demographic profile. The majority of them are dependent on the tea plantation economy and work as labourers at tea gardens. It makes them vulnerable in two ways: plantation is dependent on climate and the declining condition of the tea garden. According to PLA,1951, the labourers are supposed to get the residential house, health care facilities, water, ration at subsidized prices, and many other facilities. But, due to sickness in the tea industry, now most of these facilities are withdrawn. Some tribal groups are associated with agriculture. But these are all subsistence forms of agriculture. The tribal communities which are found in agricultural villages are marginally better than the tea garden labourers as they are cultivating their own land and are not answerable to anyone. On the other hand, forest dwellers are also self-reliant, but their occupation is not permanent. The study further points out the poor healthcare infrastructure, dependence on public or natural water sources, and food insecurity due to low income. This condition makes them more vulnerable to natural hazards, which are heavy rainfall and flood-like situations almost every year.

Empirical studies on household vulnerability to climate change among the tribal communities in the Dooars region are limited. Hence, this micro-regional study will help policymakers getting acquainted with the tribal vulnerability scenario of this region. The present study shows the susceptibility of the tribal communities to climate extremities due to their impoverished socio-demographic profile, livelihood strategy, health condition, access to water, and food. The policymakers could be able to implement policies accordingly. It could minimize the risk of climate-induced natural hazards and uplift the local economy and widen the source of livelihood opportunities. Besides, the policymakers can evaluate their existing policies and modify them accordingly. Besides, the development of literacy status among the tribes, the extension of adult literacy to make them aware, and the revival of the tea industry to enhance the income opportunities are the necessary needs here. In this way, the adaptive capacity of the people may increase, which in turn could be able to reduce the level of risk among the residing tribal peoples.

References

Adu DT et al (2018) Application of livelihood vulnerability index in assessing small holder maize farming households' vulnerability to climate change in Brong-Ahafo region of Ghana. Kasetsart J Soc Sci 39:22–32

Ahsan N, Warner J (2014) The socioeconomic vulnerability index: A pragmatic approach for assessing climate change led risks–A case study in the south-western coastal Bangladesh. Int J Disaster Risk Reduction 8:32–49

Alexander D (1991) Natural disasters: a framework for research and teaching. Disasters 15:209–226

Antwi-Agyei P et al (2013) Characterising the nature of household vulnerability to climate variability: empirical evidence from two regions of Ghana. Environ Dev Sustain 15:903–926

Barrett CB et al (2001) Nonfarm income diversification and household livelihood strategies in rural Africa: concepts, dynamics, and policy implications. Food Policy 26:315–331. https://doi.org/10.1016/s0306-9192(01)00014-8

Bhawal P (2016) Uponibeshik porbe Duars er cha shromikder jibon somossya (in Bengali). In: Ghosh AG, Biswas S (eds) Poribortoner dharai Teesta bonger Jalpaiguri jela. Progotisheel Prokashok, Kolkata, pp 243–260

Bhowmik SK (1996) Tea plantation workers in West Bengal. In: Bhowmik SK, Xaxa V, Kalam MA (eds) Tea plantation labour in India. Friedrich Ebert Stiftung. New Delhi, pp 43–80

Bhowmik SK (2016) Tribal labour in the tea plantations of West Bengal: problems of migration and settlement. In: Radhakrishna M (ed) First citizens: studies on adivasis, tribals and indigenous peoples in India. Oxford University Press, New Delhi, pp 207–227

Census of India (2011) West Bengal census handbook. Directorate of Census Operations, Kolkata, West Bengal

Chakraborty PK (1987) Jalpaiguri jelar bhougolik porichoy (In Bengali). In: Bhattacharya A (ed) Madhuparni, bishes Jalpaiguri jela sonkha. Sahitya Porishad, Kolkata, pp 22-29

Cutter SL et al (2000) Revealing the vulnerability of people and places: a case study of georgetown county, south carolina. Ann Assoc Am Geogr 90(4):713–737

Das BK (2009) Flood disasters and forest villagers in Sub Himalayan Bengal. Econ Polit Wkly XLIV(4):24–30

Deb P (2015) Changing habitat, economy and society of the major tribes in Jalpaiguri District. Unpublished PhD thesis, Department of Geography, Visva-Bharati, Santiniketan

District wise meteorological data. Retrived from https://www.indiawaterportal.org/. Accessed 26 Sept 2020

Eakin H, Bojorquez-Tapia LA (2008) Insight into the composition of household vulnerability from multicriteria decision analysis. Glob Environ Chang 18:112–127

Eakin H, Luers AL (2006) Assessing the vulnerability of social-environmental systems. Annu Rev Environ Resour 31:365–394

Electoral roll of Jalpaiguri district. Retrieved from http://ceowestbengal.nic.in/ACName?DCID=2, Accessed 5 Apr 2014

Ghosh M, Ghosal S (2020) Climate change vulnerability of rural households in flood-prone areas of Himalayan foothills, West Bengal, India. Environ Dev Sustain. https://doi.org/10.1007/s10668-020-00687-0

Gruning JF (1911) Eastern Bengal and Assam District Gazetteers, 'Jalpaiguri'. Pioneer Press, Allahabad

Hahn MB et al (2009) The livelihood vulnerability index: a pragmatic approach to assessing risks from climate variability and change—A case study in Mozambique. Glob Environ Chang 19:74–88

IPCC (2001) Climate change 2001: Synthesis report. Contribution of working groups I, II and III to the third assessment report of the intergovernmental panel on climate change. Cambridge, Cambridge University Press, UK

IPCC (2007) Climate change 2007: Synthesis report. Contribution of working groups I. II and III to the fourth assessment report of the intergovernmental panel on climate change. IPCC, p 104

Irrigation and Waterways Department (2004) Annual Flood Report for The Year 2004. Govt. of West Bengal, Kolkata
Irrigation and Waterways Department (2014) Annual Flood Report for The Year 2014. Govt. of West Bengal, Kolkata
Irrigation and Waterways Department (2020) Annual Flood Report for The Year 2019. Govt. of West Bengal, Kolkata
Kelly PM, Adger WN (2000) Theory and practice in assessing vulnerability to climate change and facilitating adaptation. Clim Change 47:325–352
Lieten GK, Datta P (1995) Panchayats in dealing with natural disasters: 1993 floods in Jalpaiguri. Econ Pol Wkly 30(15):828–832
Lindell MK et al (1997) Adoption and Implementation of Hazard Adjustments. Int J Mass Emerg Disasters 15:323–453
Mahalanobis PC (1927) Rainfall and floods in North Bengal 1870–1922. Bengal Secretariat Press, Kolkata
Nandi R (2015) Political Ecology of a Multi-ethnic Setting: Environmental History of Dooars, West Bengal. In: Barman B, Barman KC (eds) History and culture of North Bengal. Pragatishil Prakashak, Kolkata, pp 78–142
Palm R, Carroll J (1998) Illusions of safety: culture and earthquake hazard response in California and Japan. Westview Press CO, Boulder
Pandey R, Jha SK (2012) Climate vulnerability index - measure of climate change vulnerability to communities: a case of rural Lower Himalaya, India. Mitig Adapt Strateg Glob Change 17:487–506. https://doi.org/10.1007/s11027-011-9338-2
Rakshit SK (2003) Forest Resource: its problems and prospects a study of Darjeeling and Jalpaiguri District, West Bengal. Unpublished PhD Thesis. Department of Geography and Applied Geography, University of North Bengal
Roy S (2011) Flood Hazards in Jalpaiguri District. Unpublished Ph.D. Thesis, Department of Applied Geography, University of North Bengal, Siliguri
Sam AS et al (2016) Quantifying household vulnerability triggered by drought: evidence from rural India. Climate Dev. https://doi.org/10.1080/17565529.2016.1193461
Sam AS et al (2017) Quantifying household vulnerability triggered by drought: evidence from rural India. Climate Dev 9(7):618–633. https://doi.org/10.1080/17565529.2016.1193461
Sarkar S (2008) Flood hazard in the Sub-Himalayan North Bengal, India. In: Singh S et al (eds) Environmental changes and geomorphic hazard. Bookwell, Shillong, pp 247–262
Torry W (1979) Hazards, hazes, and holes: a critique of the environment as hazard and general reflections on disaster research. Can Geogr 23:368–383
UNDP (2007) Human development report 2007/2008: fighting climate change: human solidarity in a divided world. Palgrave Macmillan, New York
Van den Berg M (2010) Household income strategies and natural disasters: dynamic livelihoods in rural Nicaragua. Ecol Econ 69(3):592–602

Dr. Pamela Deb completed her Ph.D. from the Department of Geography, Visva-Bharati, Santiniketan, India. Previously she taught in Salesian College, Siliguri as an Assistant Professor for more than 4 years. Her areas of interest are Social Geography with special focus on tribes, Political Geography and Urban Geography. She has published several articles in various reputed national and international journals and chapters in edited books.

Dr. Rameswar Mukherjee is an Assistant Professor in the Department of Geography of Samsi College, Malda. He has completed his Ph.D. from Banaras Hindu University, Varanasi, India. His areas of interests are: digital geomorphometry, quaternary geomorphology, channel dynamics, impact of climate change on human, flood studies. He has published 17 articles in various reputed national and international journals and chapters in edited books.

Chapter 16
Exploring the Options for Sustainable Livelihood in the Indian Sunderbans: An Attempt Through Contingent Valuation Method

Nilendu Chatterjee, Bappaditya Koley, and Anindita Nath

Abstract The Sundarban is a unique ecosystem—the largest delta and estuarine mangrove forest in India. The people in the Sunderbans of West Bengal are primarily dependent on the extraction of renewable resources like fishery, forestry to maintain their livelihood in a sustained manner. The degree of dependency, the awareness about sustainability of renewable resources, and the knowledge-base for a sustainable livelihood by protecting the ecosystem of Sunderbans can best be examined through a valuation exercise to conserve the overall ecosystem of this area. It will again help to conserve the renewable resources of the region concerned. It is rightly assumed that if and only if the ecosystem of the area can be protected, it will generate and conserve renewable resources such as fishery and forestry; again, it will attract visitors All these things together will generate earnings for the people of this area. The study finds that the *willingness to pay* (WTP) for conserving the ecosystem in the Sunderbans is reasonably high, given that the people of this area are economically impoverished and socially backward. This finding confirms that the people of the Sunderbans are pretty aware of the necessity of ecosystem conservation in their residing areas for a long-run benefit.

Keywords Sustainability · Ecosystem · Contingent valuation method · Willingness to pay

16.1 Introduction

West Bengal, one of the few states in India, is the land of various weather zones, habitats, and biomes. There is vast dryland in the west of the state, whereas we also find the tide-fed Sunderban in the south. It envisages a broad spectrum of dependency on natural resources as the state hosts different types of natural resources at its different

N. Chatterjee
Department of Economics, Bankim Sardar College, South 24 Parganas, West Bengal, India

B. Koley (✉) · A. Nath
Department of Geography, Bankim Sardar College, South 24 Parganas, West Bengal, India

M. Mishra et al. (eds.), *Regional Development Planning and Practice*,
Advances in Geographical and Environmental Sciences,
https://doi.org/10.1007/978-981-16-5681-1_16

parts, leading to a difference in livelihood earning among the people of different parts. Sunderban is the only mangrove forest in India with its unique ecosystem. Therefore, it demands greater significance for its maintenance from every aspect. Sunderban has an abundance of two natural resources—mangrove forests and fish. While the mangroves provide substantial ecosystem services, local households earn cash income by selling different ecosystem products (such as fish) at local markets that support the livelihood of residing communities.

Along with these two natural resources, Sunderban is also famous for its natural beauty and the existence of the Royal Bengal Tiger. All these factors are very much integrated with each other and with the livelihood of people of this backward (from the socio-economic aspect, yet abundant with natural resources) area. Sunderban is a famous tourist spot, and thousands of tourists visit it across the world throughout the year. This feature contributes to the livelihood earning of the people residing here and is very much associated with this area's natural resources, which brings the tourists (Datta et al. 2011). Apart from these features, Sunderban has a mangrove forest that enhances the natural beauty of this place and makes this place unique. Apart from being a source of earning, Mangrove provides safety to the people of Sunderban from numerous natural disasters Beginning from the 'Aila' cyclone in 2009 to the latest 'Amphan' in 2020, all the cyclones and other forms of natural disasters have badly hit West Bengal. The meteorologists have observed that the effects of such disasters could have been worse without the existence of Mangrove forest in Sunderban. So, the existence of mangroves provides safety not only to the people of this area but to the whole of West Bengal (Ghosh et al. 2015). That's why the government has felt the importance of increasing the area of mangrove forest cover in Sunderban and has taken positive initiatives for that purpose.

The area of Sunderban is actually shaped like a delta. The Sunderban is an intricate web of tidal waterways, seawater, rivers, creeks, and mudflats, formed by the gradual deposition of alluvial silt, at the merger of the Ganga and Brahmaputra, and Meghna rivers in the Bay of Bengal (Roy Chowdhury and Vyas 2007). The area of Sundrerban has been named after the *Sundari* (*Heritierafomes*) and the *bani* (*Avicenniaofficinalis*) mangroves. The Indian part of Sunderbans consists of 4,263 sq. km (WBFD 2003) area, but the entire Sunderbans consist of around 10,000 sq. km area. The remaining part belongs to Bangladesh. The Indian Sunderban is divided into two parts. One is Sunderban Tiger Reserve consisting of around 2600 sq. km area, and the second one is Reserve Forest area covering the area of 1600 sq. km. (WBFD 2003). Situated in the north-east coast of India, Indian Sundrrban has 106 islands, out of which 56 are inhabited and ranges over 19 blocks, covering the two districts of North 24 Parganas and South 24 Parganas, having a population of approximately 4.7 millions (Sel 2016). All these islands are separated from each other by a network of small rivers, popularly known as '*khari*' in Bengali. The main occupational groups in the Sundarbans are fishermen, the *bowalis* (wood cutters or '*golpata*' leaf collectors), and the crab and shell collectors (UNDP 2002). Fishing and collection of non-timber forest products—these two have been the main source of livelihood for the people of Sunderbans because very few people have access to agricultural land in this region, and the nature of soil prevents extensive agricultural activity.

This evergreen land of mangroves has been recognized as a 'UNESCO World Heritage Site' (Chaudhuri et al. 1994). It is an essential and distinctive natural resource endowment that is famous for its brackish water fish production and marine fisheries. The National Park of Sunderban is surrounded by a buffer zone of 885 square kilometers (Bairagya and Bairagya 2017). This also mainly consists of mangrove forests. The core area of the park has its own natural boundaries with the river Matla on its west, the river Haribhanga on its East, with Netidhopani and Gosba in the North. Sunderbans has been the nursery for almost 90% of the aquatic species of the eastern coast of India. Jhingran (1977) recorded a total of 172 species from a variety of sources and also mentioned that the diversity of the Hooghly-Matlah estuary increases along an increasing salinity gradient. The government of West Bengal has also been helping the fishermen of Sunderban with various infrastructural facilities. For example, five fishing harbors have been built in Frazerganj, Diamond Harbour, Kakdwip, Sagar, and PatharPratima (Chandra and Sagar 2003). The mangrove forest of Sundarban is valuable because of its rich biodiversity. The Non-Timber Forest Produce (NTFPs) is one of the epitomes for the livelihoods of many forest fringe dwellers (Bhattacharya and Hayat 2004). The NTFPs collected from the mangrove forest of Sunderban include tannin bark, Nypafruticans (*Golpata*), natural honey and bee wax, fuelwood, and small poles (Singh et al. 2010). However, it is an utter disappointment that the commercial collection attitude exploits a significant portion of these livelihood-support resources of the commoners of the Sundarbans.

From the above discussion, it is evident that for the sake of sustainable livelihood of the people of this region, we need to protect its natural resources with manifold benefits. It generates earnings from various sources as well as protects this region along with the entire state. Hence the necessity to protect this unique ecosystem and look towards its further improvement comes into question.

If we look at the subset of vast existing literature on CVM that emphasizes the ecosystem, we would see that authors such as Qian et al. (2011) have applied CVM for environmental valuation in the process of mineral resource development in China. Mineral resource development causes both positive and negative externalities. Therefore, we have applied both the willingness to pay as well as the willingness to accept approaches in this paper and have found significant outcomes with respect to the sustainable development of the society. Al-Assal (2015) has applied CVM for the valuation of forest services by the following *Willingness to Pay* (WTP) approach in Northern Jordan. He has found a relation between Willingness to Pay and income of the households. He believes that CVM can be a very useful technique for the valuation of the environment in Middle East nations. Resende et al. (2017) have used CVM for the valuation of services provided by a protected area in Brazilian Cerrado. They have interviewed 514 households about their willingness to pay and obtained quite a good figure of the amount as the mean willingness to pay for conservation of services of Cerrado. Zhang et al. (2012) have applied CVM for the evaluation of forest ecosystem services of HeshuiWaterdhed in China. Their approach used on 200 households indicated the fact that the respondents were willing to pay a significant amount of money for the protection of forest services. Now the fact that natural resources are mainly common property resources (CPR) makes the question of conservation and sustainable use even more important. We find that authors like Angelson and Wunder (2003) have searched for the reasons for the poor economic condition of forest-dependent people. Cavendish (2000) has tried to link CPR with a sustainable livelihood. At the national level, the important works regarding common property rights (CPR), poverty, and forest degradation are by Jodha (1986), Reddy and Chakraborty (1999), Adhikary (2002; 2005), etc. Significant works on mangrove-fishery linkages at the international level are done by Ruitenbeek (1994), Grasso (1998), Gupta (2005), etc. Most of these works focus on the benefits and costs associated with mangroves.

In spite of the existence of vast literature on CPR and CVM, works related to valuation as well as conservation of the ecosystem of Sunderbans are very few. This lacuna in the literature of economic analysis has been the biggest motivation behind this work. It naturally becomes very interesting to see how the people who earn their livelihoods from these natural resources have been valuing these resources and how far they are able to make a contribution for protection as well as further improvement of these resources as well as the entire ecosystem.

The remaining part of the present chapter is organized in the following manner. Section 16.2 deals with the objectives and hypotheses of the study. In the next section, that is, Sect. 16.3 considers the database and survey design of the study. Methodology, econometric specification, and the results of the study are shown in Sect. 16.4. Finally, the concluding remarks are made in Sect. 16.5.

16.2 Objectives and the Hypotheses of the Study

The primary objectives of the present study are—Firstly, to see how the people of the Sunderban are valuing the ecosystem of Sunderban as a whole which includes conservation of significant natural renewable resources on which they are mainly dependent for their livelihood; Secondly, to examine the determinants of willingness to pay (WTP) for the conservation of ecosystem through the conservation of natural renewable resources in the Sunderban and Lastly, to link the valuation analysis with that of sustainable livelihood of the stakeholders in the study region.

The following presumptions have been considered in the present study to achieve the above objectives:

- Here we have considered the bids of the respondents regarding Willingness to Pay (WTP) by applying 'single-bounded dichotomous choice closed-ended referendum', as generally followed in the case of dichotomous choice random utility models, with the help of dice throwing.
- Conservation of ecosystem and its uses in our study region implies overall conservation of ecosystem and renewable resources in the backward areas not only as a 'use value' but also as a 'non-use value'.

- We have estimated WTP by applying both the logit model as well as OLS model. Logit model has been applied with the help of a single-bound closed-ended referendum (for choosing the bids), and OLS has been applied with the help of an open-ended referendum. So two WTPs following two different methods will be estimated.

Here we have considered the following hypotheses for the present study.

- In the case of the closed-ended referendum, we have followed the literature of CVM, and for the purpose of estimation of WTP, we have mainly used various socio-economic variables and dummy variables.
- Here we have compared the two mean WTPs. These are obtained from two different formats. This comparison helps us to know the existence of anchoring bias (Chatterjee 2019).

16.3 Database and Survey Design

This study is based on primary data, collected from different regions of Sunderbans. Partly stratified and partly random sampling techniques have been followed for this purpose. For the selection of blocks or villages, we have followed stratified sampling. The stratification is done in a manner so that we have blocks of South 24 Parganas district where the stakeholders are dependent on using resources of ecosystem, that is, fishery or forestry or both.[1] First, to conduct the study on the valuation of the ecosystem, we have selected the blocks based on their dependency on the use of products or resources generated by the ecosystem. Then we have selected villages from the blocks. We have focused on those activities that require the use of the ecosystem. However, it does not mean that we have surveyed people associated with these activities only. Since we have already mentioned that the ecosystem of this region, as a whole, provides various safety measures to the people, it is to be noted that people in the same family of the Indian Sunderbans are dependent not only on forestry but also on fishery and tourism, which provide direct and indirect benefits to many people. Although, for the Sunderbans as a whole, the fishery is the primary occupation, followed by forestry and tourism. There are some blocks in this region where the majority of the people are dependent on the fishery. We have selected villages like Jhorkhali (of Basanti block) and Pakhirala (of Gosaba block) for our purpose based on consultation with the local Panchayat members of the two blocks. We have selected 300 families each from the two villages. We have surveyed in Namkhana block. So for valuation of the ecosystem, we have considered the sample from the blocks of Gosaba (300 families from village Pakhirala) and Basanti (300 families from village Jhorkhali) and 400 more families from Namkhana block (consisting of 200 families from Bijoybati village, 100 families from Amarabati (east) village, 100 families from Amarabati (west) village). From the literature of CVM,

[1] They may be associated with tourism also, thus becoming a dependent on eco-system.

Table 16.1 Division of households of sample survey in case of the Sunderbans

District	Block	No. of households used for sample survey
South 24 pgs	Basanti	300
South 24 pgs	Gosaba	300
South 24 pgs	Namkhana	400

Source Field survey by the authors

one can see that most of the works on CVM are based on a sample size of 200–400. But, we all know that a larger sample means more reliability and predictability and a better reflection of the view of the population. In order to capture the view of the population in a better way, we have increased the sample size far more than what other works on CVM normally do. Since CVM involves direct views of the people, we believe that the sample size of 1000 for a district is good enough, and it gives a better reflection of the views of the people. The household survey was done between December 2019 and February 2020. The villages in Namkhana block are also selected based on discussion with the Panchayat members, and also, the stratification within a village is done based on consultation with them during the pilot survey. The families from each village are selected randomly. So, our total sample size is 1000. We have also considered this sample for our socio-economic study.

We have selected three blocks for the survey in the Sunderbans, namely Basanti, Gosaba, and Namkhana (Table 16.1). Basanti and Gosaba are very close to the forestry and rivers of Sunderban, and people of these blocks are dependent on forestry and fishery, and tourism for their livelihood. However, in Namkhana block, we have witnessed dependency on fishery and tourism only as forests are far from this block. But, overall, we have seen good awareness among people regarding the conservation of the ecosystem for the sake of sustainable livelihood.

We have conducted a household survey to get data on various socio-economic aspects of the respondents' families (like the number of family members, age–sex composition, educational status, income earned from different sources, landholding, livestock holding, etc.).[2]

In our sample, we find that almost 58% of the respondents are Male, and the rests are female (based on Primary Survey). Next, we have categorized the respondents into different income groups for both areas. Finally, we have classified the entire income range in various groups, frequency of people falling in each group with their percentages in Table 16.2.

As reflected in Table 16.2, a considerable proportion of respondents in all three blocks belong to the income group of less than INR 10,000 per month, while Namkhana is comparatively better than the other two. However, households with a monthly income of more than INR 10,000 are rare.

[2] These socio-economic indicators have been used as variables in the CVM study.

Table 16.2 Different income groups of respondents with frequency and percentage in the Sunderbans

Blocks	Basanti (N = 300)		Gosaba (N = 300)		Namkhana (N = 400)	
Income groups (INR)	Frequency	Percentage (%)	Frequency	Percentage (%)	Frequency	Percentage (%)
0–2500	70	23.33	75	25	68	17
2501–5000	82	27.33	91	30.33	95	23.75
5001–10,000	118	39.33	123	41	217	54.25
10,001–20,000	22	7.33	7	2.33	11	2.75
More than 20,000	8	2.66	4	1.33	9	2.25

Source Field survey by the authors

16.4 Valuation of the Ecosystem in the Study Area: Methodology and Results

The present study focuses on the valuation of the ecosystem in the Indian Sunderbans. While doing so, it indirectly seeks to evaluate various services provided by the ecosystem, which obviously includes extraction of renewable resources, their uses in the Sunderbans, and their linkages with sustainable livelihoods. To understand the implications of the importance of the ecosystem in Sunderbans, one should focus on the services provided by the ecosystem, that is, the valuation of resources associated with the ecosystem. For the valuation of the ecosystem, here we have applied the same procedure of CVM as has been done by Chatterjee (2019), Gupta and Chatterjee (2021) but the bids offered are different, and the area of the works, i.e., study-area, are also different. CVM has a long and vast history of application. So, the theoretical part followed here has been the same as that applied by many researchers following Haraou et al. (1998).

We have applied the widely used Contingent Valuation Method (CVM) to evaluate the ecosystem and services connected with it in our study area. Our study, along with its findings, has significant meaning in the sense that a high value of willingness to pay (WTP) to conserve the ecosystem means that the stakeholders or the people of that area have the knowledge base to conserve the services and resources.

In the Sunderbans, the leading renewable resources of the ecosystem that are extracted and also on which the stakeholders are dependent for their livelihood are fishery and forestry, and the natural beauty of the ecosystem brings tourists from all over the world and influences the livelihood of the people of different professions, different socio-economic activities in manifold ways. We have considered in this paper the valuation of the whole ecosystem, which incorporates these services in a combined manner. One should remember that though fishery is the primary occupation in the Sunderbans, forestry conservation helps to promote fishery in the long run. This is because the forests here are in the form of mangrove swamps, and they are the nursery grounds for various fish stock.

16.4.1 Valuation of Ecosystem: Methodology

The sampling procedure has already been discussed in Sect. 16.3. However, we need to mention that we have followed stratified random sampling for our study. It has been observed that the response rate is 100% which is high. A higher percent of response rate in our study can be considered an unconventional but good outcome in a developing country.[3]

The Contingent Valuation Method (CVM) is based on a closed-ended bidding game of dichotomous choice. For this kind of analysis, it is crucial to decide the bid first and then determine how these bids are shown to the respondents.[4] The bids that we have considered for the ecosystem of the Sunderbans are INR 10, INR 20, INR 40, INR 50, INR 80, and INR 100 (per month). Here the bids are determined after discussing with the local people through pilot surveys which gave us an idea of the maximum and minimum amounts that we should put forward to the respondents as bid amounts. The next step is to identify the 'valid' responses out of 1000 respondents. For this, we have followed the strategy adopted by Chatterjee (2014) and Gupta and Chatterjee (2021). We have categorized the respondents into three bid groups—'*low*', '*medium*', and '*high*'. Then, we have applied the single-bound dichotomous choice CV method. The *low* bid group implies bids of INR 10 and INR 20 per month. For the *medium* bid group, the bid amounts are INR 40 and INR 50 per month. For the *high* bid group, the bid amounts are INR 80 and INR 100 per month. This process of offering bids, amounts of bids, and accordingly grouping of respondents in three bid groups, just described, is the same for all the surveyed areas or blocks.

Now, we need to explain how bids are selected in a closed-ended referendum under single-bound dichotomous choice CVM. *First* of all, a dice has been thrown before every respondent. Based on the outcomes, the procdure of finalizing the bid amount or *protest bidder*[5] is presented in Fig. 16.1. The respondents have responded to the offered bids in terms of either '*YES*' or '*NO*' to a concerned bid. In this way, a particular bid has been shown to a particular respondent from different events when the events are mutually exclusive, equally likely, and independent. So, for a particular respondent, we have thrown a dice twice, firstly, for randomly selecting the bid-group for each respondent and, secondly, for randomly selecting the amount of bid that was

---

[3] We attribute this high response rate to the '*face-to-face in-person interviews*' that we have conducted, as suggested by the Arrow et al. (1993) panel. One more reason for such high response rate could be the fact that from Chap. 3 we have seen the degree of forest dependency is very high, almost 100%, for the people of our surveyed areas. So, regarding uptiftment of forestry, they might be very eager.

[4] One can refer to Saha (2015) and Chatterjee (2017) for a similar type of bidding procedure. However, the present bidding strategy is different from the bidding strategy followed by Saha (2015), rather it is close to the work of Chatterjee (2017). In our case and also in Chatterjee (2017) anchoring has been done in offering bids in terms of a dice throwing first and then the bids are offered again by throwing a dice for the second time. On the other hand in Saha (2015) first a dice has been thrown and then the bids are offered in terms of tossing a coin.

[5] Protest bidders are those who do not prefer the stated programme and therefore provide zero WTP value.

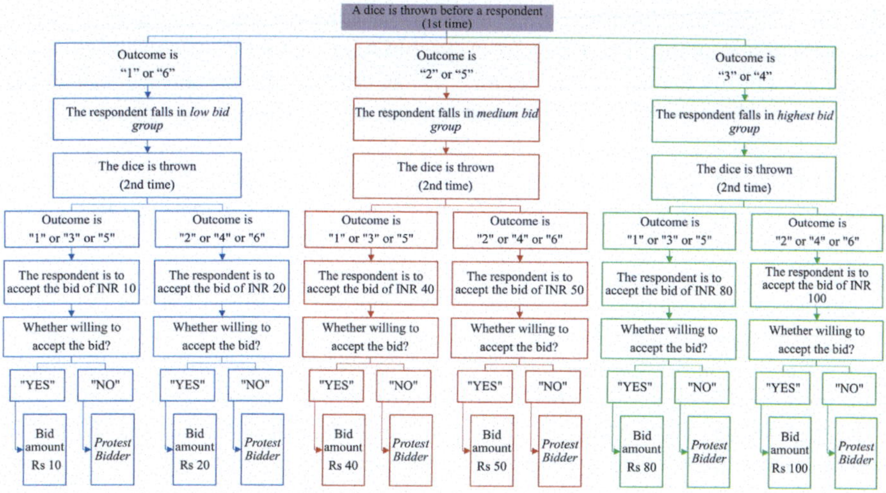

Fig. 16.1 Procedure of finalizing the bid amount and selection of the *protest bidders* (*Source* Drawn based on Chatteerjee 2017)

offered to the respondent.[6] Finally, we have directly asked the respondents about their maximum willingness to pay (Max WTP) for the open-ended segment of the study.

The present survey reflects that out of 1000 respondents, 371 respondents are not willing to accept the bids. Thus, we will consider 629 respondents as '*willing*' respondents and 371 respondents as '*non-willing*' for further analysis. Here 371 '*non-willing*' participants are considered as '*protest bidders*' (Bateman et al. 2002). It is seen that these bidders give '*zero WTP values*' but still have a preference to participate in the program and enjoy the benefits of it. So we cannot omit those respondents.

The bidding process as described above is a 'single bounded dichotomous choice' bidding process. The theoretical part for this type of bidding process can be explained briefly with the help of a random utility model. In case of Random Utility Model (RUM), the surveyor describes an imaginary situation of improved ecosystem in front of the respondent which might take place if most of the people or all the respondents agree to accept the bid. This improved scenario could be presented during pilot survey or during meeting with the panchayat bodies where most of the villagers remain present. In our case, we did it during meetings held with the panchayat bodies. Then, during the survey, the respondent is offered a bid of a certain amount (say X) and asked whether he or she agrees to accept this bid, that is, he or she is ready to pay this amount. Since we have applied a single-bounded close-ended referendum, the respondent is offered a single bid amount after the random process of offering the bid is completed and asked to accept or reject that. If the respondent feels that this offered

[6] This procedure of throwing a dice twice before offering a particular bid to the respondent was followed for bringing simplicity in the survey process.

bid is more significant than his utility after the said improvement in the ecosystem, he or she will accept the bid, else he or she would reject it. Obviously, the respondent compares his utility between two scenarios—*first*, utility in the present scenario of the ecosystem at which the bid is offered and; *second*, utility in the described scenario by a surveyor or in the post-improvement scenario. If the utility in the post-improvement scenario surpasses that of the present scenario, then the respondent generally accepts the bid. Obviously, this utility depends on many socio-economic aspects along with the described scenario by the surveyor, in which the respondent feels that he or she would enjoy a higher utility level by accepting the offered bid. So, along with socio-economic aspects, the knowledge about understanding the utility model or described post-improvement scenario is important. Lack of knowledge about understanding the described scenario might lead to the rejection of the offered bid. Since we are using logit model, hence it is the log-odds ratio, that is, the log value of the ratio of the probability of accepting the bid and the probability of rejecting the bid is the dependent variable, and it depends on numerous independent variables, most of which are socio-economic in nature along with the offered bid amount.

The above-described procedure verbally describes the theoretical methodology of Haraou et al. (1998). The mathematical expression of the methodology, developed by Chatterjee (2019) and Gupta and Chatterjee (2021), based on Haraou et al. (1998), is given in Annexure to help readers getting the methodological context better. The bidding process, that is, the process by which an individual respondent is offered an amount of bid randomly, is shown with the help of a schematic diagram (Fig. 16.1) and The independent variables used in this model are described in Table 16.3.

The independent variables are the socio-economic variables and are more or less common for CVM studies. We have used several socio-economic aspects as independent variables by using a dummy. In a poverty-stricken, backward area, it is expected that these aspects can play an important role in the respondents' response. The estimated results of the logit model along with Marginal effects are shown in Table 16.4.

Table 16.3 Description of independent variables of the model

DC bid	Bids vector of INR 10, INR 20, INR 40, INR 50, INR 80 and INR 100
Income	Total monthly income from all sources
Family size	Household size
Age	Age of the respondent
Eduyrs	Total years of education of the respondent
Sex	Dummy variable: 0 for males and 1 for females
Caste	Dummy variable: 0 for general caste, 1 for OBC, 2 for SC and 3 for ST
Dom animals	Dummy variable: 0 for having no animal and 1 for having any

Source Authors

Table 16.4 Results of the Logit model for ecosystem (Indian Sunderbans)

Variable	Coefficient	Marginal effects (dY/dX)
DC bid/close-ended bid	−0.000543*** (−9.12)	−0.0000321*** (−7.98)
Income	0.000345*** (7.98)	0.0000303*** (4.73)
Family size	0.003987*** (11.21)	0.0028355*** (9.41)
Age	0.321445 (0.87)	0.004590 (0.69)
Eduyrs	0.760999** (2.12)	0.201307** (2.05)
Sex	0.765444 (0.57)	0.081876 (1.05)
Caste	−0.022575*** (−5.51)	−0.000345*** (−7.41)
Dom animals	0.345980** (1.91)	0.217654** (1.97)
Constant	−8.560987*** (−5.72)	
Log-likelihood	−210.7622	The terms in the parentheses for both coefficient and marginal effects are the t-values
LR chi-square	304.78	
Prob > chi-square	0.000	
Pseudo R^2	0.5541	
Total no. of observations	1000	

** denotes significance at 5% levels
*** denotes significant at 1% levels
Source Computed by the authors

16.4.2 Valuation of Ecosystem: Results

The results related to the valuation of the ecosystem in terms of the logit model are reported in Table 16.4. We find that age and sex are insignificant, but others are significant. Among the significant variables, the dichotomous choice bid has a negative sign before it, which proves that for one unit of change in the offered bid, the probability of accepting that bid or the probability of willingness to pay or the log-odds ratio falls. The negative coefficient for the bid-vector is because as the bid value rises, the probability of *YES* (or acceptance) decreases. The negative sign of caste signifies that people of the general categories are more willing to contribute, which implies their good economic condition, but when we move to OBC, SC, and ST categories of residents, the willingness to contribute decreases. It is evident that a larger portion of the people belonging to OBC, SC, and ST categories in this region are in impoverished economic conditions. Along with an increase in income, willingness to pay is expected to go up, that is why people's income has a positive impact on WTP. This result is irrespective of all types of income groups.

In Table 16.4, we find that the larger the family size, the higher is the number of working members, dependent on the use of various ecosystem facilities.[7] So, the

[7] This dependency could well be in the form of working not only as foresters but also the fisher-men (if any in the family) are dependent on forests as mangrove forests of the Sunderbans are nursery grounds for fish. The honey collectors are also very much dependent on forests. Again, people associated with different tourism activities are also benefited from the conservation and development

Table 16.5 Estimation of mean WTP for forestry in Indian Sunderbans (DC model under closed-ended referendum)

Measure	WTP	LB	UB
Mean	28.85	11.25	57.27
Achieved significance level for testing H_0: WTP $\leq$ 0 versus H_1: WTP > 0			
LB: lower bound; UB: upper bound			

Source Computed by the authors

variable family size has a positive sign before its coefficient. The coefficients of dc bid, income, and family size are highly significant (at 1% level of significance). Education has a positive sign and is significant at 5% level. It is also expected because educated people are more aware of the protection of the ecosystem. We have used domestic animals as a dummy of an economic asset. The coefficient with a positive sign indicates that the households with domestic animals are in a better economic condition than others. Hence, their WTP is relatively high, which justifies a high value of the coefficient for domestic animals, which is highly significant with a positive sign. The value of pseudo-R^2 usually judges the goodness of fit of the model used. In our model, the value of pseudo-R^2 is 0.5541. So, our model gives a good fit. It is vital to examine whether some degree of multicollinearity is present in the model as it is common in any cross-section data. The variance inflationary factor (VIF)[8] and the tolerance[9] for the model show that the models do not suffer from severe multicollinearity problems as the value of 'mean VIF' is 2.59, which is well below the value of 5.[10] We have also estimated the marginal effects for the logit equation. It shows the rate of change in the probability of willingness to pay due to a change in the value of an independent variable X_j (j = 1, 2… n) (Table 16.4). From Table 16.4, we find that as income changes by one unit, holding other factors constant, the probability of WTP also rises; the same explanation applies in the case of variables like years of education, family size, and domestic animals. The opposite explanation applies to the variables like dichotomous choice bid and caste. For these variables, one unit of change causes the probability to accept an assigned bid fall.

Then, we have estimated the WTP in the case of a closed-ended referendum under the dichotomous choice model. The mean WTP of our model is INR 28.85 per month, with the lower bound of INR 11.25 and upper bound of INR 57.27 These are shown in Table 16.5.

---

of the eco-system. So it is expected that all the stakeholders will opt for the conservation of the eco-system in the Sunderbans.

[8] Variance Inflationary Factor (VIF) measures how much the variance of the estimated regression coefficients are inflated as compared to when the predictor variables are not linearly related. This is used to describe how much multicollinearity (correlation between predictors) exists in a regression/logit analysis. When there is no collinearity, VIF will be 1.

[9] 1/VIF is known as tolerance.

[10] As 'the rule of thumb', if 1 < VIF < 5, it implies variables are moderately correlated and if 5 < VIF < 10, then the variables are highly correlated.

Table 16.6 Regression results of open-ended referendum (the ecosystem of the Indian Sunderbans)

Variables	Coefficients	t values
Constant	−7.90876***	−11.39
Income	0.089655***	10.32
Family size	−0.650210 ***	−4.45
Age	0.098345	1.07
Eduyrs	0.420369**	2.09
Sex	0.548987	0.81
Caste	−0.459620***	−9.23
Dom animals	0.854416***	8.41
Dependent variable: max WTP (open ended maximum WTP)		
N = 1000, F = 147.55, Prob > F = 0.000, adjusted $R^2 = 0.5129$		

** denotes significant at 5% level
*** denotes significant at 1% level
Source Computed by the authors

Though the mean WTP, in general, appears to be low, given the backwardness of the study area and a large portion of the respondents living below the poverty line, the mean WTP figure of INR 28.85 per month to conserve the ecosystem in the Sunderbans can be considered quite reasonable. We can thus say that the closed-ended WTP is INR 28.85 per month. Thus, one can say that the people of Sunderbans and communities dependent upon various aspects of the ecosystem are willing enough to conserve and further develop the ecosystem and its services in the Indian Sunderbans. Therefore, the Median WTP in our study appears to be INR 32.51 per month.

We now want to examine the WTP under an open-ended referendum. Here we consider WTP for ecosystem conservation when the respondents are offered open-ended bids, that is, the respondents are directly asked how much they are willing to pay. In this case, we use OLS regression techniques to determine the WTP, and the regression result is summarized in Table 16.6.

From Table 16.6, we find that the same variables are significant in both the logit model and the open-ended model, with the same signs (except for family size) before their coefficients.[11] We have used simple OLS estimation for the open-ended part. The value of adjusted R^2 implies that more than 51% of the dependent variable variation is explained by the independent variables included in the model (adjusted to degrees of freedom). So, in terms of cross-section data, one can say that it is a well-fitted model. Also, the t-values suggest that most of the parameter estimates are significant at 1% level. It is only years of education which is significant at 5% level. From Table 16.6 we can easily calculate the open-ended (maximum) mean WTP, and it has been calculated to be INR 25.73 per month. By considering the

[11] The only possible reason for having a negative sign before the family size may be due to the fact that large family size can use relatively less per-capita income and hence reduces the maximum WTP to conserve eco-system in the Sunderbans.

average of dichotomous choice WTP and open-ended WTP, we find that the value of '*true WTP*' for the conservation of ecosystem in the Indian Sunderbans is INR 27.29 (approximately INR 27) per month.

We have performed the convergent validity test to test the presence of anchoring bias. For this purpose, we have conducted the paired t-test by comparing the two mean WTPs obtained from two different models. Here, we have assumed the null hypothesis that the difference between the two WTPs is zero and the alternative hypothesis states that this difference, not zero. Here the difference between 'two WTPs' is -3.12, and the value of paired t-statistic is -7.57 with 999° of freedom. It is significant at 1% level. So, the two mean WTPs are significantly different. It implies that anchoring bias exists in our CVM exercise for the ecosystem in the Sunderbans. This anchoring bias exists maybe because of the fact that people of the Sunderbans are not highly educated enough, and also, they do not have the proper perception about the issue of sustainability of the ecosystem. Our study finds that in the Sunderbans, the '*true WTP*' for conserving the ecosystem is approximately INR 27 per month.

16.5 Concluding Remarks

This chapter has started with a socio-economic analysis for some selected areas of the Indian Sunderbans to analyze the backwardness. The people in the Sunderbans are primarily dependent on the extraction of renewable resources like fishery and forestry as well as on tourism to maintain their livelihood in a sustained manner. So, their dependency on the ecosystem is very high. The degree of dependency, the awareness about sustainability of the ecosystem, and the knowledge base for sustainable livelihood can best be examined by conducting a valuation exercise to conserve the various components of the ecosystem.

In this study, we have observed that the people of the Sunderbans are willing to conserve the ecosystem base of the region. This is reflected in the high willingness to pay by the people. The fishery is one of the primary sources of livelihood in this region. It can be in the form of employment in the aquaculture shrimp farms or the form of shrimp fry collector (mainly by the women) or the estuary/marine fishermen. Conventional wisdom suggests that the expansion of aquaculture fish farms destroys mangrove forests. However, our analysis suggests that the Sunderbans people are pretty aware of the conservation of forests and the expansion of tourism (as reflected in terms of high WTP for the ecosystem). In this study, we could have applied the Hedonic Pricing Method instead of CVM. Again, here we have applied a single bounded dichotomous choice bidding process. But, we could have applied a double bounded dichotomous choice bidding process also. A different bidding process would have given us a different result. These facts open up the scope for further study.

CVM can be used for the valuation of all those environmental goods having either use-value or non-use value. However, it is mainly used to capture the non-use value of environmental goods. From that sense, CVM can be used to capture the view of people of any region regarding any environmental aspect. For example, one can investigate the options of reducing pollution in Sunderbans using the CVM.

By using its regional bodies, such as Panchayats, Joint Forest Management groups, or even NGOs, the government can organize or design a successful mechanism by which people will be informed about the proposed change along with its benefits. Then, by organizing a meeting among themselves, they can fix the willingness to pay, and in this way, they could proceed to apply CVM successfully.

Thus, the government should take more initiative to improve the condition of the stakeholders in this region by promoting conservation of various aspects and facilities generated by the ecosystem. Such a policy would help protect and expand the forestry in Sunderbans. Moreover, it would also help in the growth of fishery, in the long run, providing environmental safety and the tourist influx hand in hand. Growth of these three aspects—forestry, fishery, and tourists-would undoubtedly provide economic benefits and provide them with sustainable earning sources. In a nutshell, it will help in the socio-economic prosperity of the Sunderbans and their people.

Acknowledgements We are immensely grateful to the anonymous reviewer and Dr. Mukunda Mishra for his many insightful contributions and valuable suggestions for our manuscript.

Annexure 1

Random Utility Model

The random utility model follows or copies all those choices which an individual generally faces in the presence of a real market. The surveyor presents a specific monetary value (e.g., INR X) to the respondent and proposes a policy change, and he/she is asked to make a judgment of accepting or rejecting the offer. The size of X or the offered bid is randomly varied across the sample of a study.

The DC elicitation method provides only a limited amount of information about the respondents' WTP value, namely, 'YES' or 'NO' answers to a particular bid.

$$\text{If, Bid Amount (X)} > \text{WTP, then the response is 'No'}$$
$$\text{If, Bid Amount (X)} \leq \text{WTP, then the response is 'YES'}$$

Our next focus is on the Random Utility Version of the model. A respondent will respond by saying '*YES*' or will accept the offered bid if and only if his/her utility from the conservation and improvement in ecosystem's measure is larger than or equal to her utility compared to status quo position; and '*NO*', otherwise.

$$(U_1 - U_0) \geq 0, \text{ the individual will accept to pay the bid X} \quad (16.1)$$

$$(U_1 - U_0) < 0, \text{ the individual will reject to pay the bid X} \quad (16.2)$$

Since the utility U of the individual is not directly observable (hence the differences are also not directly observable). However, its determinants are observable. Therefore, under the two different scenarios, one with the acceptance and the other with the rejection of the bid, the following specification of the utility function can be put forward:[12]

$$U_1(1, y - X; S) = V(1, y - X; S) + e_1 \quad (16.3)$$

$$U_0(0, y; S) = V(0, y; S) + e_0 \quad (16.4)$$

where $V(.)$ is the utility function without random elements, and $U(.)$ is the utility function with a random element. It is to be noted that in Eqs. (16.3) and (16.4) we find y = total income; 1 = acceptance of the bid; 0 = rejection of the bid; S = other socio-economic features; e = random error component due to the limited knowledge of the utility model of the individual by the analyst.

From Eqs. (16.3) and (16.4) we can write:

$$\Delta U = \Delta V - e \quad (16.5)$$

where $(U_1 - U_0) = \Delta U$,

$(e_0 - e_1) = e$, and

$[V(1, y{-}X; S){-}V(0, y; S)] = \Delta V$, given Eq. (16.5), the inequalties in Eqs. (16.1) and (16.2) can be written as,

$$\Delta V \geq e \rightarrow \text{Acceptance of X} \quad (16.6)$$

$$\Delta V < e \rightarrow \text{Rejection of X} \quad (16.7)$$

There are two types of models for estimating the mean WTP value from the DC bids—the probit and the logit model. Here we have considered a logit model (logistic distribution of the error term) for our purpose.[13]

The probability that the individual agrees to accept the bid is, therefore:

$$P(acceptX) = P(Y = 1) = P(e \leq \Delta V) = F(\Delta V) \quad (16.8)$$

where Y is the observed dichotomous variable,

---

[12] This part briefly describes the theoretical methodology of Harou, Markandya, Bellu and Cistulli (1998) and naturally the methodology part is similar to the work of Saha (2015).

[13] The choice of the model depends on the probability distribution of the error term where probit is used if the error term follows a normal distribution and logit is used if the error term follows a logistic distribution. However, most of the studies that used DC format follow the logit model since the difference between the two is minor and the logistic function is simpler to deal with.

acceptance = 1,
refusal = 0.

Assuming that the random variable *e* follows a logistic probability distribution, we can write:

$$P(acceptX) = F(\Delta V) = 1/[1 + exp(-\Delta V)] \quad (16.9)$$

When the individual accepts to pay the proposed bid X, its means that the maximum Willingness to Pay (WTP) is greater than the proposed bid X. The probability of acceptance, given a bid X, is the probability of individual $WTP \geq X$.

Therefore, we can write:

$$P(acceptX) = P(WTP > X) = 1/[1 + exp(-\Delta V)] \quad (16.9.1)$$

This means that the probability the WTP is less than or equal to X is:

$$P(WTP \leq X) = G(X) = 1 - 1/[1 + exp(-\Delta V)] \quad (16.10)$$

where G(X) is the probability distribution of the WTP.

The mean of the WTP distribution is commonly assumed to be indicators of the individual WTP.

The mean of the maximum WTP can be calculated using the formula that relates the mean of a random variable to its probability distribution:

$$E(WTP) = \int_0^\infty \{1 - G(X)\}dX \quad (16.11)$$

Let us consider the econometric specification of the DC model for a closed-ended referendum. Here we want to derive the mean WTP for the ecosystem of the Sunderbans. To estimate the WTP, we have used a logit model, and we have derived the values for the DC bids used for the respondents.

The logit model[14] used for the study contains the following variables:

$$\text{Dependent Variable: } \ln \frac{P_i}{1 - P_i}$$

where, P_i is the probability of WTP amount greater than or equal to an assigned bid.

$\ln \frac{P_i}{1-P_i}$ is the log odds ratio.[15]

[14] Most of the variables used in this model have been selected after going through the literature on CV technique.

[15] The ratio of probability of willingness to pay (P_i) and non-willingness to pay ($1 - P_i$). It is to be noted that as P_i increases the log-odds ratio increases.

References

Adhikary B (2005) Poverty, property rights and collective action: understanding the distributive aspects of common property resource management. Environ Dev Econ 10(1):7–31

Adhikary B (2002) Household characteristics and common property forest use: complementarities and contradictions. J For Livelihood 2(1)

Al-Assal A (2015) Applying contingent valuation to measure the economic value of forest services: a case study in Northern Jordan. Int J Sust Dev World 22(3):242–250

Angelson A, Wunder S (2003) Exploring the forest-poverty link: key concepts, issues and research implications. CIFOR Occasional Paper 40. https://doi.org/10.17528/cifor/001211

Arrow K, Solow R, Portney PR, Leamer EE, Radner R and Schuman H (1993) Report of the NOAA panel on contingent valuation. Federal Register, 58(10): 4601–4614

Bairagya H, Bairagya S (2017) Influence of geographical phenomenon on education: a case study on sundarbans. J Educ Dev 7(13):184–203

Bateman I, Carson RT, Day B, Hanemann WM, Hanley N, Hett T, Jones A, Loomes G, Mourato S, Ozdemiroglu E, Pearce DW, Sugden R and Swanson J (2002) Economic valuation with stated preference techniques: A manual. Edward Elgar, Cheltenham

Bhattacharya P, Hayat SF (2004) Sustainable NTFP management for rural development: a case of Madhya Pradesh, India. Int For Rev 6(2):161–168

Cavendish W (2000) Empirical regularities in the poverty environment relationship of rural households: evidence from Zimbabwe. World Dev 28(11):1979–2003

Chandra G, Sagar R (2003) Fisheries in sundarbans: problems and prospects. 1–6. https://doi.org/10.2139/ssrn.2084014

Chatterjee N (2019) Valuation of forestry in selected dryland areas of West Bengal: a contingent valuation approach. Econ Aff 64(1):167–177. https://doi.org/10.30954/0424-2513.1.2019.21

Chatterjee N (2014) Willingness to pay for drinking water in some selected dryland areas of West Bengal: a contingent valuation approach. Vidyasagar Univ J Econ XIX(2014–15):14–37

Chatterjee N (2017) Dependency on natural resources in dryland areas of West Bengal. Unpublished PhD dissertation, Department of Economics, Rabindra Bharati University, Kolkata, India

Chaudhuri AB, Choudhury A, Hussain MZ, Acharya G (1994) Mangroves of the Sunderbans: India, vol 1. IUCN, Bangkok, Thailand, p 247. ISBN-13: 9782831702094

Datta D, Chattopadhyay RN, Deb S (2011) Prospective livelihood opportunities from the mangroves of the sunderbans, India. Res J Environ Sci 5:536–543. https://doi.org/10.3923/rjes.2011.536.543

Ghosh A, Schmidt S, Fickert T, Nüsser M (2015) The Indian sundarban mangrove forests: history, utilization, conservation strategies and local perception. Diversity 7:149–169. https://doi.org/10.3390/d7020149

Grasso M (1998) Ecological-economic model for optimal mangrove tradeoff between forestry and fishery production: comparing dynamic optimization and a simulation model. Ecol Model 112:131–150

Gupta K (2005) Mangrove-fishery linkages under alternative policy regimes. In: Sengupta N, Bandyopadhyay J (eds) Biodiversity and quality of life. Mc. Milan India Ltd., New Delhi, pp 272–293

Gupta AC, Chatterjee N (2021) Economic values for the environment with special reference to the contingent valuation method. In: Sikdar PK (ed) environmental management: issues and concerns in developing countries. Springer, Cham. https://doi.org/10.1007/978-3-030-62529-0_14

Haraou P, Markandya A, Bellu L, Cistulli V (1998) Environmental economics and environmental policy: a workbook. Economic Development Institute (EDI) of World Bank

Jhingran VG (1977) Fish and fisheries of India. Hindustan Publishers, New Delhi

Jodha NS (1986) Common property resources and rural poor in dry regions of India. Econ Pol Wkly 21(27):1169–1181

Qian Z, Yaoji J, Qun W (2011) Research on value of eco-environment and CVM in the process of mineral resources development. In: International conference on business management and electronic information. https://doi.org/10.1109/ICBMEI.2011.5917934

Reddy SRC, Chakraborty SP (1999) Forest dependence and income distribution in subsistence economy: evidences from India. World Dev 27(7):1141–1149

Resende FM, Fernandes GW, Andrade DC, Néder HD (2017) Economic valuation of the ecosystem services provided by a protected area in the Brazilian Cerrado: application of the contingent valuation method. Braz J Biol 77(4):762–773. https://doi.org/10.1590/1519-6984.21215

Roy Chowdhury B and Vyas P (2007) Sunderbans- The mystic mangroves. Niyogi Books

Ruitenbeek HJ (1994) Modelling economy-ecology linkages in mangroves: economic evidence for promoting conservation in Bintuni Bay, Indonesia. Ecol Econ 10:233–247

Saha D (2015) Some economic aspects of forestry in the sunderbans. Unpublished PhD dissertation, Department of Economics, Rabindra Bharati University, Kolkata, India

Sel A (2016) Reaching the unreached in sunderbans. Community Eye Health J 29(95):S10–S12. https://www.ncbi.nlm.nih.gov/pmc/articles/PMC5340112/pdf/jceh_29_95_S10.pdf

Singh A, Bhattacharya P, Vyas P, Roy S (2010) Contribution of NTFPs in the livelihood of mangrove forest dwellers of Sundarban. J Hum Ecol 29(3):191–200. https://doi.org/10.1080/09709274.2010.11906263

United Nations Development Programme (2002) Strengthening sustainable livelihoods for biodiversity conservation in Sundarbans. Report of United Nations Development Programme (UNDP), New Delhi

West Bengal Forest Department (2003) Annual report. Sundarban Tiger Reserve, West Bengal Forest Department (WBFD), India

Zhang T, Haiming Y, Jinyan Z (2012) Economic valuation of forest ecosystem services in Heshui watershed using contingent valuation method. Procedia Environ Sci 13:2445–2450

Dr. Nilendu Chatterjee is an Assistant Professor in the Department of Economics, Bankim Sardar College, West Bengal, India. He has a research interest in Resource Economics, General Equilibrium, Development Economics, etc. He has published research papers in various National and Internationally renowned journals, including the International Journal of Sustainable Economies Management, Economic Affairs, and Foreign Trade Review.

Dr. Bappaditya Koley is presently the Head of the Department in Geography, Bankim Sardar College, South 24 Parganas, West Bengal, India. He is involved in investigating in the field of Applied Geomorphology and Landslide Disaster Mitigation Studies. His areas of research interest cover Landslide Hazard, Vulnerability and Risk Assessment; Landslide Early Warning System; Coastal Risk Assessment, Coastal Planning and Management; Climate Change; and Environmental issues, etc. He has published research papers in various National and International reputed journals.

Anindita Nath has completed the degree of M.Sc in Geography and presently working as a State Aided College Teacher in the Department of Geography, Bankim Sardar College, South 24 Parganas, West Bengal, India. She does research in Coastal Geomorphology and Hazards Studies. Her areas of research interest cover Coastal Hazard, Vulnerability and Risk Assessment; Coastal Planning and Management; Landside Hazards Studies, and Environmental issues, etc. She has published research papers in various National and International reputed journals.

The manufacturer's authorised representative in the EU is Springer Nature Customer Service Centre GmbH, Europaplatz 3, 69115 Heidelberg, Germany. If you have any concerns regarding our products, please contact ProductSafety@springernature.com

Printed and bound by CPI Group (UK) Ltd, Croydon, CR0 4YY
30/09/2026
02225241-0013